Molecular Plant Biology
Volume Two

The Practical Approach Series

Related **Practical Approach** Series Titles

Essential Molecular Biology V1 2/e
Radioisotopes 2/e
Plant Cell Biology 2/e
Bioinformatics: sequence, structure
  and databanks
Functional Genomics
Essential Molecular Biology V1 2/e
Differential Display
Gene Targeting 2e
DNA Microarray Technology
In Situ Hybridization 2/e
Mutation Detection
Molecular Genetic Analysis of
  Populations 2/e

PCR3: PCR In Situ Hybridization
Antisense Technology
Genome Mapping
DNA and Protein Sequence Analysis
DNA Cloning 3: Complex Genomes
Gene Probes 2
Gene Probes 1
Non-isotopic Methods in Molecular
  Biology
DNA Cloning 2: Expression Systems
DNA Cloning 1: Core Techniques
Plant Cell Culture 2/e
PCR 1

Please see the **Practical Approach** series website at

http://www.oup.com/pas

for full contents lists of all Practical Approach titles.

No. 259

Molecular Plant Biology
Volume Two
A Practical Approach

Edited by

Philip M. Gilmartin
Centre for Plant Sciences,
Faculty of Biological Science,
University of Leeds, UK.

and

Chris Bowler
Laboratory of Molecular Plant Biology
Stazione Zoologica, Naples, Italy.

OXFORD
UNIVERSITY PRESS

OXFORD

UNIVERSITY PRESS

Great Clarendon Street, Oxford OX2 6DP

Oxford University Press is a department of the University of Oxford.
It furthers the University's objective of excellence in research, scholarship, and
education by publishing worldwide in

Oxford New York

Auckland Bangkok Buenos Aires Cape Town Chennai Dar es Salaam Delhi
Hong Kong Istanbul Karachi Kolkata Kuala Lumpur Madrid Melbourne
Mexico City Mumbai Nairobi São Paulo Shanghai Taipei Tokyo Toronto

with an associated company in Berlin

Oxford is a registered trade mark of Oxford University Press in the UK and in
certain other countries

Published in the United States by Oxford University Press Inc., New York

A catalogue record for this title is available from
the British Library

Library of Congress Cataloguing-in-Publication Data
(Data available)

ISBN 0 19 963819 5 (Hbk)
ISBN 0 19 963818 7 (Pbk)

Typeset in Swift by Footnote Graphics, Warminster, Wilts
Printed in Great Britain on acid-free paper
by The Bath Press, Avon

Preface

Nearly fourteen years separates the publication of these volumes from the original *Plant Molecular Biology – A Practical Approach*. The original book, edited by Charlie Shaw, represented a milestone in the Practical Approach series as the first book in the series dedicated to the application of molecular biology to plant biology. In 1988, the field of Plant Molecular Biology was burgeoning. A number of recombinant DNA techniques had been adapted for the analysis of plant genes, and several dozen genes had been isolated and characterised from plants. The techniques for transformation of a few species using *Agrobacterium tumefaciens* mediated gene transfer had been developed although there were comparatively few laboratories with practical experience of producing transgenic plants. The methods for using endogenous plant transposons for gene tagging had been established and the first steps towards using transposons as tools for heterologous gene tagging had been taken. Several techniques for the analysis of plant organelles were also available, including the post-translational transport of proteins into chloroplasts. Approaches for analysis of the subcellular localization of macromolecules included light, fluorescent and electron microscopy. In addition to the growing use of recombinant DNA approaches to the study of higher plants, these techniques were also being applied to plant viruses as well as other models including *Chlamydomonas* and Cyanobacteria. All these approaches were featured in *Plant Molecular Biology – A Practical Approach*, and an appendix reported the use of the 'recently developed reporter gene' β-glucuronidase.

The advances since 1988 are astounding. There have been an unimaginable number of technical developments that have revolutionised the tools available and the scale of experiments that are possible. A large number of Practical Approach Books have been published since 1988 and several of these have focused on aspects of plant science or included approaches applicable to the analysis of plants, but there was never a second edition of *Plant Molecular Biology – A Practical Approach*. In compiling this all-new two-volume book, *Molecular Plant Biology*, we were mindful that this was not a second edition, but a sequel. Many of the approaches and applications described in the initial book are as relevant and important today as they were in 1988. However, as technologies have advanced there are now a myriad of approaches that build on the original techniques, as well as entirely new approaches based on technologies that did not exist at the time of publication of the original book.

The development and wide-spread use of PCR, the advent of automated DNA sequencing, the availability of entire plant genome sequences, the discovery of new reporter genes, the technological advances in bio-robotics and imaging that have resulted in DNA micro-arrays, and the ability to routinely transform a wide range of plant species, including all our major crops, highlight just a few of these developments. Not only have these advances revolutionised the way that fundamental experimental biology can be done, but they have also been applied commercially with the appearance, and in some cases subsequent removal, of products from transgenic plants on our supermarket shelves.

These tremendous advances have occurred in a relatively short time period, and we have both been fortunate enough to have witnessed all these advances during our own research careers. As editors of this new two-volume book, *Molecular Plant Biology* we havedrawn on our combined research experiences and contacts to identify and persuade an international array of authors with particular expertise to contribute the chapters that we feel represent the breadth and scope of the field today. We would like to express our extreme gratitude to these authors for contributing to these books. The range of approaches covered in these two volumes is immense. Not only would it be impossible to find an individual with hands-on experience and expertise in all the topics covered in these two books, but there will surely be few laboratories anywhere where all these approaches are in routine use. We would like to thank all the contributors for sharing their expertise and making these books possible.

In editing these books we were keen to incorporate practical approaches as applied to a range of plant species, and not just the common models. Although much can be gained from the focused analysis of one species, there are many aspects of biology that are unique to non-model organisms. Some approaches covered in the original book are not covered in these new volumes. This is not meant to signify that they are no longer relevant or important, it is merely a reflection of the breadth of the field and a lack of space to incorporate and update approaches for specific organisms such as *Chlamydomonas* and Cyanobacteria. An exception to this has been the inclusion of a new chapter dealing with the moss *Physcomitrella patens*, which is a relatively new molecular model and represents the only plant in which homologous recombination is routine.

The two volumes are used to divide the topics into two themes, gene identification and isolation (Volume 1) and gene expression and gene product analysis (Volume 2). Each volume is divided into three sections and the rationale for the organisation of topics within these Sections is described below. However, many of the approaches have multiple applications, and others may therefore see alternative logic.

In Volume 1, the chapters in Section 1 (Gene Identification) provide methods and considerations for gene identification by classical mutagenesis. Also included here is a chapter on plant transformation, because it is an essential tool for the following two chapters which cover gene tagging strategies in transgenic plants. The final chapter in Section 1 focuses on genomic subtraction for gene identification. In Section 2 (Gene Organization) we have included a chapter on approaches for gene mapping, and a further chapter on the techniques for construction and screening of YAC, BAC and cosmid libraries. Also covered are approaches for chromosome *in situ* analysis. In Section 3 (Library Screening and cDNA Isolation) there are three chapters that describe PCR cloning strategies, western and South-western library screens and complementation cloning.

In Volume 2, the first Section (Gene Expression) covers a range of methodologies for the analysis of gene expression, with chapters on transcript analysis, *in situ* RNA hybridisation and DNA micro-arrays. Approaches for the *in vitro* analysis of DNA:protein interactions are also covered, along with a final chapter on inducible gene expression in plants. In Section 2 (Gene Product Analysis) there are six chapters covering heterologous expression of recombinant proteins, analysis of proteins and protein targeting to chloroplasts, and biochemical approaches for the analysis of plant tissues. The application of the yeast two-hybrid system for analysis of protein:protein interactions is also included with the final two chapters in this Section devoted to antibody techniques and the construction and application of phage display libraries. The final Section of the book is devoted to functional analysis *in vivo*. In this Section we have included three chapters where the approaches primarily permit analyses in living cells. The first of these focuses on calcium imaging, the second describes a range of approaches using various reporter genes and the third is devoted to *Physcomitrella* patens as a model organism in which homologous recombination provides an exciting new *in vivo* approach. As the final chapter in the two volumes, this brings us full circle back to the first chapter in Volume 1 that is dedicated to mutagenesis.

We hope that this organisation illustrates that the approaches described in these volumes are not a linear progression through genetic, biochemical and molecular techniques, but that integration of these approaches and reiteration of analyses are essential to generate a global perspective of plant biology. In compiling the indexes for these two volumes we have provided one combined index for both books to facilitate full cross-referencing between the two volumes and we hope that this proves to be useful.

In 1988 when we first used *Plant Molecular Biology – A Practical Approach*, one of us had just started a PhD, the other had just started a postdoc. We hope that the readers of these new books will find them as useful as we found the original book during our own time at the bench.

P. G. and C. B.
February 2002

Contents

Protocol list

Assessment of cellular fractions and the use of marker enzymes

Working with yeast—the techniques for the yeast two-hybrid system

Performing two-hybrid screens

Making antibodies

Using antibodies

Naïve, synthetic, and immune scFv phage display libraries

Affinity selection (panning) of scFvs from a phage display library

Isolation and analysis of monoclonal phage antibodies

Abbreviations

2-CBSU	2-chlorobenzene sulfonamide
AcMNPV	*Autographa californica* nuclear polyhedrosis virus
AM	acetoxymethyl (as in AM-ester)
AOI	area of interest
AP	alkaline phosphatase
APS	ammonium persulphate
BCIP	5-bromo-4-chloro-3-indolyl phosphate
BSA	bovine serum albumin
BTH	thiadiazole-7-cabothioic acid S-methyl ester
BTP	bis-Tris propane
CaM	calmodulin
CaMV	cauliflower mosaic virus
CAT	chloramphenicol acetyl transferase
CCD	charged coupled device
CCLR	luciferase cell culture lysis reagent
CDRs	complementarity determining regions
CFI	correction for infinity
CFP	cyan fluorescent protein
Chl	chlorophyll
CTAB	cetyl trimethyl ammonium bromide
DAB	chromogen 3,3'-diaminobenzidine
DBD	DNA–binding domain
DEPC	diethyl pyrocarbonate
DEX	dexamethasone
DIC	differential interference contrast
DIG	digoxigenin
DMF	dimethylformamide
DMS	dimethylsulphate
DTE	dithioerythritol
DTT	dithiothreitol
DW	dual wavelength
EB	extraction buffer
EDTA	ethylenediaminetetra-acetic acid
ELISAs	enzyme-linked immunosorbent assays
EMSA	electrophoretic mobility shift assay

ER   endoplasmic reticulum
EST   expressed sequence tag
FAE   mixture of paraformaldehyde, acetic acid, and ethanol
FCS   foetal calf serum
FISH   fluorescence *in situ* hybridization
FITC   fluoroscein isothiocyanate
FRET   Fluorescence Resonance Energy Transfer
GFP   green fluorescent protein
GR   glucocorticoid receptor
GST   glutathione S-transferase
GUS   β-glucuronidase
HAT   hypoxanthine, aminopterin, and thymidine
HBD   hormone binding domain
HBW   half bandwidth
HDSS   high density sucrose solution
HPRT   hypoxanthine-guanine phosophoribosyl transferase
HRP   horseradish peroxidase
HS   heat shock
HSE   high-salt extraction buffer
IDAs   immuno-dot assays
IMAC   immobilized metal affinity chromatography
INA   2,6-dichloroisonicotinic acid
ipt   isopentenyl transferase
IPTG   isopropyl β-D-thiogalactopyranoside
KLH   keyhole limpet haemocyanin
LacZ   β-galactosidase
LDSS   low density sucrose solution
LSCM   laser scanning confocal microscopy
LUC   luciferase
MBP   maltose binding protein
MES   2-(*N*-morpholino)ethanesulfonic acid
MOPS   (3-[N-Morpholino]propane sulphonic acid)
NA   numerical aperture
NBT   nitroblue tetrazolium
NDPase   nucleotide diphosphatase
NIB   nuclei isolation buffer
PCR   polymerase chain reaction
PEG   polyethylene glycol
PEP   *Physcomitrella* EST programme
Ph   phase contrast
PMSF   phenylmethylsulfonylfluoride
PNPG   *p*-nitrophenyl glucuronide
PPase   pyrophosphatase
PPEB   periplasmic protein extraction buffer
PR   pathogenesis related
PSF   point spread function
PVA   polyvinyl alcohol

PVP	polyvinyl pyrrolidone
PVPr	polyvinyl propylene
SA	salicylic acid
SAR	systemic acquired resistance
SB	sonication buffer
scFv	single-chain Fv
SD	synthetic dropout
SDS	sodium dodecyl sulphate
SIP	selectively infective phage
SNR	signal-to-noise ratio
SW	single wavelength
tc	tetracycline
T-DNA	transfer DNA
TEMED	N,N,N′,N′-tetramethylethylenediamine
TES	*N*-tris(hydroxymethyl)methyl-2-aminoethanesulfonic acid
TetR	*tet* repressor
UTR	untranslated region
UV	ultraviolet
wtGFP	wild-type GFP
YFP	Yellow fluorescent protein

Section 1 Gene expression

Chapter 1
Transcript analysis

Graham R. Teakle
Horticulture Research International, Wellesbourne, Warwick CV35 9RZ, UK

Charles P. Scutt
Reproduction et Développement des Plantes, École Normale Supérieure de Lyon, 46 Allée d'Italie, 69364 Lyon cedex 07, France

Philip M. Gilmartin
Centre for Plant Sciences, University of Leeds, Leeds LS2 9JT, UK

1 Introduction

The isolation and characterization of messenger RNA is fundamental to a wide range of gene expression analyses. In this chapter, we provide a collection of experimental approaches for RNA extraction, purification, quantitation, and analysis. New techniques for the analysis of gene expression profiles have developed rapidly over the past few years. Two such approaches, *in situ* hybridization (see Chapter 2) and micro-array analysis (see Chapter 3) are dealt with in detail in dedicated chapters. However, there is still a need for access to techniques that allow the broad expression analysis of individual transcripts. Transcript analysis may be required to determine the expression level of a particular gene, for example, by northern analysis or may be required to map mRNA molecules relative to their genomic sequences, using RNase or S1 transcript mapping, or primer extension analysis. In this chapter we have attempted to cover a range of approaches for transcript analysis that complement those described in other chapters of this book.

A number of precautions are required when working with RNA, both to preserve the integrity of the RNA and minimize the risk of degradation by RNase, but also to protect the researcher. Many of the reagents used in the experimental procedures are toxic or dangerous, and individual researchers should familiarize themselves with the hazardous nature of these reagents. Although we have endeavoured to highlight specific hazards associated with particular procedures, it is important to follow local safety procedures to minimize any potential health risks not highlighted in this chapter. We have included a section highlighting specific precautions required to minimize contamination of the RNA samples with RNases and general guidance for working with RNA. Accurate quantitation of RNA is essential for many of the procedures described; this is covered in Section 3 of this chapter. The isolation of RNA from different plants, or from different tissues, may require different experimental procedures. In Section 4 we describe two different protocols that have proved to be reliable in our own laboratory for the extraction of RNA form a range of plant species and tissues. Also in this section, we describe the isolation of messenger RNA from total RNA. Section 5 covers a range of approaches for transcript abundance analysis by blotting and filter hybridization. The exact procedures used will be dictated by the abundance of the transcript under analysis or the specific expression information required. In Sections 6 and 7, we describe approaches for the analysis of transcript structure, although the techniques of RNase protection, S1 mapping and primer extension can also be used for monitoring transcript abundance. Again, the specific approach taken will be determined by the information required and the abundance of the transcript under analysis. We hope that, in combination with other chapters in this book, ours will provide useful information for a wide range of transcript analysis applications.

2 Handling RNA

Extensive precautions are required when handling RNA as RNases are extremely stable and have been used as models for studying the refolding of proteins following denaturation. RNase is not destroyed by autoclaving; indeed, boiling RNase is a method used to remove DNase contamination from RNase. RNases are also prevalent, being present in skin secretions and are released from cells when they are broken open to extract the RNA. If much RNA work is to be performed on a regular basis, many laboratories set aside bench space, chemicals, and equipment designated for RNA use only. Gloves should be used at all times, and particular care should be taken not to contaminate electrophoresis equipment, tubes, pipettes, and reagents, etc. RNases may be removed from solutions by treatment with diethyl pyro-carbonate, as described below. Equipment such as certain glassware and metal items can be baked overnight at 200 °C. In addition, many of the reagents used in the analysis and manipulation of RNA are extremely toxic and suitable precautions must be taken

We have found that prepacked sterile plastic tubes and pipette tips are essentially free from RNase contamination and need no treatment. If pipette tips need to be racked by hand then gloves should be worn and it is best to take the tips from a fresh bag. Alternatively, tubes and pipette tips, and solutions can be treated with diethyl pyrocarbonate (DEPC). DEPC reacts with amino groups and chemically deactivates RNases and can subsequently be removed by autoclaving (*Protocol 1*). Solutions of reagents with amino groups, such as Tris, should not be treated with DEPC, but instead made up from bottles of reagent reserved for RNA use and dissolved in DEPC-treated water. It is important to remove DEPC since it will carboxy-methylate amino groups in RNA.

During the initial stages of RNA extraction, when the plant tissue is being disrupted, the use of decontaminated equipment is less important, since the extraction buffer used will be denaturing and designed to inhibit any RNase activity. However, during subsequent steps when the denaturant has been removed it is essential to work in an RNase-free environment. In addition to the above precautions, placental ribonuclease inhibitor, which is available commercially from a number of companies, may be added to reactions containing RNA. This is a protein and so will be removed during any further purification steps, such as phenol: chloroform extraction, and so may need to be added again. RNA can be precipitated from solution using ethanol. This procedure requires the addition of 2.5 vols of ethanol to the combined volume of RNA and salt. The salt used in ethanol precipitation of RNA is most commonly 0.3 M sodium acetate.

Protocol 1

DEPC treatment of plasticware and solutions

Reagent

- Diethylpyrocarbonate (DEPC) solution, 0.1% (v/v) in water[a]

Method

1 Soak plasticware or glassware in 0.1% (v/v) DEPC at 37 °C for at least 2 h or overnight.
2 Autoclave the plasticware or glassware to sterilize and remove DEPC before use.
3 Make solutions with water containing 0.1% (v/v) DEPC.
4 Autoclave solutions to sterilize and remove DEPC before use.

[a] DEPC decomposes to produce CO_2 and care should be taken when handling sealed containers as these may be under pressure from DEPC decomposition.

3 Quantification of RNA

3.1 Spectrophotometric quantification

Quantification of RNA can be accurately achieved using an ultraviolet (UV) spectrophoto-meter and quartz cuvettes (*Protocol 2*). Dilution of a small sample of RNA for quantification is preferable to the measurement of the entire sample. However, under some circumstances, such as during the analysis of small amounts of poly A^+ RNA, the entire sample may need to be placed in a cuvette and recovered after quantification. In such circumstances, every precaution must be taken to ensure that the integrity of the RNA sample is maintained. RNA samples can also be quantified by comparison to known concentration standards after staining with ethidium bromide (*Protocol 3*). This procedure is rapid and can be useful for quantification of large numbers of samples, but is not as accurate as spectrophotometric quantification. The advantage of this protocol over quantitation of RNA fractionated by electrophoresis is that the RNA sample will usually be composed of a population of molecules with a range of molecular weights. This method is suitable for measuring small amounts of mRNA.

Protocol 2

Spectrophotometric quantification of RNA

Equipment and reagent

- UV spectrophotometer
- Quartz cuvettes
- methanol:HCl (1:1)

Method

1 Soak clean quartz cuvettes in methanol:HCl (1:1) for 1 h at room temperature.
2 Rinse the cuvettes thoroughly with DEPC-treated sterile water.

Protocol 2 continued

3   Dilute the RNA sample or use undiluted if concentration of RNA is very low.[a]

4   Measure OD_{260}, OD_{280} and OD_{310} in the spectrophotometer.[b]

5   Calculate RNA concentration using OD_{260} reading. A 40 μg/ml solution of RNA has an absorption at 260 nm of 1.

[a]   If sufficient RNA is available the diluted sample can be discarded after quantification. For limited amounts of RNA, such as poly A^+ preparations, quantification of the whole sample may be necessary to obtain reliable absorbance values. In these case, it is imperative that all precautions are taken to minimize the risk of RNase contamination.

[b]   A pure solution of RNA has OD_{260}/OD_{280} ratio of 2.0. Its absorbance at 310 nm should be minimal by comparison to its absorbance at 260 nm. If this is not the case, the RNA solution contains impurities and its absorbance at 310 nm should be subtracted from the 260 nm value to obtain a more accurate estimate of RNA concentration.

Protocol 3

Quantification by ethidium bromide staining

Equipment and reagents

- 1% agarose (w/v) in water and autoclave to dissolve the agarose
- 0.25 mg/ml ethidium bromide solution[a]
- Plastic Petri dish
- RNA standard solutions (a range of standards will be required and can be prepared by dilution of a stock that has been quantified by spectrophotometry)[b]

Method

1   Melt the 1% agarose in a glass bottle and when cool enough to handle add ethidium bromide to 0.25 μg/ml.

2   Pour 1% agarose containing ethidium bromide into a plastic Petri dish and allow to set.

3   Spot 1 μl volumes of different dilutions of RNA of a known concentration[b] onto the agarose.

4   Spot 1 μl volumes of several dilutions of the RNA solution to be quantified onto the agarose.

5   View the agarose plate on UV transilluminator[c] and compare spot intensities either by eye to estimate the concentration as compared to the known standards or using a digital imaging system with image quantification analysis software to gain a more accurate estimation.[d]

[a]   Ethidium bromide is toxic and carcinogenic, necessary precautions should be taken.

[b]   The range of dilutions required will depend on the concentration of the unknown samples to be quantified.

[c]   The plastic of some Petri dishes may absorb UV light, in this case illuminate the sample from above.

[d]   This method is suitable for measuring small amounts of mRNA and is better than quantification following gel electrophoresis, as RNA is usually composed of a population of molecules with a range of molecular weights.

4 Isolation of RNA

When isolating RNA from plant tissues, a number of factors may influence the yield and quality of the RNA. Some tissues may be available in only very small quantities, whereas others may be available in abundance. Extraction of RNA from some samples, such as leaf tissue, may give higher yields than root or stem tissue. Furthermore, the presence of compounds such as starch, phenolics, anthocyanins, or other metabolites can dramatically reduce the yield or quality of RNA samples. For common laboratory plants, such as *Arabidopsis* and tobacco, isolation of RNA is routine and uncomplicated. For certain plants, it may be necessary to evaluate the effectiveness of a number of RNA isolation protocols to identify one that gives the required yield and quality of RNA from the plant or tissue under investigation. A number of protocols are provided for comparison.

4.1 Isolation of total RNA using LiCl precipitation

Protocol 4 can be conveniently scaled from 10 g of fresh weight plant tissue, where the extraction is performed in two 50 ml disposable plastic centrifuge tubes, to 100 mg plant tissue, where the extraction is performed in 1.5 ml microcentrifuge tubes. We have successfully used this protocol on a range of tissues including flowers, leaves, stems, and roots. The volumes are for 10 g of plant tissue, but for smaller amounts the volumes can be adjusted accordingly. This method yields several mg RNA from 10 g plant tissue and greater than 100 μg from 200 mg plant tissue, which is enough for several lanes on northern blots.

Protocol 4

Isolation of total RNA using LiCl precipitation

Equipment and reagents

- 1 M Tris pH 7.6[a]
- 8 M LiCl[b]
- 0.25 M EDTA[b]
- 10% SDS[b]
- DEPC treated sterile water
- Extraction buffer (EB): 50 mM Tris pH 7.6, 150 mM LiCl, 5 mM EDTA, 1% SDS
- Phenol:chloroform:isoamyl alcohol (25:24:1)
- 3 M sodium acetate pH 5.2

- 100% ethanol at −20 °C
- 70% ethanol at −20 °C
- Liquid nitrogen
- Chilled pestle and mortar
- 50 ml disposable sterile plastic tubes with conical bottoms that are resistant to phenol (e.g. Falcon product number 2070)
- RNase-free 30 ml Corex tubes

Method

1 Freeze 10 g plant tissue in liquid nitrogen.

2 Grind plant tissue to a fine powder in liquid nitrogen in a pestle and mortar.

3 Aliquot 15 ml extraction buffer into two 50 ml disposable sterile plastic tubes.[c]

4 Divide the frozen ground powder evenly between the two tubes.[d]

5 Add 15 ml of phenol:chloroform:isoamyl alcohol (25:24:1) to the tubes and agitate gently so as to mix the organic and aqueous phases for several min.[e]

6 Centrifuge[f] the tubes in a bench-top centrifuge, cooling is not necessary.

Protocol 4 continued

7 Carefully remove the supernatant to a new tube using a pipette. It is better to leave some supernatant behind than to transfer debris from the interface.

8 Extract twice more with phenol:chloroform:isoamyl alcohol as described in steps 6–8.

9 Add 15 ml chloroform:isoamyl alcohol (24:1) to the supernatant, vortex, and centrifuge for 5 min in a bench-top centrifuge, cooling is not necessary.

10 Remove the supernatant to a clean sterile 30 ml Corex tube.

11 Estimate the volume of supernatant from the gradations on the side of the disposable tube or by transferring the sample to the Corex tube by pipette.

12 Add 1/3 volume of 8 M LiCl to the supernatant from step 10 for a final concentration of 2 M.

13 Precipitate the RNA by incubation overnight at 4 °C.[g]

14 Pellet the RNA by centrifugation at 12 000 g for 20 min at 4 °C.

15 Remove the supernatant and re-dissolve the pellet in 4 ml DEPC-treated sterile H_2O.

16 Add 0.1 vol (0.4 ml) of 3 M sodium acetate pH 5.2 and 2.5 vols (11 ml) of −20 °C 100% ethanol, and mix by vortexing.

17 Incubate at −20 °C for 1 h.

18 Pellet the RNA by centrifugation at 12 000 g for 20 min at 4 °C.

19 Remove all the supernatant taking care not to disrupt the pellet.

20 Add 5 ml of −20 °C 70% ethanol and vortex.

21 Pellet the RNA by centrifugation at 12 000 g for 20 min at 4 °C.

22 Remove all the supernatant taking care not to disrupt the pellet.

23 Dry the pellet of RNA in a vacuum dessicator.

24 Re-dissolve the pellet in 0.5 ml DEPC-treated H_2O and transfer to a microcentrifuge tube for storage.

25 Quantify the RNA (see *Protocols 2 and 3*)

26 Store RNA at −20 °C or −80 °C.

[a] DEPC treated water should only be used after removal of DEPC by autoclaving as DEPC will react with amine groups in the Tris.

[b] Solutions can be made up in water containing DEPC and then autoclaved to remove the DEPC.

[c] Use a ratio of 3 ml extraction buffer to 1 g plant tissue.

[d] Make sure that no liquid nitrogen is remaining in the tube when the lid is put on as this could be extremely hazardous.

[e] Phenol is extremely toxic by all routes of uptake and is also caustic, causing severe burns on skin contact. Wear hand and face protection for the use and preparation of phenol, and all reagents containing it. All manipulations involving phenol must be conducted in a fume hood.

[f] All centrifugation steps should be performed at a speed of least 2800 g, unless otherwise stated. This corresponds to approximately 4000 rpm in a typical bench top centrifuge. If a micro-centrifuge is used, the maximum centrifugation speed (up to 20,000 g) can be selected.

[g] This step selectively precipitates the RNA and not the DNA. Because the RNA is precipitated it is stable at this stage and can be left for longer than overnight without any significant detrimental effects.

4.2 RNA extraction using guanidinium hydrochloride

Protocol 5 is based on that of Logemann *et al.* (1), which was originally used for RNA preparation from potato tubers. It incorporates washing steps using concentrated sodium acetate to remove starch from RNA samples and may be particularly useful for plant tissues that are rich in carbohydrates. The method uses guanidinium hydrochloride, a potent inhibitor of RNases, and is also well adapted to tissues such as the roots of soil-grown plants, where RNA degradation during preparation can be a problem. Furthermore, it is a convenient method of RNA preparation to perform on a small scale and so is useful for plant tissues that cannot easily be obtained in large quantities. The volumes and quantities given in the following protocol can be varied according to sample size.

Protocol 5

RNA extraction using guanidinium hydrochloride

Equipment and reagents

- Pestle and mortar
- Liquid nitrogen
- 15 ml or 50 ml disposable sterile plastic tubes with conical bottoms that are resistant to phenol (e.g. Falcon product numbers 2069 and 2070)[a]
- Polypropylene mortars made to fit 1.5 ml centrifuge tubes (e.g. product number 749520, Kontes, Vineland, New Jersey)
- RNase-free 15 ml or 30 ml Corex tubes[a]
- 0.25 M EDTA
- 1 M MES pH 7.0

- Extraction buffer: 8 M guanidinium hydrochloride, 20 mM MES pH 7.0, 20 mM EDTA, 50 mM β-mercaptoethanol. Dilute the MES buffer and EDTA stocks, dissolve the guanidinium hydrochloride and add β-mercaptoethanol. Do not autoclave
- Phenol/chloroform/isoamylalcohol (25:24:1)
- 100% ethanol at −20 °C
- 1 M acetic acid
- 3 M sodium acetate (pH 5.2)
- DEPC treated sterile water (see *Protocol 1*)

Method

1 Freeze plant tissue in liquid nitrogen.

2 Grind frozen tissue to a fine powder in a mortar and pestle pre-cooled with liquid nitrogen. For samples of less than 200 mg tissue, grind 1.5 ml centrifuge tubes using disposable polypropylene mortars.

3 Transfer tissue powder to a disposable polypropylene tube[a], pre-cooled in liquid nitrogen.[b]

4 Add 2 vols of extraction buffer and mix thoroughly.

5 Pellet cell debris by centrifugation at 13 000 g for 5 min in 1.5 ml microcentrifuge tubes or at 2800 g for 15 min in 15 or 50 ml plastic tubes.

6 Transfer supernatant to a clean tube.[a]

7 Add an equal volume of phenol/chloroform/isoamylalcohol (25:24:1) and mix thoroughly by vortexing.[c]

8 Centrifuge the sample to separate the phases as in step 5.

Protocol 5 continued

9 Transfer the aqueous (upper) phase to a clean microcentrifuge tube or Corex tube,[a] and precipitate the RNA by addition of 0.7 vols of ethanol and 0.2 vols of 1 M acetic acid.

10 Incubate at $-20\,^{\circ}$C for at least 1 h.

11 Pellet the RNA by centrifugation. Samples in 1.5 ml microcentrifuge tubes should be centrifuged at 13 000 g for 10 min. Samples in Corex tubes should be centrifuges at 4000 g for 20 min.

12 Carefully remove the supernatant and add a volume of 3 M sodium acetate (pH 5.2) equal to the volume of extraction buffer used in step 4 at room temperature and gently vortex.

13 Centrifuge the samples as described in step 11.

14 Carefully remove the supernatant and wash the pellet in ice-cold 70% ethanol.

15 Centrifuge the samples as described in step 11.

16 Carefully remove the supernatant taking care not to dislodge the pellet and resuspend the RNA in DEPC-treated, sterile distilled water.

17 Quantitate the RNA (see *Protocols 2* and *3*).

18 Store the RNA frozen, preferably at $-80\,^{\circ}$C.

[a] The size of the tube used will depend on the size of the tissue sample.

[b] Make sure that no liquid nitrogen is remaining in the tube when the lid is put on as this could be extremely hazardous.

[c] Phenol is extremely toxic by all routes of uptake and is also caustic, causing severe burns on skin contact. Wear hand and face protection for the use and preparation of phenol and all reagents containing it. All manipulations involving phenol must be conducted in a fume hood.

4.3 Isolation of mRNA

Messenger RNA (mRNA) represents only about 1–5% of total cellular RNA. The remainder is mostly tRNA and ribosomal RNA. Messenger RNA is purified from the other RNAs by making use of the fact that it possesses a poly A tail. This poly A tail will anneal to oligo-dT$_n$, where n is approximately 15 residues, coupled to a solid support. Types of support include cellulose columns, paramagnetic beads, or even modified paper. The rationale is to bind the mRNA in a high salt buffer, wash to remove the unwanted RNA species, and then elute the mRNA in lower salt buffer or water. There are many commercial kits available for the purification of mRNA and for the purification of small quantities of mRNA we have found that those using paramagnetic beads were extremely quick and convenient. Detailed instructions are provided with such kits and these will not be repeated here.

Protocol 6 below is adapted from Apel and Kloppstech (2) and does not involve the purchase of expensive kits. It is well suited to larger scale isolation of poly A$^+$ RNA, but can also be scaled down to 1.5 ml microcentrifuge tubes. In *Protocol 6*, the mRNA is directly isolated from a partially purified cellular homogenate using oligo(dT)-cellulose. Other methods are available based on purification of mRNA from total RNA preparations.

Protocol 6

Isolation of mRNA

The protocol below is for 10 g of plant tissue, for smaller amounts of tissue, scale all volumes proportionally.

Equipment and reagents

- Extraction buffer: 100 mM NaCl, 50 mM Tris.HCl pH 9.0, 10 mM EDTA, 2% SDS 0.1 mg/ml proteinase-K (added just before use)
- Binding buffer: 0.5 M NaCl, 10 mm Tris.HCl pH 7.5, 1 mM EDTA, 0.2% SDS
- Elution buffer: 10 mM Tris.HCl pH 7.5, 1 mM EDTA, 0.05% SDS
- 5 M NaCl
- Liquid nitrogen
- Pestle and mortar
- Polytron homogenizer (Philip Harris Scientific or similar)
- 250 ml centrifuge tubes that are resistant to phenol

- 50 ml disposable sterile plastic tubes with conical bottoms that are resistant to phenol (e.g. Falcon product number 2070)
- Phenol:chloroform:isoamyl alcohol (25:24:1)
- Chloroform:isoamyl alcohol (24:1)
- Oligo(dT)-cellulose
- Baked 10 ml glass pipette fitted with a siliconized glass wool plug and a DEPC-treated rubber tube at the bottom to which a tube clamp can be attached. Alternatively, an RNase-free small water-jacketed column could be used

Method

1 Freeze plant tissue in liquid nitrogen.
2 Grind tissue to a powder in liquid nitrogen in a pestle and mortar.
3 Transfer powder to an approximately 200 ml tube containing 30 ml extraction buffer.[a]
4 Homogenize using a Polytron homogenizer (or equivalent) at full power for 1 min.
5 Pour the homogenate into a 250 ml centrifuge tube and add 30 ml phenol:chloroform:isoamyl alcohol (25:24:1).[b]
6 Shake the tubes vigorously for 15 min.
7 Separate the phases by centrifugation at 4000 g for 5 min.
8 Transfer the aqueous upper phases to two clean 50 ml polypropylene tubes and add 15 ml phenol:chloroform:isoamyl alcohol (25:24:1) to each.
9 Mix thoroughly by vortexing.
10 Separate the phases by centrifugation at 1500 g for 5 min in a bench-top centrifuge.
11 Repeat steps 8 and 9.
12 Remove the aqueous upper phases to new 50 ml polypropylene tubes and add 15 ml chloroform:isoamyl alcohol (24:1).
13 Mix thoroughly by vortexing and centrifuge at 1500 g for 5 min as in step 10.
14 Remove the upper aqueous phases and pool the two samples in a clean 50 ml polypropylene tube.
15 Estimate the volume and add 5 M NaCl to give a final NaCl concentration of 0.5 M.
16 Add 0.25 g oligo(dT)-cellulose to 5 ml binding buffer and equilibrate for 5 min.
17 Add equilibrated oligo (dT)-cellulose to sample from step 15.
18 Mix on an orbital shaker at 200 rpm for 15 min.

Protocol 6 continued

19 Collect the oligo(dT)-cellulose by centrifugation at 3000 g for 2 min.

20 Remove the supernatant and resuspend the oligo(dT)-cellulose pellet in sufficient binding buffer to make a slurry (approximately 1 ml).

21 Transfer the slurry to a small column prepared from a baked 10 ml glass pipette with a small plug of RNase free siliconized glass wool and a rubber tube fitted to the bottom.

22 Wash the oligo(dT)-cellulose column slowly with several column volumes of binding buffer until it has returned to its white colour.

23 Check the eluate has no absorption at 260 nm.

24 Allow the buffer to drain to the surface of the column and stop the flow, by clamping the rubber tube at the bottom of the column.

25 If the column has a water jacket then connect this to a 45 °C water bath. If a water-jacketed column is not available run column at room temperature.

26 Add elution buffer pre-warmed to 65 °C to the top of the column.

27 Collect 0.2 ml fractions, assay them for RNA content and pool the most concentrated fractions.[c]

28 For a second round of purification adjust the NaCl concentration of the eluted RNA to 0.5 M.

29 Denature the sample by heating to 65 °C for 5 min.

30 Reload the sample onto the oligo(dT)-cellulose column.

31 Wash the oligo(dT)-cellulose column with three column volumes of binding buffer.

32 Elute the mRNA by following step 23–step 27.

33 Concentrate the poly A$^+$ RNA by adding NaCl to 0.1 M and 2.5 vols of cold 100% ethanol. Incubate at −20 °C for 1 h.

34 Pellet the RNA by centrifugation in a microcentrifuge at 13 000 g for 10 min.

35 Wash the RNA pellet twice with 70% ethanol, centrifuging as in step 34 after each wash.

36 Dry the pellet and re-dissolve in 200 μl DEPC-treated H_2O.

37 Quantify the RNA either by UV absorption at 260 nm or by ethidium bromide fluorescence.

Note: Alternatively, the mRNA may be purified from the total RNA by first preparing the oligo(dT)-cellulose column equilibrated with binding buffer and adjusting the salt concentration of the total RNA to that of the binding buffer before loading the sample onto the column.

[a] A 200 ml tube is used to contain the volume of froth produced during homogenization.

[b] Phenol is extremely toxic by all routes of uptake and is also caustic, causing severe burns on skin contact. Wear hand and face protection for the use and preparation of phenol and all reagents containing it. All manipulations involving phenol must be conducted in a fume hood.

[c] After a single selection using oligo dT cellulose, the poly A$^+$ RNA is about 50% pure.

5 Blotting and hybridization analysis of RNA

Transcript analysis by blotting and subsequent hybridization can be undertaken in one of three ways. RNA can be fractionated by agarose gel electrophoresis, transferred by blotting onto a membrane, and then used in a hybridization assay using radiolabelled or non-radioactive probes. This procedure of northern blotting is described in Section 5.1. The process of reverse northern analysis involves the immobilization of cDNAs onto a membrane and screening with a probe derived from RNA by reverse transcription. Procedures for such analyses are described in Section 5.2. The process of virtual northern blotting relies on agarose gel fractionation of full length cDNA sequences derived from RNA by reverse transcription and subsequent PCR analysis. Analysis of these amplified cDNA pools is referred to as virtual northern blotting and is described in Section 5.4.

5.1 Northern analysis

Northern analysis can be used to measure steady state transcript abundance, as well as determine the size of specific RNA molecules. The RNA is separated on a gel (*Protocol 7*) followed by blotting it onto a membrane (*Protocol 8*). The membrane is then hybridized (*Protocol 10*) with a labelled RNA or DNA probe (*Protocol 9*). RNA:RNA hybrids are stronger than RNA:DNA hybrids so greater sensitivity is obtained using an RNA probe. For most purposes Northern analysis can be undertaken using total RNA or it can be done using mRNA for the detection of low abundance transcripts. Northern analysis using mRNA provides a 10–100-fold increase in sensitivity as compared to total RNA. Typically, 10 µg of total RNA is fractionated per lane on an agarose gel, but up to 30 µg may be loaded if necessary. If mRNA is used then typically this would be in the range of 0.5–3 µg. Since RNA is single-stranded it can anneal to itself or to other RNA molecules, which would give anomalous sizes on a gel. Therefore, before loading the RNA onto the gel it needs to be denatured. It is possible to stably denature RNA by treating it with the alkylating reagent methylmercuric hydroxide. However, this is very toxic, so a simple solvent-denaturing method is given. Once the RNA sample has been denatured it is fractionated on a denaturing gel containing formaldehyde (*Protocol 7*).

Protocol 7

Gel and sample preparation

Equipment and materials

- DEPC treated sterile water (see *Protocol 1*)
- 5× MOPS buffer: prepare 800 ml 40 mM Na-acetate, 5 mM EDTA, DEPC treat, and autoclave[a]. Add 20.6 g MOPS, which will give a final concentration of 100 mM MOPS. Adjust the pH to 7.0 with NaOH and make up to 1 l. Sterilize through a 0.2 µm filter and store in the dark at 4 °C. MOPS should not be autoclaved. MOPS buffers will yellow with age
- 10% SDS
- Formaldehyde[b]
- Formamide[c]
- 1.2% Formaldehyde agarose gel: to prepare a 100 ml denaturing gel, melt 1.2 g agarose in 62.5 ml H_2O, add 18 ml formaldehyde, and 19.5 ml 5× MOPS
- Formaldehyde gel loading dye: 50% glycerol, 1 mM EDTA pH 8.0, 0.25% bromophenol blue, 0.25% xylene cyanol FF[a]
- Gel electrophoresis apparatus: soak the gel tray and comb in 10% SDS then wash thoroughly with sterile DEPC treated water.

Method

1 Make each RNA sample up to 4.5 µl RNA using DEPC treated H_2O.

2 Add 2 µl 5× MOPS buffer.

3 Add 10 µl formamide.

4 Add 3.5 µl formaldehyde.

5 Heat to 65 °C for 15 min.

6 Cool samples on ice and pulse spin in a microcentrifuge to collect any condensation.

7 Add 1 µl formaldehyde gel loading dye and mix.[d]

8 Place the formaldehyde denaturing gel in the electrophoresis tank in the 1× MOPS buffer.

Protocol 7 continued

9 Pre-run gel at 5 V/cm for 5 min.

10 Load the samples onto the formaldehyde denaturing gel.

11 RNA size markers should also be loaded. Commercially prepared markers are available.

12 Run the gel at 3–4 V/cm. Recirculation of the buffer is not necessary, but it is a good idea to mix buffer half way through.[e]

13 Run the gel until bromophenol blue has migrated about three-quarters of the length of the gel.

14 If you wish to visualize the RNA, stain the gel in 0.5 μg/ml ethidium bromide in 0.1 M ammonium acetate for 30 min.[d]

[a] Solutions can be made up in water containing DEPC and then autoclaved to remove the DEPC.

[b] If formaldehyde is <pH 4.0 a new stock should be used or the sample deionized. Note: Formaldehyde is toxic and should be handled in a fume hood.

[c] If the formamide appears yellow it should be deionized by equilibration with Amberlite monobed resin (1 g per 500 ml formamide) with stirring for 1 h.

[d] If required, a small amount of ethidium bromide (1 μl of 0.5 mg/ml) can be added to the RNA at this stage to prevent the need for later staining of the gel. However, the presence of ethidium bromide may reduce hybridization detection sensitivity. We have found it useful to prepare twice the required amount of RNA, add ethidium bromide to half the sample, and run this aliquot in separate lanes on the gel. This section of the gel is not blotted, but instead is used to visualize the quality of the RNA. If the RNA is of good quality, distinct well-resolved ribosomal RNA bands should be visible. If the RNA is degraded, the ribosomal RNA bands will have a smear trailing below them or the whole lane will be a smear. If this degradation is observed, new RNA should be prepared. mRNA is too small a proportion of the total RNA to be seen as a smear on the gel.

[e] Some laboratories routinely use buffer recirculation for formaldehyde gels. However, this is not essential and can be replaced by mixing the buffer during running. For safety reasons, the gel should be switched off during this procedure. An alternative is to place the gel tank on two magnetic stirrers during electrophoresis with a revolving magnetic flea at each end of the tank.

Protocol 8

Transfer of RNA to membrane

Equipment and reagents

- Nylon or nitrocellulose membrane
- 20× SSC: 3 M NaCl, 0.3 M tri-sodium citrate
- Whatman 3MM paper
- Absorbent tissues
- 0.5 M sodium acetate pH 5.2
- Methylene blue

Method

1 Wet the nylon or nitrocellulose membrane in H_2O.

2 Equilibrate the membrane in 20× SSC for several min.

3 Set up the capillary blot ensuring that there are no trapped air bubbles between the agarose gel, the membrane, and the Whatman 3MM paper soaked in 20× SSC (see *Volume 1, Chapter 6*).

4 Blot overnight.

Protocol 8 continued

5 Dismantle the blot and allow the membrane to air dry.

6 Fix the RNA to the filter either by baking at 80 °C for 2 h[a] or cross-link with UV light (e.g. in a Stratagene Stratalinker).

7 Cut the marker lane off the filter for staining.[b]

8 Stain the marker lane in 0.5 M sodium acetate pH 5.2 containing 0.04% methylene blue for 5–10 min at room temperature.

9 Destain in H_2O as required.

[a] If using nitrocellulose, the filter should be baked under vacuum.

[b] Staining the marker lane enables visualization of the size markers and provides confirmation of transfer of RNA from the gel to the membrane.

Protocol 9

Preparation of radiolabelled probes

Adapted from Feinberg and Vogelstein (3). (See also *Volume 1, Chapter 6*)

Equipment and reagents

- Sterile distilled water
- Random hexameric oligonucleotide mix (1.6 mg/ml)
- 1× TE buffer: 10 mM Tris-HCl and 1 mM EDTA, pH 7.6
- Empty disposable spin column units (e.g. Bio Rad product number 732–6204)
- Sephadex G50: prepare as a slurry by mixing 3 g of Sephadex-G50 with 50 ml of 1× TE and either incubate the mixture overnight at 4 °C

  or autoclave it for 10 min. Mix again immediately before loading the Sephadex G-50 into spin columns
- 10× labelling buffer: 0.5 M Tris-HCl pH 7.6, 0.1 M magnesium chloride
- Deoxynucleotide mix: 1 mM each of dATP, dGTP, and dTTP[a]
- [α-^{32}P]dCTP 10 μCi/μl[a]
- Klenow DNA polymerase, 2 units/μl.

Method

1 Add between 50 and 250 ng of DNA to be labelled to a 1.5 ml microcentrifuge tube.

2 Add 10 μl of random sequence hexameric oligonucleotide mix and sterile distilled water to a total volume of 33 μl.

3 Incubate the tubes at 100 °C for 5 min[b], chill immediately on ice for 5 min, and spin briefly at full speed in a microcentrifuge to collect the condensate.

4 Add 5 μl of 10× labelling buffer, 5 μl of deoxynucleotide mix, and 5 μl of the [α-^{32}P]dCTP to each tube.

5 Add 2 μl (4 units) of Klenow DNA polymerase.

6 Incubate at 37 °C for 30 min.

7 Purify probes from unincorporated nucleotides using spin columns.

8 Mix the Sephadex G50 slurry and add 600 μl aliquots to empty disposable spin columns. Place spin columns in 2 ml centrifuge tubes and centrifuge these at 600 g for 1 min in a micro-centrifuge.

Protocol 9 continued

9 Immediately transfer the spin columns to 1.5 ml centrifuge tubes and add the reaction from step 6 to the centre of the Sephadex G-50 matrix.

10 Centrifuge the spin columns at 600 g for 2 min. Radiolabelled probe is present in the excluded fraction collected from the spin column.

11 Quantitate the probe by scintillation counting.

a If a radiolabelled nucleotide triphosphate other than $[\alpha\text{-}^{32}P]dCTP$ is to be used, the deoxynucleotide mix should be modified accordingly.

b Either use screw cap tubes or tubes with lid locks to prevent the lid popping open during the incubation.

Protocol 10

Hybridization

Equipment and reagents

- Formamide
- 10 mg/ml denatured sonicated salmon sperm DNA
- 20× SSPE: 0.2 M sodium phosphate, 2.98 M sodium chloride, 20 mM EDTA, pH 7.4
- 20× SSC: 3 M NaCl, 0.3 M tri-sodium citrate
- 50× Denhardts reagent: 1% (w/v) Ficoll, 1% (w/v) polyvinylpyrrolidone, 1% (w/v) BSA

- 10% SDS
- Hybridization buffer: 5 ml formamide, 100 μl 10 mg/ml denatured sonicated salmon sperm DNA, 3 ml 20× SSPE, 0.5 ml 100× Denhardts reagent, 0.5 ml 10% SDS, 0.9 ml H_2O. This provides 10 ml of hybridization buffer
- Rotating hybridization oven[a]

Method

1 Wet the nylon or nitrocellulose filter from *Protocol 8* in 2× SSC and transfer to hybridization bottle.

2 Prehybridize the filter at 42 °C for at least 1 h in hybridization buffer.

3 Add denatured probe[b] and hybridize at 42 °C overnight.

4 Wash the blot three times for 15 min each time in 2× SSC, 0.1% SDS prewarmed to 42 °C.

5 Wash once with 0.1× SSC, 0.1% SDS at 60 °C for 10 min for a final high stringency wash.

6 Seal the wet membrane in a plastic bag and autoradiograph at −80 °C with intensifying screens for appropriate times.[c]

a A number of manufactures supply rotating hybridization ovens. Alternatively, filters can be prehybridized and hybridized in sealed flat plastic pouches.

b Denature the probe by placing the tube in a boiling water bath for 5 min, chill on ice, and spin briefly in a microcentrifuge to recover any condensation. Use a screw cap microcentrifuge tube or one with a lid lock to prevent the tube popping open during incubation.

c By keeping the filter damp, it may be rewashed at a higher stringency at a later date. If the filter is allowed to dry the probe will become fixed to the filter and cannot be removed. Filters may be reused a number of times after stripping the probe by adding with boiling 0.1% SDS, allowing to cool for 15 min and repeating.

5.2 Reverse Northern blotting

Northern blotting is the standard procedure for comparing the quantities of a transcript in different tissues. However, such data may be required for a large number of different mRNAs and reverse northern blotting provides an approach for such analyses. Micro-array analysis (*Chapter 3*) is based on the same principle as reverse northern blotting. Both approaches are performed by immobilizing sets of cDNA sequences representing defined transcripts onto hybridization membranes for reverse northern blotting and glass slides for micro-array analysis. Although micro-arrays provide the opportunity for the analysis of large numbers of transcripts, protocols for reverse northern blotting are included here (*Protocols 11* and *12*) as many laboratories do not yet have access to the necessary equipment for micro-array analysis, or may wish to analyse only a small subset of sequences. Reverse northern blotting offers a quick, cheap and reliable approach for multiple transcript analysis. The cDNAs may either be arrayed as dot blots using a dot-blot manifold, or capillary blotted from agarose gels. The reverse northern blots are then differentially screened with labelled, first-strand cDNA probes prepared from mRNA samples from the tissues of interest (*Protocols 13* and *14*). Some cDNAs from messages that are deemed to be equally abundant in the tissues under examination should be included on reverse northern blots to control the exposure times of autoradiographs. Typically, cDNAs for ATPase, actin, or tubulin can be used for this purpose.

In a reverse northern blot experiment, the cloned cDNAs arrayed on the membrane will be in large excess over their homologues within the labelled first-strand cDNA probes. For this reason, the exact amount of DNA present in each cDNA dot or band on a reverse northern membrane is not a critical factor. For the same reason, the activity of a first-strand cDNA probe, rather than its specific activity, will determine the signal strength achieved in a reverse northern filter experiment. This is the opposite of conventional northern hybridizations.

The protocols below give two options for the construction of reverse northern blots: cDNAs can be arrayed as duplicate sets of DNA dot-blots (*Protocol 11*) or can be blotted bi-directionally from agarose gels onto two hybridization membranes simultaneously (*Protocol 12*). The method for dot-blotting given is based on that of Marsluff and Huang (4), though any standard plasmid dot-blotting procedure could be used. The alternative bi-directional blotting procedure detailed below is based on that of Smith and Summers (5). The cDNAs to be blotted may be prepared from recombinant plasmids by their linearization using appropriate restriction endonucleases. Alternatively, the DNA samples to be dot-blotted may be prepared in PCR reactions using either recombinant λ bacteriophage or recombinant plasmids as templates. PCR reactions for this purpose are performed using pairs of standard primers such as M13 sequencing primers, SP6, T3, or T7 promoter primers that have sites flanking the cDNA insertion site in the cloning vector in use. One microlitre of bacterial culture or λ bacteriophage suspension generated from a single plaque is normally sufficient as a template in such PCR reactions without any DNA extraction procedure being necessary.

Protocol 11

Preparation of reverse Northern dot blots

Equipment and reagents

- 20× SSC: 3 M NaCl and 0.3 M trisodium citrate, pH 7.0
- Sterile distilled water
- Nylon hybridization membrane
- Denaturation solution: 0.2 M NH_4OH (1% by volume from '880' ammonia), 2 M NaCl
- Dot blot manifold
- UV cross-linker

Protocol 11 continued

Method

1 Cut the required number of nylon hybridization membranes to size, pre-wet them in distilled water, and soak in 20× SSC for at least 10 min.

2 Dilute aliquots of linearized plasmids to 100 μl with distilled water.[a] Use multiples of these volumes for the number of replicate dot-blots to be produced.

3 Add 20 μl of denaturation solution per 100 μl of DNA solution to each sample and mix.

4 Heat the samples to 70 °C for 2 min and place directly on ice.

5 Clamp the first nylon membrane into the dot-blot manifold. Adjust a water (or other) pump to give a sample flow rate through the membrane of approximately 20 μl/s. Apply each 100 μl sample to the dot-blot manifold, followed by 200 μl 2× SSC. (Repeat this procedure for each replicate dot-blot membrane to be produced.)

6 Allow the dot-blots to dry, dismantle the dot blot manifold, and UV cross-link the cDNAs to the membrane using UV doses recommended for the nylon membrane type in use (e.g. 70 000 J/cm^2 for Hybond-N, Amersham).

[a] Typically, 1/10 of the yield of a high copy-number plasmid DNA mini-preparation from 1.5 ml of bacterial culture, or 10 μl of a PCR reaction product.

Protocol 12

Bi-directional reverse Northern capillary blots

Equipment and reagents

- 1× TAE buffer: 1 l of 50× TAE buffer stock contains 242 g Tris base, 57 ml glacial acetic acid, and 100 ml from a stock solution of 0.5 M EDTA (pH 8.0)
- Agarose
- 0.5 μg/ml ethidium bromide
- Denaturing solution: 1.5 M NaCl, 0.5 M NaOH
- Neutralizing solution: 1.5 M NaCl, 0.5 M Tris–HCl pH 7.0
- Electrophoresis equipment

Method

1 Excise cDNA inserts from plasmids with appropriate restriction endonucleases, or prepare PCR products from cloned cDNAs in bacterial or phage stocks.

2 Prepare a 1% agarose gel in 1× TAE buffer containing 0.5 μg/ml ethidium bromide.

3 Adjust the volume of the DNA samples to be analysed to approximately 3/4 of the capacity of the wells.[a]

4 Fractionate the samples by electrophoresis at 2–4 V/cm and photograph the gel.

5 Denature the cDNA samples by soaking the gel twice for 20 min each time in several gel volumes of denaturing solution. Re-neutralize the gel by soaking it twice for 20 min each time in several gel volumes of neutralizing solution.

6 Set up a bi-directional capillary blot by first placing a layer of paper towels of approximately 2.5 cm thickness on a flat, level, non-porous surface.

7 On top of the paper towels place four layers of Whatman 3MM chromatography paper cut to the size on the gel and soaked in 20× SSC.

Protocol 12 continued

8 Place a piece of nylon hybridization membrane (e.g. Hybond-N, Amersham) that has been cut to the size of the gel, pre-wetted in distilled water and soaked in 20× SSC for 10 min on top of the Whatman 3MM paper.

9 Place the gel on top of the nylon membrane taking care to exclude air bubbles.

10 Place a second hybridization membrane treated as described in step 8 on top of the gel.

11 Place a further four sheets of 3MM paper, soaked in 20× SSC followed by a second layer of paper towels on top of the assembled blot.

12 Place a glass plate and a weight of approximately 0.5 kg/100 cm^2 of gel area on top of the assembly and allow transfer to take place for at least 6 h, or overnight if convenient.

13 Dismantle the assembly and fix the cDNAs to the reverse Northern blots by UV exposure, as described in *Protocol 11*.

[a] It is essential that the samples loaded into the wells of a gel that will be bi-directionally blotted are all of the same volume and that the wells are filled almost to capacity to ensure that each sample is blotted equally in both directions.

Protocol 13

cDNA synthesis and purification for reverse northern probes

Equipment and reagents

- Poly A^+ RNA for probe preparation
- 0.5 μg/μl oligo $dT_{(12-18)}$ primer (40 μM)
- Sterile distilled water
- 20 mM dithiothreitol (DTT)
- 5× reverse transcriptase buffer (e.g. Life Technologies Ltd)
- dNTP mixture: 10 mM each of dATP, dCTP, dGTP and dTTP
- Superscript II reverse transcriptase, 200 units/μl (Life Technologies Ltd)

- 1× TE buffer: 10 mM Tris-HCl and 1 mM EDTA, pH 7.6
- Sephadex G50: prepare as a slurry by mixing 3 g of Sephadex-G50 with 50 ml of 1× TE, and either incubate the mixture overnight at 4 °C or autoclave it for 10 min. Mix again immediately before loading the Sephadex G-50 into spin columns.
- Empty disposable spin column units (e.g. Bio Rad product number 732–6204)

Method

1 For each RNA sample, add 1 μg of poly A^+ RNA, 1 μl of oligo $dT_{(12-18)}$ primer and sterile distilled water to a total volume of 11 μl to a sterile microcentrifuge tube.

2 Incubate at 70 °C for 2 min and place immediately on ice for 2 min then spin briefly in a microcentrifuge to collect the condensation.

3 To each tube add 4 μl of 5× reverse transcriptase buffer (Life technologies Ltd), 2 μl 20 mM DTT, 2 μl dNTP mixture and 1 μl (200u) Superscript II reverse transcriptase.

4 Mix gently and centrifuge briefly to ensure all components are at the bottom of the tube.

5 Incubate the cDNA synthesis reactions in an air incubator at 42 °C for 1 h.

6 Separate the unincorporated nucleotides from the first strand cDNA using Sephadex G50 spin columns.[a]

Protocol 13 continued

7 Mix the Sephadex G50 slurry and add 600 μl aliquots to empty disposable spin columns.

8 Place spin columns in 2 ml centrifuge tubes and centrifuge them at 3000 rpm for 1 min in a microcentrifuge.

9 Immediately transfer the spin columns to 1.5 ml centrifuge tubes and add the reaction from step 6 to the centre of the Sephadex G-50 matrix.

10 Centrifuge the spin columns at 600 g for 2 min. The cDNAs will be found in the excluded fractions collected below the spin columns.

11 Check an aliquot of 2 μl of each purified first-strand cDNA on a 1% agarose gel.

12 Radiolabel the first strand cDNA as outlined in *Protocol 9* using between 50 and 250 ng of the products of first-strand cDNA synthesis.[b]

[a] cDNA must be purified from unincorporated nucleotides following cDNA synthesis as these will otherwise reduce the incorporation of labelled nucleotides in the subsequent labelling reaction.

[b] Five microlitres of the eluate from a spin column will typically contain approximately 50–250 ng of cDNA.

Protocol 14

Reverse northern blot hybridization

Equipment and reagents

- 20× SSC: 3 M NaCl, 0.3 M trisodium citrate, pH 7.0
- 50× Denhardt's reagent: 1% (w/v) bovine serum albumen (fraction V), 1% (w/v) polyvinyl pyrrolidone, and 1% (w/v) Ficoll type 400
- 10% SDS
- 0.5% (w/v) tetrasodium pyrophosphate

- 10 mg/ml denatured, sonicated herring testes DNA
- Hybridization solution: 5× SSC, 5× Denhardt's Reagent, 0.1% (w/v) SDS, 0.05% (w/v) tetrasodium pyrophosphate, 20 μl/ml denatured, sonicated herring testes DNA
- Rotary hybridization oven[a]

Method

1 Pre-hybridize reverse Northern blots in hybridization solution for at least 2 h at 65 °C in a rotary hybridization oven or in heat-sealed polythene bags within sealed containers in a shaking water bath.

2 Denature probes from *Protocol 13* by incubation at 100 °C for 5 min, place these on ice for 1 min and centrifuge briefly in a microcentrifuge to collect condensation.

3 Add the probes directly to the pre-hybridization solutions.

4 Hybridize[b] with rotation (or shaking) at 65 °C overnight.[c]

5 Remove the hybridization solution containing the probe and wash membranes in 50 ml volumes, of 2× SSC, 0.1% SDS at 65 °C for 15 min.

6 Repeat four times.

7 Wash the membranes with a high-stringency wash in 0.1× SSC, 0.1% SDS at 65 °C for 10 min.

8 Seal the wet membrane in a plastic bag[d] and autoradiograph at −80 °C with intensifying screens for appropriate times.[e]

[a] For blots of up to 400 cm², 25 ml of pre-hybridization solution is sufficient.

Protocol 14 continued

b Hybridization should be for at least 15 h.

c A number of manufactures supply rotating hybridization ovens; alternatively filters can be prehybridized and hybridized in sealed flat plastic pouches.

d By keeping the filter damp, it may be rewashed at a higher stringency at a later date. If the filter is allowed to dry the probe will become fixed to the filter and cannot be removed. Filters may be reused a number of times after stripping the probe by adding boiling 0.1% SDS, allowing to cool for 15 min and repeating.

e Due to their very high sensitivity, Kodak Biomax MS-1 films used in conjunction with Kodak Biomax MS intensifying screens are highly recommended for reverse Northern blotting.

5.3 Virtual northern blotting

Virtual northern blotting is useful when the relative amounts of a transcript in different tissues are to be compared, but total RNA or poly A+ RNA samples cannot easily be obtained in sufficient quantities for a conventional northern blot (16). The procedure for a virtual northern blot is to extract small quantities of total RNA from the tissues to be compared, make full length first-strand cDNA from these (*Protocol 15*), and then amplify each cDNA sample in its entirety by PCR (*Protocol 16*). Full length cDNA for this procedure can be produced using the BD Biosciences Clontech SMART™ cDNA synthesis system. Equal quantities of each PCR product are then fractionated on agarose gels, transferred to nylon membranes, and probed with a radiolabelled DNA probe corresponding to the mRNA of interest (*Protocol 17*). The resulting virtual northern blot is, in reality, a Southern blot, as the immobilized nucleic acid is DNA, rather than RNA. However, this type of blot resembles a true northern blot in two respects. First, it shows bands corresponding to the sizes of mRNA transcripts and, secondly, the relative intensities of those bands in different tracks reflects mRNA abundance in the tissues from which the RNA samples were prepared. It is possible that first-strand cDNA synthesis and PCR amplification can introduce a degree of bias into virtual northern experiments. However, in our laboratory we have investigated a number of constitutive and differentially-expressed genes, and found comparable results between virtual and conventional northern blotting experiments.

RNA can be isolated for virtual northern blotting by any standard procedure, but if virtual northern blot analysis is required it is likely that small-scale RNA preparations will be performed. Suitable methods include that of Logemann *et al.* (1), see *Protocol 5* and various kit-based methods. It is not necessary to perform poly A+ purification for virtual northern blotting.

The amplification of first-strand cDNAs for use in virtual northern analysis is facilitated by the incorporation of two primer sites at the termini of cDNA molecules during first-strand cDNA synthesis (*Protocol 15*). At the 3′-terminus, this oligonuleotide consists of an oligo-dT sequence, which anneals to the poly A tails of mRNA molecules and primes first-strand cDNA synthesis. This cDNA synthesis primer contains a 5′-terminal extension of 15 bases to facilitate subsequent PCR amplification of the cDNAs and two degenerate bases at its 3′-end to anchor it at the first base of the poly A tail of mRNAs. During first strand cDNA synthesis, a terminal transferase activity associated with MMLV reverse transcriptase catalyses the addition of a small number of dC residues to the 3′-termini of first strand cDNA molecules. These residues are only added at the 5′ end of the cDNA once the terminal transferase has reached the end of the transcript template. This is the basis of the BD Biosciences Clontech SMART™ cDNA synthesis system[a]. Incorporation of a template-switching oligonucleotide that contains three rG or dG residues at its 3′ terminus in the reverse transcription reaction leads to incorporation of the complementary sequence of this oligonucleotide (6, 7).This procedure results in the production of full length cDNA sequences containing specific oligonucleotide primer sequences at their 3′ and 5′ termini. BD Biosciences Clontech market the kits and reagents for this procedure as SMART™ cDNA synthesis system.

To amplify first-strand cDNA samples in their entirety, PCR reactions are assembled containing primers that anneal to the incorporated primer sequences of the cDNA molecules (*Protocol 16*). The PCR products formed are mixtures representing all of the mRNA molecules expressed in each tissue under analysis. We have found that if sufficient PCR cycles are carried out, typically 30–35, the total yields of PCR products from different reactions appear very similar following agarose gel analysis. There is a possibility that the PCR reactions may become exhausted before the last cycle, in which case the complex mixtures of PCR products formed may be single-stranded as a result of high temperature incubation steps during the last few rounds of PCR. This will not affect the relative abundances of the different cDNAs present, but will cause them to run at different rates to those of double-stranded molecules on non-denaturing agarose gels. For this reason, we use denaturing alkaline agarose gels (8) for virtual northern analyses (*Protocol 17*), modified from the method given by Sambrook *et al.* (9). On these gels, all DNA samples run in their single-stranded state. Size markers, such as phage lambda DNA restriction fragments or commercially available DNA ladders, can be used on alkaline agarose gels. These can be visualized either by ethidium bromide staining of the appropriate part of the gel following neutralization or, more conveniently, after transfer of the gel to a hybridization membrane, by incorporating a small amount of size-marker DNA in labelling reactions together with the hybridization probe.

[a]SMART™ technology is covered by US Patents #5, 962, 271 and 5,962,272

Protocol 15

cDNA synthesis for virtual Northern blotting

The oligonucleotides used to prime first-strand cDNA synthesis and template-switching in the virtual northern blot protocol given here contain an oligo-dT stretch and three 3′-terminal dG residues, respectively. The remaining nucleotide stretches of these oligonucleotides were designed to contain sequences that could be amplified using available primers. These sequences differ from those offered in the BD Biosciences Clontech SMART™ cDNA system and we have not compared their efficacy directly. It is possible that other convenient sequences could also be used as PCR primers for the subsequent PCR amplification step; these could be constructed to contain restriction sites, which could be useful in other applications of the general technique. In our hands, the oligonucleotide sequences given below worked well, however, if this is your first time doing cDNA synthesis or virtual northern blots we recommend to use commercially available primers and reagents from the BD Biosciences Clontech SMART™ cDNA synthesis kit.

Equipment and reagents

- cDNA synthesis primer (10 μM) 5′-GTAACAACGCAGAG(T)$_{18}$VN, where V indicates a mixture of bases A, C, and G, and N indicates a mixture of all four bases
- Template-switching primer (10 μM) 5′-TCCAATGACTACGGCATCCTACAACGCGGG
- Nuclease-free sterile distilled water
- 5× reverse transcriptase buffer)
- 20 mM DTT
- Superscript II reverse transcriptase, 200 units/μl (Gibco-BRL).
- dNTP mixture: 10 mM each of dATP, dCTP, dGTP, and dTTP

Protocol 15 continued

Method

1 For each RNA sample, mix 1µg of total RNA, 1 µl of cDNA synthesis primer and 1 µl of template-switching primer, and nuclease-free sterile distilled water to a final volume of 5 µl in a sterile 0.5 ml centrifuge tube.

2 Incubate the tubes at 70 °C for 2 min, place on ice for 2 min, and centrifuge briefly to collect condensation.

3 Add to each tube 2 µl 5× reverse transcriptase buffer, 1 µl 20 mM DTT, 1 µl dNTP mixture, and 1 µl (200 units) Superscript II reverse transcriptase.

4 Gently mix and spin briefly to ensure all the reaction components are at the bottom of the tube.

5 Incubate the tubes in an air incubator at 42 °C for 1 h.

6 Dilute the samples to 50 µl volumes with sterile distilled water.

7 Store the samples at −20 °C until required.

Protocol 16

Virtual northern PCR amplification

Equipment and reagents

- 10× Taq polymerase buffer: 500 mM KCl, 100 mM Tris.Cl, pH 8.3 at room temperature, 15 mM $MgCl_2$, 0.1% gelatin
- dNTP mixture: 10 mM each of dATP, dCTP, dGTP, and dTTP
- Virtual northern PCR primer A[a] 5′-GTAACAACGCAGAGT, 2 µM
- Virtual northern PCR primer B[b] 5′-GAGATC-TAGAATTCAATGACTACGGCATCCTACA, 2 µM
- Taq DNA polymerase, 0.5 u/µl

- Agarose
- 1× TAE buffer: 1 l of 50× TAE buffer stock contains 242 g Tris base, 57 ml glacial acetic acid, and 100 ml from a stock solution of 0.5 M EDTA, pH 8.0
- 3 M sodium acetate, pH 5.2
- Absolute ethanol
- 70% (v/v) ethanol

Method

1 Mix 1 µl first strand cDNA (*Protocol 15*) with 5 µl Taq polymerase buffer, 1 µl dNTP mixture, 5 µl each of virtual northern PCR primers A and B, 1 µl (0.5 units) Taq DNA polymerase in a total volume of 50 µl.[c]

2 Perform PCR amplifications for 30–35 cycles of: 94 °C for 30 s; 55 °C for 30 s; 72 °C for 3min 30 s.[d]

3 Check aliquots (5–10 µl) of each sample on a non-denaturing 1% agarose 1× TAE gel to verify that smears of PCR product are of similar appearance and concentration from all samples.[e]

4 Precipitate equal amounts of each PCR product by addition of 1/10 volume 3 M sodium acetate and 2.5 vols ethanol.

5 Incubate at −20 °C for 30 min.

6 Collect the DNA by centrifugation in a microcentrifuge at 13 000 g for 15 min.

7 Wash the DNA pellet in 70% ethanol and recover the pellet by centrifugation in a microcentrifuge at 13 000 g for 15 min.

Protocol 16 continued

8 Dry the DNA pellets under vacuum and resuspend in 20 μl sterile distilled water.

9 Quantify the DNA concentrations by spectrophotometry.

[a] This primer anneals to 3′ end of cDNA.

[b] This primer anneals to 5′ end of cDNA.

[c] Do not use hot-start PCR methods.

[d] The extension time of 3 min 30 s can be reduced to 2 min 30 s if the transcript of interest is known to be under 3 kb in length.

[e] If the yield from a single reaction is insufficient for virtual northern blotting, multiple reactions of 50 μl or 100 μl volumes can be performed and their products pooled following amplification.

[f] Between 1 and 5 μg of each PCR product will be needed for each lane for the virtual northern gel.

Protocol 17

Virtual northern blot preparation and hybridization

Equipment and reagents

- Sterile distilled water
- 1× Alkaline agarose gel running buffer: 50 mM NaOH, 1 mM EDTA
- 6× alkaline agarose gel loading buffer: 300 mM NaOH, 6 mM EDTA, 18% (w/v) Ficoll type 400, 0.15% (w/v) bromocresol green
- Neutralizing solution: 1.5 M NaCl and 0.5 M Tris/HCl, pH 7.0, 20× SSC: 3 M NaCl, 0.3 M trisodium citrate, pH 7.0

Method

1 Prepare a 1% agarose gel in distilled water.[a]

2 Soak the gel twice for 15 min each time in five gel volumes of 1× alkaline running buffer.

3 Dilute 1–5 μg of PCR products (*Protocol 16*) to 20 μl with sterile distilled water.

4 Add 5 μl of 6× alkaline agarose gel loading buffer and mix.

5 Load the samples onto the gel and fractionate at 1.5 V/cm in 1× alkaline running buffer until the bromocresol green tracking dye has migrated 2/3 of the length of the gel.

6 Neutralize the gel by soaking in neutralizing solution for 30 min. If required, cut marker tracks from gel for ethidium bromide staining and visualization under UV light.

7 Set up a Southern capillary blot in 20× SSC to transfer the DNA onto a nylon membrane (such as Hybond-N; Amersham) (see *Volume 1, Chapter 6*).

8 Immobilize the DNA on the membrane by UV cross-linking.

9 Prepare a radioactively labelled DNA probe (*Protocol 9*).

10 Perform blot pre-hybridization, hybridization to the radio-labelled probe, post-hybridization washes, and autoradiography as described in *Protocol 14*.

[a] Use a 1.5% gel if the mRNA/cDNA to be analysed is shorter than 2 kb.

6 Nuclease protection assays

The various northern blotting approaches described in Section 5 are relatively quick and easy ways of detecting the expression levels of mRNAs. However, these techniques do not discriminate between closely-related sequences. Thus, unless sufficient information is available to enable a gene-specific probe to be designed, there is a risk that the probe will cross-hybridize with other related transcripts. If the cross-hybridizing transcript is same size as the authentic one, an erroneously high expression level will be observed, if it is a different size, multiple signals will be seen on the blot. A further possible limitation of northern analysis is sensitivity for the detection of low abundance transcripts. The maximum loading capacity of agarose gels is 20–30 μg RNA. Although problems of detection can be overcome in some cases by the use of poly A$^+$ northern blots or by virtual northern analysis, the potential problems of cross-hybridization remain. Two alternative methods that provide sensitive and gene specific transcript detection are RNase and S1 nuclease protection assays.

In both RNase and S1 protection assays, a single stranded probe, either RNA or DNA, respectively, is designed that will hybridize to the desired transcript. When this probe is hybridized to a population of RNA molecules it will anneal to its homologous transcript, but may also hybridize to similar, but non-identical sequences. In this latter scenario, there will be residues that do not base-pair between the probe and transcript resulting in regions of the probe that remain single-stranded. This single stranded sequence is a template for S1 nuclease if single-stranded DNA probe is used or RNase A if a single-stranded RNA probe is used. The portions of the probe that anneal to form a double stranded molecule are protected from nuclease digestion. If the sequences of the transcribed genes that cross-hybridize to the probe are known, then the sizes of the resulting RNase or S1 nuclease digested products can be predicted. If the sequence of the gene of interest is known the protected size of the gene-specific probe can be defined and it will be unlikely that annealing of the probe to another mRNA will give the same protected fragment size. In addition, as the hybridization of the probe to the RNA is done in solution, up to 100 μg of RNA can be assayed. These techniques, therefore, have advantages over northern analysis in terms of their sensitivity and gene specificity.

A good example of where the use of S1 nuclease protection assays has revealed the differing expression patterns of different gene family members is that of Millar and Kay (10). The *CAB* gene family in *Arabidopsis* consists of *CAB1*, *2*, and *3*, and RNA slot blots probed with a sequence that detected all three genes showed only minor changes in expression over a 24-h period. However, by using S1 nuclease protection assays with a probe that could distinguish *CAB2/CAB3* from *CAB1* they found that the expression of *CAB1* changed little over this time course, whereas *CAB2/CAB3* showed a dramatic circadian expression pattern.

6.1 RNase protection assays

For RNase protection analysis an antisense riboprobe must be synthesized (*Protocols 18–21*). Probe synthesis is typically achieved from a plasmid clone in which the insert is flanked by binding sites for the RNA polymerases T3, T7, or SP6. Plasmid constructs containing either cDNA sequences or fragments of genomic clones can be used to generate probes that ideally should be about 200–300 nucleotides long. If genomic fragments containing introns are used, the mismatch between intron and transcript will provide regions susceptible to RNaseA digestion that will enable a size difference to be observed between the probe and protected products. If a cDNA is used as a template for the probe, a sufficient length of polylinker sequence should be incorporated into the probe to provide a detectable difference between the size of the probe before and after digestion with RNaseA. This is a useful control as it

25

demonstrates that RNase digestion has worked. The production of probes corresponding to the 3′ end of the gene of interest can be achieved by in vitro transcription of an antisense transcript containing the 3′ untranslated sequence including some vector sequence downstream of the poly A tail. For production of 5′ gene specific probes, a 200–300 bp fragment from the 5′ end of the cDNA should be subcloned into an in vitro transcription vector. The internal region of the cDNA should be ligated adjacent to the T3, T7, or SP6 promoter such that in vitro transcription will proceed through the 5′ end of the cDNA and into vector sequence beyond.

Although 200–300 bases is optimal, the size of the probe will likely be limited by the choice of restriction sites and we have successfully used probes over 600 nucleotides long. We have found that in the following reaction conditions for probe preparation, T3 polymerase provides better in vitro transcription than T7 polymerase. We have found that SP6 polymerase does not work well with the following conditions. Since the probe must contain sequence complementary to the RNA transcript, the choice of polymerase will probably be determined by the orientation of the insert within the vector. One consequence of producing an antisense probe from the 3′ end of a cDNA is that if the clone contains a 3′ untranslated region (UTR), this will be incorporated into the probe. The presence of this 3′ UTR will most likely act as a gene-specific probe to enable discrimination between transcripts from different gene family members. To make the probe it is also important to linearize the template to give a transcript of a defined size. The restriction enzyme used should give either blunt ends or recessed 3′ ends. The linearization of plasmids with restriction enzymes that leave recessed 5′ ends is best avoided. There are reports that 3′ overhangs inhibit the RNA polymerases, and also that RNA polymerase can misprime off a 3′ overhang and make top strand contaminants that will anneal to the probe and produce strong background protected bands. Protocols 18–23 describe the preparation of riboprobes, hybridization, digestion and analysis of products.

Protocol 18
Riboprobe template preparation

Equipment and reagents
- Sterile DEPC treated water (Protocol 1)
- Appropriate restriction enzymes
- Appropriate restriction enzyme buffers
- 10 mg/ml proteinase K
- Agarose
- 10 mg/ml ethidium bromide solution[a]
- 3 M sodium acetate, pH 5.2, (DEPC treated, Protocol 1)
- Phenol:chloroform:isoamyl alcohol (25:24:1)
- 1× TAE buffer: 1 l of 50× TAE buffer stock contains 242 g Tris base, 57 ml glacial acetic acid, and 100 ml from a stock solution of 0.5 M EDTA (pH 8.0)
- Absolute ethanol
- 70% (v/v) Ethanol

Method
1 Linearize 10 μg of plasmid template in 50 μl restriction enzyme reaction.[b]
2 Add 1 μl 10 mg/ml proteinase K.
3 Incubate at 37 °C for 15 min.
4 Remove 1 μl of the digest and check that it is linearized on a 1% TAE agarose gel stained with ethidium bromide.[a]
5 Extract the linearized plasmid DNA twice with an equal volume of phenol/chloroform/isoamyl alcohol separating the phases each time by centrifugation at 13 000 g for 5 min.

Protocol 18 continued

6 Precipitate the DNA from the aqueous phase by addition of 1/10 volume 3 M sodium acetate and 2 vols ethanol.

7 Incubate at −20 °C for 30 min.

8 Collect the DNA by centrifugation in a microcentrifuge at 13 000 g for 15 min.

9 Wash the DNA pellet in 70% ethanol and recover the pellet by centrifugation in a microcentrifuge at 13 000 g for 15 min.

10 Dry the DNA pellets under vacuum and resuspend in 10 μl DEPC H$_2$O.

11 Quantitate the DNA concentrations by comparison with DNA standards on an agarose gel.

[a] Ethidium bromide is toxic and carcinogenic, necessary precautions should be taken.

[b] The restriction enzyme used should provide a probe of between 200–300 nucleotides in length and should leave either recessed 3′ or blunt ends as protruding 3′ ends lead to problems with probe preparation.

Protocol 19

Test transcription using linearized template

Equipment and reagents

- 1 μg/μl linearized template (*Protocol 18*)
- 5× transcription buffer: 200 mM Tris pH 8.0, 40 mM MgCl$_2$, 10 mM spermidine, 250 mm NaCl
- ATP stock: 10 mM ATP
- GTP stock: 10 mM GTP
- CTP stock: 10 mM CTP
- UTP stock: 10 mM UTP
- 0.75 M DTT
- Sterile DEPC treated H$_2$O (*Protocol 1*)
- RNA polymerase T3, T7 or SP6 as appropriate (10 U/μl)

- RNase-free DNase (10 U/μl)
- Phenol:chloroform:isoamyl alcohol (25:24:1)
- 3 M sodium acetate, pH 5.2, (DEPC treated, *Protocol 1*)
- Absolute ethanol
- 70% (v/v) ethanol
- Agarose
- 1× TAE buffer: 1 l of 50× TAE buffer stock contains 242 g Tris base, 57 ml glacial acetic acid and 100 ml from a stock solution of 0.5 M EDTA (pH 8.0)
- 10 mg/ml ethidium bromide solution[a]

Method

1 Mix 1 μl (1 μg) linearized template with 5 μl 5× transcription buffer, 1μl each ATP, GTP, CTP, and UTP stocks, 1 μl 0.75 M DTT, and 13 μl DEPC treated H$_2$O.

2 Mix all components and briefly spin in a microcentrifuge to collect all reagents in the bottom of the tube.

3 Add 1 μl (10 U) appropriate RNA polymerase.

4 Incubate at 37 °C for 30 min.

5 Add 1 μl (10 U) RNase free DNase.

6 Incubate at 37 °C for 15 min.

7 Extract the transcription reaction with an equal volume of phenol/chloroform/isoamyl alcohol and separate the phases by centrifugation at 13 000 g for 5 min.

Protocol 19 continued

8 Precipitate the RNA from the aqueous phase by addition of 1/10 volume 3 M sodium acetate and 2.5 times the new volume of ethanol.

9 Incubate at −20 °C for 30 min.

10 Collect the RNA by centrifugation in a microcentrifuge at 13 000 g for 15 min.

11 Wash the RNA pellet in 70% ethanol and recover the pellet by centrifugation in a micro-centrifuge at 13 000 g for 15 min.

12 Dry the RNA pellet under vacuum and resuspend in 20 μl DEPC H_2O.

13 Check 1 μl of the *in vitro* produced transcript on a 1% agarose TAE gel containing 0.25μg/ml ethidium bromide.[a] If the transcription has worked, a band of the expected size should be seen on the gel with no smearing.

[a] Ethidium bromide is toxic and carcinogenic, necessary precautions should be taken.

Protocol 20
Preparation of riboprobes

Equipment and reagents

- RNase-free screw-cap 1.5 ml microcentrifuge tubes
- 5× transcription buffer: 200 mM Tris pH 8.0, 40 mM $MgCl_2$, 10 mM spermidine, 250 mm NaCl
- 100 mM DTT
- 10× nucleotide mix: ATP 8 mM, GTP 8 mM, CTP 8mM, UTP 0.12 mM
- DEPC H_2O
- [α^{32}P]UTP (800 Ci/mmol)
- RNA polymerase T3, T7 or SP6 as appropriate (10 U/μl)
- Sterile DEPC treated H_2O (*Protocol 1*)
- RNase-free DNase (10 U/μl)
- Phenol:chloroform:isoamyl alcohol (25:24:1)
- 3 M sodium acetate, pH 5.2, (DEPC treated, *Protocol 1*)
- Absolute ethanol
- 70% (v/v) Ethanol
- 10 U/μl placental ribonuclease inhibitor (e.g. Stratagene RNase Block)

Method

1 Add the following components together in an a RNase-free screw-cap 1.5 ml microcentrifuge tube: 4 μl 5× transcription buffer, 2 μl 100 mM DTT, 1 μl (1 μg) linearized template[a], 2 μl 10× nucleotide mix. 4 μl DEPC H_2O, 5 μl [α^{32}P]UTP (800 Ci/mmol), 1 μl (10U) placental ribonuclease inhibitor.

2 Mix and recover the solution to the bottom of the centrifuge tube by briefly spinning in a microcentrifuge.

3 Add 1μl (10 U) of the appropriate RNA polymerase.

4 Incubate at 37 °C for 1 h.[b]

5 Add 1 μl RNase-free DNase.[c]

6 Incubate at 37 °C for 15 min.

7 Precipitate the probe by addition of 2.1 μl 3 M sodium acetate and 2.5 vols ethanol. Incubate at −20 °C for 30 min.

Protocol 20 continued

8 Collect the RNA by centrifugation in a microcentrifuge at 13 000 g for 15 min.

9 Dry the RNA pellet under vacuum.

10 If the probe is to be used directly, resuspend in 20 μl DEPC H_2O.[d]

11 If the probe is to be gel purified, proceed to *Protocol 21*.

[a] It may be necessary to add more than 1 μl of linearized template depending on amount obtained from *Protocol 18*. If more than 1 μl is needed, reduce the volume of H_2O accordingly.

[b] Incubation at 10 °C has been reported to give a higher proportion of full-length probe.

[c] It is important to remove template DNA before hybridization as this could hybridize to the probe.

[d] This type of probe may also be used for probing RNA or DNA blotted onto filters, and as such there is no need for any further purification.

Protocol 21

Gel purification of radiolabelled probe

An optional step after synthesis of a labelled riboprobe as outlined in *Protocol 20*, is to gel purify the probe. This procedure serves three purposes. It enables a check on probe synthesis, it eliminates any products that are not the correct size that could give spurious bands when analysing the protection products and unincorporated radiolabel is also removed. Particular caution should be exercised while performing this technique as the reactions contain radiolabelled nucleotides and a number of manipulations are required.

Equipment and reagents

- Urea
- 40% acrylamide:bis acrylamide (19:1)[a]
- 5× TBE: 450 mM Tris–HCl , 450 mM boric acid, 10 mM EDTA
- 10% Ammonium persulphate (freshly prepared)

- TEMED
- Denaturing gel loading buffer: 80% formamide, 1 mM EDTA, bromophenol blue, and xylene cyanol to 1 mg/ml each
- Gel elution buffer: 2 M NH_4OAc, 1% SDS, 25 μg/ml tRNA

Method

1 Mix the following components for a 6% polyacrylamide denaturing gel: 20g urea (8.3 M final), 11.3 ml H_2O, 6 ml 40% acrylamide:bis acrylamide (19:1), 8 ml 5× TBE. The volume occupied by 20 g of urea makes this 40 ml.

2 Add 0.33 ml 10% ammonium persulphate and 23 μl TEMED.

3 Pour a vertical polyacrylamide gel between glass plates.

4 Once the gel has polymerized assemble in a vertical gel apparatus with 1× TBE running buffer.

5 Redissolve the precipitated labelled RNA probe from *Protocol 20* in 10 μl denaturing gel loading buffer.[b]

6 Heat denature the probe at 95 °C for 5 min. Use a screw cap microcentrifuge tube or one with a locking lid to prevent the tube popping open during incubation.

7 Before loading the sample onto the gel, flush wells with 1× TBE gel running buffer using a pipette to flush out urea from the wells, which is denser than the sample.

Protocol 21 continued

8 Load the probe onto the gel.

9 Run the gel at 200 V until the bromophenol blue is about two thirds of the way down the gel

10 Working with a radiation shield, split the plates apart and cut off the gel just above the bromophenol blue band with a sharp blade and carefully dispose of the lower part of the gel, which contains unincorporated radiolabel.[b]

11 Wrap gel on the glass plate in Saran wrap and place a second glass plate on top to shield the radiation.[b]

12 Working in a dark room behind a radiation shield with sides, remove the top protective glass plate

13 Working with a safe light, place a sheet of X-ray film on top of the gel making sure to align the edges of the glass plate with the X-ray film.[b]

14 Expose the film for 30 s.

15 Develop the X-ray film

16 Align the X-ray film underneath the glass plate with the gel behind a radiation shield.

17 Cut the probe bands out[c] and check that the correct region has been excised using a Geiger counter.

18 Place the gel slice containing the radiolabelled probe in screw-cap microcentrifuge tube containing 0.4 ml gel elution buffer.

19 Incubate the sample at 37 °C for 4–16 h.

20 Transfer the radioactive eluate to a clean screw-cap tube and dispose of the remaining gel slice, which will still be very radioactive.

21 Precipitate the RNA probe by addition of 1 ml 100% ethanol.

22 Collect the precipitate by centrifugation at 13 000 g in a microcentrifuge for 15 min.

23 Remove the supernatant from the radioactive pellet.[b]

24 Air-dry the pellet behind a radiation shield.

25 The probe is now ready for use. Alternatively a 70% ethanol wash of the pellet may be performed before the pellet is dried and resuspended.

[a] Caution, acrylamide is a neurotoxin, appropriate safety precautions must be taken.

[b] Caution, appropriate safety precautions must be taken when using radioisotopes.

[c] The bands should appear as a single band with no signs of probe degradation such as a smeary band.

Protocol 22

RNase protection analysis hybridization

The optimal ratio of RNA probe to the mRNA under analysis is 10:1. Under these conditions all the mRNA of interest should be being driven into mRNA:probe hybrids and so the band intensity after gel electrophoresis will reflect the relative expression level. The use of a larger excess of probe may produce additional background bands. The amount of probe to be added should be determined in a preliminary experiment by performing a probe titration against a fixed amount of RNA. The optimal amount of probe is determined by the lowest concentration that gives maximum signal strength and after which additional probe does not increase in the intensity of the protected band.

Protocol 22 continued

Equipment and reagents

- RNase-free sterile screw cap microcentrifuge tubes
- Total RNA samples
- 10 μg/μl tRNA control
- RNA–DNA hybridization buffer: 40 mM PIPES pH 6.4, 0.4 M NaCl, 1 mM EDTA, 80%

formamide. It is convenient to prepare a 5× stock of the buffer without formamide and dilute with 4 parts formamide just prior to use

Method

1 Lyophylize 10 μg sample of total RNA to be analysed[a] and 10 μg tRNA as a control in separate screw cap microcentrifuge tubes.[b]

2 Dissolve the RNA pellets in 15 μl RNA–DNA hybridization buffer.

3 Dissolve the purified probe in sufficient RNA–DNA hybridization buffer for addition of 15 μl to each sample RNA tube.[c]

4 Add 15 μl probe to each RNA sample and mix.

5 Collect the solution to the bottom of the tube by a brief spin in a microcentrifuge.

6 Denature the RNA at 85 °C for 10 min.

7 Incubate the hybridization reactions at 50 °C overnight in a hybridization oven.[d]

[a] Up to 100 μg total RNA can be used for low abundance transcript analysis.

[b] Do not dry to completion as the RNA is more difficult to re-dissolve if too dry.

[c] Save sufficient probe for use as the undigested control.

[d] It may be necessary to determine the optimal hybridization temperature by setting up parallel reactions and incubating between 37 °C and 50 °C.

Protocol 23

RNase digestion and analysis

Equipment and reagents

- RNase digestion buffer: 10 mM Tris pH 7.5, 300 mM NaCl, 5 mM EDTA
- 40 mg/ml RNase A[a]
- 2 mg/ml RNase T1[b]
- 10% SDS
- 20 mg/ml proteinase K, prepared just before use
- Phenol:chloroform:isoamyl alcohol (25:24:1)
- 10 μg/μl carrier tRNA
- Reagents for preparation of 6% polyacrylamide gel (*Protocol 21*)
- Denaturing gel loading buffer: 80% formamide, 1 mM EDTA, bromophenol blue and xylene cyanol to 1 mg/ml each

Method

1 Prepare a mixture of RNase digestion buffer containing 40 μg/ml RNase A and 2 μg/ml RNase T1[b] just before use, allowing 350 μl for each digestion reaction.

2 Add 350 μl RNase digestion buffer containing RNase to each RNA:probe hybridization reaction.

3 Incubate the samples at 30 °C for 40 min.

Protocol 23 continued

4 Add 20 μl 10% SDS.

5 Add 2.5 μl freshly made 20 mg/ml proteinase K.

6 Incubate the reactions at 37 °C for 15 min.

7 Extract the digestion reactions once with phenol:chloroform:isoamyl alcohol and separate the phases by centrifugation in a microcentrifuge for 5 min.

8 Add 1 μl (10 μg) carrier tRNA.

9 Precipitate the nucleic acids by addition of 1 ml absolute ethanol.

10 Incubate at −20 °C for 30 min.

11 Collect the precipitated RNA by centrifugation in a microcentrifuge for 15 min.

12 Remove the supernatant to leave a radioactive pellet.

13 Allow samples to air dry at room temperature on a bench or in 37 °C heat block for 15 min.

14 To analyse the samples run them on a 6% acrylamide denaturing gel prepared as described in *Protocol 21*.

15 Dissolve the dried pellets in 20 μl denaturing gel loading buffer and mix well to ensure all pellet has dissolved.[c]

16 Denature the samples at 100 °C for 3 min.

17 Flush the wells of the polyacrylamide gel with 1× TBE running buffer and load the samples onto the gel including appropriate controls[d] and radiolabelled size markers.[e]

18 Run the gel until the bromophenol blue is about 2/3 down the gel.

19 Dry the gel and autoradiograph.

[a] RNase A cleaves single-stranded RNA 3′ of C and U residues, RNase T1 cleaves on the 3′ sides of G residues.

[b] Some protocols omit T1 RNase.

[c] There should be no radioactive counts remaining in the tube after the liquid has been removed.

[d] Suitable controls should include: (1) undigested probe diluted appropriately to give a similar exposure to the sample—this sample should produce a larger molecular weight band on the gel than any of the protected bands; (2) RNase treated probe—this sample should give no protected fragments; (3) the probe hybridized to tRNA followed by RNase treatment. This sample should also give no protected fragments. Any bands produced by the above two controls will identify background effects, rather than true protected bands. Such bands may be due to the probe hybridizing to itself.

[e] Radioactive markers may be prepared by filling the end of commercial DNA size markers with a radionucleotide using Klenow polymerase. Depending on the gel conditions, RNA runs about 5–10% slower than DNA.

6.2 S1 nuclease protection assays

S1 nuclease is a single strand specific nuclease that degrades single-stranded DNA and RNA, and as such can be used for transcript mapping in an approach similar to that described for RNase protection analysis. However, S1 nuclease is more active on DNA than RNA and the probes used for S1 transcript analysis are DNA, rather than RNA (see ref. 9, for comments regarding the potential pitfalls of using S1 nuclease protection). There are a number of ways to prepare probes for S1 nuclease protection assays. dsDNA probes may be used and hybridization performed under conditions optimized to minimize the formation of DNA:DNA hybrids, but maximize the formation of RNA:DNA hybrids (11). Alternatively, ssDNA probes can be generated by linearizing a plasmid containing the probe sequence, followed by end-labelling the restriction site. A second restriction endonuclease digestion then releases the labelled probe fragment and the probe strand is obtained by purification from a strand separating gel (12). Alternatively, a primer extension reaction using a radioactively labelled oligonucleotide can be used to generate a single stranded probe followed by purification on a denaturing gel (13). The method presented below (*Protocol 24*) uses a linear amplification of a cloned gene to produce an evenly labelled single stranded probe (adapted from Ito *et al.*, ref. 14). Probe hybridization and S1 analysis are described in *Protocol 25*.

Protocol 24

Preparation of probe by primer extension for S1 nuclease analysis

Either cDNA or genomic clones may be used for the probe, but as for the RNase protection assay (Section 6.1) the probe needs to contain non-complementary sequence so that fragments protected from S1 nuclease digestion can be distinguished from undigested probe. The probe needs to be complementary to the mRNA so the appropriate vector-priming site, for example, the M13 forward or reverse primers, should be used to generate the non coding strand. Alternatively, a non-coding strand gene-specific primer may be used, but it is important to ensure that non-transcribed sequence will be included in the probe.

Equipment and reagents

- Linearized plasmid template (see *Protocol 18*)[a]
- 50 pmol/μl primer[b]
- 10$\times$ nucleotide mix: 2 mM, dGTP, 2 mM dTTP, 2 mM dATP, 25 μM dCTP
- 10$\times$ Taq polymerase buffer: 500 mM KCl, 100 mM Tris.Cl, pH 8.3 at room temperature, 15 mM MgCl$_2$, 0.1% gelatin
- 50 mM MgCl$_2$
- [α^{32}P]dCTP (800 Ci/mmol)
- 5 units/μl Taq DNA polymerase
- 3 M sodium acetate, pH 5.2, (DEPC treated, *Protocol 1*)

- Absolute ethanol
- 70% (v/v) Ethanol
- Denaturing gel loading buffer: 80% formamide, 1 mM EDTA, bromophenol blue and xylene cyanol to 1 mg/ml each
- RNA–DNA hybridization buffer: 40 mM PIPES pH 6.4, 0.4 M NaCl, 1 mM EDTA, 80% formamide. It is convenient to prepare a 5$\times$ stock of the buffer without formamide and dilute with 4 parts formamide just prior to use

Method

1 Add the following components together in a microcentrifuge tube: 200 ng linearized template dissolved in H$_2$O, 1 μl 50 pmol/μl primer, 2 μl 10$\times$ nucleotide mix, 2 μl 10$\times$ Taq polymerase buffer without MgCl$_2$, 1 μl 50 mM MgCl$_2$, H$_2$O to 17 μl.

2 Mix the reaction components and collect in the bottom of the tube by a brief spin in a microcentrifuge.

3 Add 2 μl [α^{32}P]dCTP (800 Ci/mmol) and gently mix.

4 Add 1 μl (5 units) Taq polymerase and gently mix.

5 If a thermal cycler with a heated lid is not available, overlay the reaction with one drop of mineral oil.

GRAHAM R. TEAKLE, CHARLES P. SCUTT AND PHILIP M. GILMARTIN

Protocol 24 continued

6 Place in a thermal cycler and amplify under the following conditions: one cycle of 94 °C for 3 min, and 15 cycles of 94 °C for 30 s, 50 °C for 5 s, and 72 °C for 1 min.

7 Add 1/10 volume 3 M sodium acetate.

8 Precipitate the DNA from the aqueous phase by addition of 1/10 volume 3 M sodium acetate and 2 vols ethanol.

9 Incubate at −20 °C for 30 min.

10 Collect the DNA by centrifugation in a microcentrifuge at 13 000 g for 15 min.

11 Wash the DNA pellet in 70% ethanol and recover the pellet by centrifugation in a microcentrifuge at 13 000 g for 15 min.

12 Dry the DNA pellet under vacuum and dissolve in 10 μl denaturing gel loading buffer.

13 Purify the probe on a denaturing acrylamide gel as for the RNase protection assay (see *Protocol 21*).

14 Redissolve the probe in RNA–DNA hybridization buffer (see *Protocol 22*).

[a] The enzyme used to linearize the probe plasmid will be determined by the probe sequence and also by the vector used. In order to generate a probe from a cDNA clone that will hybridize to the corresponding mRNA, the restriction enzyme site should be located within the cDNA sequence, upstream of the poly A tail.

[a] The primer used for primer extension of the probe should anneal to the vector sequence down stream of the poly A tail such that the 3′ end of the oligonucleotide can be extended to produce a complement of the coding strand.

Protocol 25
Probe hybridization and S1 nuclease analysis

For every new probe sequence used for S1 nuclease analysis it is best to determine the optimal probe and RNA concentrations, and the most favourable hybridization temperature by performing a series of test hybridizations and analysing them on a denaturing acrylamide gel prior to embarking on the analysis of a large number of samples. As with the RNase protection assay (Section 6.1) the probe needs to be in excess of the mRNA being detected to ensure optimal transcript detection. For the nuclease digestion step, a range of concentrations of S1 nuclease have been reported, generally between 100 and 2000 units/ml. A preliminary experiment to determine a suitable S1 nuclease concentration should be performed, usually towards the lower end of this range, for the suggested incubation temperature given below.

Equipment and reagents

- 10× S1 nuclease buffer: 300 mM sodium acetate pH 4.5, 2.8 M NaCl, 10 mM $ZnSO_4$
- S1 nuclease mix: 125–2500 units/ml S1 nuclease and 8 μg/ml denatured carrier DNA (for example, plasmid or salmon sperm DNA) prepared just before use.
- 0.5 M EDTA pH 8.0
- 10 μg/μl carrier tRNA
- Absolute ethanol
- Reagents for preparation of 6% polyacrylamide gel (*Protocol 21*)
- Denaturing gel loading buffer: 80% formamide, 1 mM EDTA , bromophenol blue, and xylene cyanol to 1 mg/ml each

34

Protocol 25 continued

Method

1 Set up hybridization reactions and incubate as described in *Protocol 22* using single stranded probe from *Protocol 24* and RNA samples.

2 After hybridization add 30 μl 10× S1 nuclease buffer to each reaction.

3 Add 240 μl S1 nuclease mix and incubate at 25 °C for 30 min.

4 Stop the digestion by adding 3 μl 0.5 M EDTA pH 8.0.

5 Add 1 μl (10 μg) carrier tRNA.

6 Precipitate the nucleic acids by addition of 1 ml absolute ethanol

7 Incubate at −20 °C for 30 min.

8 Collect the precipitated DNA and carrier RNA by centrifugation in a microcentrifuge for 15 min.

9 Remove the supernatant to leave a radioactive pellet.

10 Allow samples to air dry at room temperature on a bench or in 37 °C heat block for 15 min.

11 To analyse the samples run them on a 6% acrylamide denaturing gel prepared as described in *Protocol 21*.

12 Dissolve the dried pellets in 20 μl denaturing gel loading buffer and mix well to ensure that all the pellet has dissolved.[a]

13 Denature the samples at 100 °C for 3 min.

14 Flush the wells of the polyacrylamide gel with 1× TBE running buffer, and load the samples onto the gel including appropriate controls[b] and radiolabelled size markers.[c]

15 Run the gel until the bromophenol blue is about 2/3 down the gel.

16 Dry the gel and autoradiograph.

[a] There should be no radioactive counts remaining in the tube after the liquid has been removed.

[b] Suitable controls should include: (1) undigested probe diluted appropriately to give a similar exposure to the sample—this sample should produce a larger molecular weight band on the gel than any of the protected bands; (2) S1 nuclease-treated probe—this sample should give no protected fragments; (3) the probe hybridized to tRNA followed by S1 nuclease treatment—this sample should also give no protected fragments. Any bands produced by the above two controls will identify background effects, rather than true protected bands. Such bands may be due to the probe hybridizing to itself.

[c] Radioactive markers may be prepared by filling the end of commercial DNA size markers with a radionucleotide using Klenow polymerase.

7 Primer extension analysis

A commonly encountered problem when characterizing cDNA sequences isolated from cDNA libraries is that they often do not contain the 5′ end of the message and so the transcription start site cannot be inferred. Several techniques can be used to map the 5′ end of the message. When a genomic clone is available the 5′ end of the corresponding transcript can be mapped by RNase protection analysis or S1 nuclease mapping. Alternative procedures to identify the 5′ end of a transcript include 5′ RACE, and the SMART™ system employed by BD Biosciences Clontech represents an efficient means of isolating missing 5′ ends of cDNA clones. An alternative method is to use primer extension and is the method given below (*Protocols 26–29*). In this method, first strand cDNA is prepared; however, instead of using oligo(dT) as the primer, a gene-specific primer is used that anneals close to the 5′ end of the message. Primers annealing too far from the 5′ terminus of the mRNA may produce prematurely aborted extension products so it is best to use a primer that anneals within 100 nucleotides of the end. A cDNA synthesis reaction is then performed extending from this primer to the end of the mRNA. After removal of the unincorporated primer, the extension products are analysed on a sequencing gel together with a sequencing reaction, preferably of a genomic clone for the gene sequenced with the primer extension oligo. However, any sequencing reaction would give the size in bp of the extension product. The use of purified poly(A)$^+$ RNA is beneficial for low abundance transcripts, however, up to 150 μg total RNA may be used. The amount required can be determined in preliminary experiments. The products are detected by radio-labelling the primer. The primer may be a synthetic oligonucleotide in the range of 20–40 nucleotides, or it may be a double-stranded restriction fragment in the range of 75–150 bp that has been labelled by filling in the ends.

Protocol 26

Probe preparation by primer labelling reaction

Equipment and reagents

- 10 pmol synthetic oligonucleotide that has not been phosphorylated
- [γ^{32}P]ATP (10 μCi/μl)
- 10× kinase buffer: 0.5 M Tris pH 7.5, 0.1 M MgCl$_2$, 1 mM EDTA, 1 mM spermidine
- Phenol:chloroform:isoamyl alcohol (25:24:1)

- 1 M DTT
- DEPC treated sterile H$_2$O (see *Protocol 1*)
- T4 polynucleotide kinase (10 U/μl)
- STE: 100 mM NaCl, 10 mM Tris pH 8.0, 1 mM EDTA.

Method

1 Add the following together in a microcentrifuge tube: 10 pmol synthetic oligonucleotide, 5 μl (50 μCi) [γ^{32}P]ATP, 1 μl 10× kinase buffer, 0.5 μl 1 M DTT, H$_2$O to 9 μl.

2 Gently mix and recover the solution to the bottom of the tube by a brief spin in a micro-centrifuge.

3 Add 1 μl (10 units) T4 polynucleotide kinase.

4 Incubate at 37 °C for 15 min.

5 Add 60 μl STE.

6 Extract with phenol:chloroform:isoamyl alcohol (25:24:1).

7 Purify by gel filtration using for example NucTrap probe purification columns (Stratagene) or ProbeQuant™ G-50 Micro Columns (Amersham Pharmacia Biotech) following the manufacturers instructions.

8 Purify by adding an equal volume of 8 M ammonium acetate and 2.5× of the resulting volume of ice-cold ethanol.

9 Incubate on ice for 30 min.

10 Pellet the DNA by centrifugation at 13,000 g in a microcentrifuge for 30 min.

11 Carefully remove the radioactive supernatant containing the unincorporated radioactivity to leave the radiolabelled oligonucleotide in the pellet.

Protocol 27

Probe preparation by restriction fragment fill-in reaction

Equipment and reagents

- 100 pmol DNA fragment[a]
- 10× Klenow buffer: 0.4 M potassium acetate pH 7.5, 66 mM MgCl$_2$, 10 mM β-mercaptoethanol (most restriction buffers work fine)
- 1 mM dGTP stock
- 1 mM dATP stock
- 1 mM dTTP stock
- DEPC treated sterile H$_2$O (see *Protocol 1*)
- [α^{32}P]dCTP (10 μCi/μl)
- Klenow DNA polymerase (I U/μl)
- 10 μg/μl carrier tRNA
- 7.5 M ammonium acetate
- Isopropanol
- 70% (w/v) ethanol

Method

1 Mix the following components in a microcentrifuge tube: 100 pmol DNA fragment, 2 μl 10× buffer, 1 μl 1 mM dGTP, 1 μl 1 mM dATP, 1 μl 1 mM dTTP, H$_2$O to 18 μl.

2 Add 1 μl (10 μCi) [α^{32}P]dCTP and carefully mix.[b]

3 Add 1 μl (1 U) Klenow DNA polymerase.

4 Incubate at room temperature for 15 min.

5 Add 29 μl H$_2$O.

6 Add 1 μl (10 μg) carrier tRNA.

7 Add 25 μl 7.5 M ammonium acetate.

8 Precipitate the nucleic acids by addition of 45 μl isopropanol.[c]

9 Incubate at room temperature for 30 min.

10 Collect the precipitated DNA by centrifugation in a microcentrifuge for 15 min.

11 Remove the supernatant to leave a radioactive pellet.

12 Wash the pellet twice in 70% ethanol recovering the pellet each time by centrifugation as in step 10.

13 After the second 70% ethanol wash, remove the supernatant and dry the pellet under vacuum.

14 Quantify the probe by scintillation counting.

[a] If the DNA fragment has been prepared with two different restriction enzymes and the site at the 5′ end has a 3′ recessed end, while the site at the 3′ end is blunt-ended or has a 3′ protruding end, the bottom strand will be more highly labelled than the top strand, and will reduce the chances of artefactual mis-primed products generated by the top strand.

[b] Ensure this nucleotide is compatible with the restriction site to be filled in.

[c] This protocol for selective precipitation of probes from unincorporated nucleotides is only suitable for probes over 100 nucleotides in length. Alternatively, follow the precipitation from step 8 in *Protocol 26*.

Protocol 28

Annealing primers to RNA for primer extension reaction

This protocol has been adapted from Grimmig and Matern (15). The products are resolved using a standard denaturing polyacrylamide sequencing gel to obtain single base pair resolution.

Equipment and reagents

- Total RNA or mRNA for analysis
- Radiolabelled oligonucleotide primer (see *Protocol 26*)
- 10 mM stocks each of dATP, dCTP, dGTP, dTTP
- 10× annealing buffer: 0.5 M Tris pH 8.3, 0.75 M KCl, 30 mM MgCl$_2$
- 0.1 M DTT
- Sterile DEPC treated H$_2$O (see *Protocol 1*)
- 10 U/μl RNase inhibitor

- 100 U/μl MMLV reverse transcriptase
- 10 μg/μl carrier tRNA
- 3 M sodium acetate, pH 5.2
- Absolute ethanol
- 70% (w/v) ethanol
- Denaturing gel loading buffer: 80% formamide, 1 mM EDTA, bromophenol blue, and xylene cyanol to 1 mg/ml each
- Sequencing gel apparatus

Method

1 Add the following components to a microcentrifuge tube: 2 μg mRNA (or total RNA), 0.8 pmol radiolabelled primer, 1 μl each of all four 10 mM dNTP stocks, 2 μl 10× annealing buffer, 2 μl 0.1 M DTT, and H$_2$O to a final volume of 18 μl.

2 Gently mix and recover the solution to the bottom of the tube by a brief spin in a micro-centrifuge.

3 Heat the mixture to 60 °C for 10 min.

4 Allow to cool slowly to below 30 °C to allow the primer to anneal.

5 Recover the solution to the bottom of the tube by a brief spin in a microcentrifuge.

6 Add 1 μl (10 U) RNase inhibitor.

7 Add 1 μl (100 U) MMLV reverse transcriptase.

8 Incubate the reaction at 37 °C for 40 min.

9 Add 1 μl (10 μg) carrier tRNA.

10 Add 1/10 vol 3 M sodium acetate.

11 Precipitate the nucleic acids by addition of 2.5 vols of absolute ethanol.

12 Incubate at −20 °C for 30 min.

13 Collect the precipitated RNA by centrifugation in a microcentrifuge for 15 min.

14 Remove the supernatant to leave a radioactive pellet.

15 Wash the pellet in 70% ethanol and recover by centrifugation in a microcentrifuge for 15 min.

16 Allow samples to air dry at room temperature.

17 Dissolve the dried pellets in 20 μl denaturing gel loading buffer and mix well to ensure all the pellet has dissolved.[a]

18 Denature the samples at 100 °C for 3 min.

19 To analyse the samples run them on a 6% acrylamide denaturing gel prepared as described in *Protocol 21*, but cast in a sequencing gel apparatus.

20 Flush the wells of the polyacrylamide gel with 1× TBE running buffer, and load the samples onto the gel including appropriate controls[b] and radiolabelled size markers.[c]

Protocol 28 continued

21 Run the gel under the usual conditions for the sequencing apparatus being used until the bromophenol blue is about 2/3 down the gel.

22 Dry the gel and autoradiograph.

a There should be no radioactive counts remaining in the tube after the liquid has been removed.

b Suitable controls should include a lane with the radiolabelled primer.

c Radioactive markers may be prepared by filling the ends of commercial DNA size markers with a radionucleotide using Klenow polymerase or ideally should be a DNA sequencing ladder produced from the same primer.

Protocol 29

Annealing restriction fragments to RNA for primer extension

This protocol is a modification of that provided by Sambrook *et al.* (9). The hybridization conditions are designed to promote RNA:DNA duplex formation while minimizing DNA:DNA duplex formation. After annealing, a cDNA synthesis reaction is performed, as for *Protocol 28*. The primer extension products are analysed on a sequencing gel.

Equipment and reagents

- Total RNA or mRNA for analysis
- Radiolabelled restriction fragment primer (see *Protocol 27*)
- 10 mM dNTP mix: 10 mM each of dATP, dCTP, dGTP, dTTP
- RNA:DNA hybridization buffer: 40 mM PIPES pH 6.4, 0.4 M NaCl, 1 mM EDTA, 80% formamide. It is convenient to prepare a 5× stock of the buffer without formamide and dilute with 4 parts formamide just prior to use
- 10× reaction buffer: 0.5 M Tris pH 8.3, 0.75 M KCl, 30 mM $MgCl_2$

- 0.1 M DTT
- Sterile DEPC treated H_2O (see *Protocol 1*)
- RNase inhibitor (10 U/μl)
- MMLV reverse transcriptase (100 U/μl)
- Carrier tRNA (10 μg/μl)
- 3 M sodium acetate, pH 5.2
- Absolute ethanol
- 70% (w/v) ethanol
- Denaturing gel loading buffer: 80% formamide, 1 mM EDTA, bromophenol blue, and xylene cyanol to 1 mg/ml each
- Sequencing gel apparatus

Method

1 Mix between 2 and 10 μg total RNA or mRNA (up to 150μg total RNA may be used if rare transcripts are being mapped) with approximately 100 cpm radiolabelled restriction fragment and add DEPC treated sterile H_2O to 25 μl.

2 Add 2.5× 3M sodium acetate pH 5.2.

3 Precipitate the nucleic acids by addition of 2 .5 vols of ethanol.

4 Incubate at -20 °C for 30 min.

5 Pellet the precipitated nucleic acids by centrifugation at 13 000 g in a microcentrifuge.

6 Remove the supernatant and air dry the pellet.

7 Resuspend the pellet in 30 μl RNA:DNA hybridization buffer.

8 Heat the sample to 85 °C for 10 min.

Protocol 29 continued

9 Transfer mixture to an annealing temperature of between 40–50 °C.[a]

10 Incubate overnight.

11 Dilute with 170 μl DEPC treated water and precipitate the nucleic acids by addition of 2 vols of ethanol.

12 Collect the precipitated nucleic acid by centrifugation at 13 000 g in a microcentrifuge for 15 min.

13 Remove the supernatant and air dry.

14 Resuspend the pellet in 10 μl 1× reaction buffer.

15 Add the following components to the microcentrifuge tube containing the re-suspended DNA/RNA hybrid: 1 μl of all four 10 mM dNTP mix, 1 μl 10× reaction buffer, 2 μl 0.1 M DTT, and DEPC treated H_2O to a final volume of 18 μl.

16 Proceed from step 5 of *Protocol 28*.

[a] The optimal annealing temperature will have to be determined for each probe.

References

Chapter 2
In situ hybridization

Sabine Zachgo

Molecular Plant Genetics Department, Max-Planck-Institut für Züchtungsfoschung, 50829 Köln, Germany

1 Introduction

The *in situ* hybridzation technique was established in the late nineteen sixties for cytological preparations (1, 2). The technique was adapted to plant research around 20 years later. A specific, labelled nucleic acid probe is hybridized to plant tissue under the appropriate conditions. The probe binds to the complementary sequence and forms a stable hybrid, which can be detected with histological techniques. *In situ* hybridization techniques in plants are used: as a diagnostic tool to detect plant pathogens (3), to analyse genomic organization of plant genes (see *Volume 1, Chapter 8*), and to analyse tissue expression patterns of mRNAs (4). The great advantage of *in situ* hybridization over other conventional nucleic acid detection techniques like northern and Southern blot hybridization (*Volume 9, Chapter 6*) is the possibility to detect the precise spatial and temporal expression of a mRNA.

This chapter focuses on *in situ* hybridization techniques using riboprobes, which are labelled antisense mRNA transcripts for detection of homologous sense mRNA target sequences in plant tissues. It is a highly sensitive technique allowing detection of low copy number mRNAs even in single cells (5), which might not be detected using a northern hybridization technique when RNA is extracted from whole tissues. Usually, the *in situ* hybridization technique is used to describe the expression pattern of steady state mRNA level. However, it has also been adapted to detect *de novo* mRNA transcription (6).

RT-PCR is considered to be more sensitive than *in situ* analysis, however, obtaining good spatial resolution depends upon the amount and size of tissue pieces used to extract RNA. RT *in situ* PCR combines the advantages of both techniques. Standard protocols as for human and animal tissue are not yet available for plant tissue. Hopefully, plant specific problems, like cell wall swelling and loss of morphology due to temperature oscillation will be overcome soon.

Initially, only radioisotopes were available for labelling, and detection was carried out by autoradiography. However, stable non-radioactive labels, like digoxigenin (DIG) and biotin, are now available and they provide the same detection sensitivity as radioisotopes using immunochemical detection systems. Furthermore, these probes allow a combination of different labels in the same experiment and therefore the co-localization of different RNAs in the same preparation. Fluorescence *in situ* hybridization (FISH) is extensively used in human cytogenetics and gene mapping (7). The technique was adapted for plant research (8) and is now used as a tool for gene mapping (see *Volume 1, Chapter 8*). Fluorescent-labelled riboprobes are so far not commonly used for expression analysis, as background signals, due to plant cell wall and chlorophyll autofluorescence often hinder specific signal detection. However, modern confocal laser microscopy techniques can overcome these problems (9) and provide the advantages of FISH in expression pattern analysis.

Table 1 summarizes the different working steps for *in situ* hybridization conducted with non-radioactive (DIG-labelled) probes and radioactive labelled (S^{35}) probes on tissue sections. It provides an overview on the time requirements for *in situ* experiments.

Until recently, information on plant RNA expression has been derived from the analysis of two-dimensional sections. Non-isotopic whole-mount *in situ* hybridization techniques allow the precise three-dimensional localization of target mRNAs. This method has been applied very successfully to study animal embryo development (10). Recently, whole mount techniques have been adapted to study more precisely mRNA expression patterns in plant tissues (9, 11–13). This method overcomes the need to conduct serial section analysis to reconstruct labelling patterns on complex three-dimensional entities like flowers. This chapter contains protocols for *in situ* hybridization to both tissue sections, as well as whole mount samples.

2 Fixation of tissue

Fixation is one of the most critical steps for successful *in situ* hybridization. Tissue fixation must give good morphology preservation together with good staining quality. Unfortunately, these two requirements are contradictory and one has to find a compromise that gives the

Table 1. Timetable for *in situ* hybridization procedure on plant sections hybridized with a DIG- or S[35] -labeled probe

Working steps		Required time	
Tissue preparation	**DIG probe**		**S[35] probe**
Fixation of tissue		2–3 days	
Embedding of tissue in paraffin		3–4 days	
Sectioning of tissue		0.5–1 day	
RNA probe preparation			
Linearization of DNA template		2 h	
Alternatively:			
Preparation of a PCR template		4 h	
Transcription from DNA template		3 h	
Hydrolysis of ssRNA		Variable, depends on starting length of RNA (for a 1 kb transcript: 2 h)	
Quantification of the transcribed mRNA		2.5 h	
Pretreatment of slides before hybridization			
Deparaffinization and dehydration		40 min	
Proteinase K digestion and washes		45 min	
Dehydration and drying		30 min	
Hybridization			
Preparation of probe and application on sections		2 h	
Incubation with probe		Overnight	
Treatment after hybridization			
Washes		2 h	
RNase digestion		30 min	
Washes		1.5 h	
Detection of DIG probes			
Blocking step and incubation with anti-DIG antibody conjugated to an enzyme	2 h		
Washes and adjusting to the staining buffer	1 h		
Incubation with staining solution containing the substrate for the enzyme	4 h to overnight		
Final washes to stop labeling	10 min		
Detection of S[35] probes			
Coating with photo-emulsion			2 h
Exposure			2–14 days
Developing and fixation			10 min
Final washes			15 min

best result for each tissue. Two types of fixatives are used, cross-linking fixatives such as formaldehyde, paraformaldehyde and glutaraldehyde, and precipitating fixatives, for example, ethanol/acetic acid.

Glutaraldehyde is the best preserver of morphology; however, its very strong cross-linking capacity prevents adequate probe penetration. The precipitating fixatives allow rapid fixation and good probe penetration, which circumvents the use of vacuum. However, morphology preservation is poor and RNA is often lost during the hybridization steps.

Choice and concentration of the fixative, as well as the fixation time need to be individually determined for different kinds of tissues like inflorescences, leaves, stems, roots, as well as for tissue from different plant species and of different age. Some plant materials, such as tissues with high starch content, leaves with thick cuticula and wax layers on the surface, wooden structures, and seeds with strong seed coats, require specifically adapted fixation procedures.

In our hands, fixation of *Antirrhinum majus* and *Arabidopsis thaliana* tissues gives best results using a paraformaldehyde fixative buffered with a sodium phosphate buffer. RNA is relatively unreactive to cross-linking reagents. Cross-linking of proteins by formation of methylene bridges produces protein lattices that retain the formed hybrids. A 4% formaldehyde solution penetrates 2 mm tissue in about 1 h at room temperature, which together with optimized protease digestion, permits good probe penetration and therefore good staining results can be obtained. This fixation also provides good results with immunohistochemistry. This dual application is advantageous since serial sections of the same tissue can be used to detect the RNA and protein localization of the gene of interest (14).

To determine the right fixation for tissue sections it is recommended to try and compare two standard fixation techniques: fixation with 4% paraformaldehyde in a sodium phosphate buffer and fixation with a mixture of paraformaldehyde, acetic acid, and ethanol (FAE). If these fixatives, described in *Protocol 1 and 2*, do not give satisfactory results, different fixation times or other fixatives (15, 16) should be tested. Fixation for whole mount *in situ* hybridization (*Protocol 3*) has been successfully applied to roots and seedlings from *Arabidopsis thaliana* (9, 11) and inflorescences of *Antirrhinum majus* (13).

Protocol 1

Paraformaldahyde fixation for tissue sections

Do not remove solutions completely during the washing steps, as the tissue might dry out.

Reagents

- Paraformaldehyde (Sigma or Merck)[a]
- 1 M NaOH
- 1 M $Na_2 HPO_4$
- 1 M $NaH_2 PO_4 \cdot H_2O$

- 1 M sodium phosphate buffer, pH 7.2: combine 68.4 ml 1 M $Na_2 HPO_4$ and 31.6 ml $NaH_2 PO_4 \cdot H_2O$, which should result in a buffer with pH 7.2
- Ethanol

Method

1 Prepare 8% paraformaldehyde stock solution by heating 80 ml H_2O to 60 °C. Add 8 g paraformaldehyde and dissolve while stirring in a fume hood. Add drop by drop 1 M NaOH until paraformaldehyde is dissolved. Adjust the volume to 100 ml and filter through filter paper to remove remaining particles. The stock solution is cooled down and should always be made fresh.

2 Make 4% fixative solution in 0.1 M sodium phosphate buffer by diluting the 8% fixative solution 1:1 with 0.2 M sodium phosphate buffer, pH 7.2.

3 Harvest plant material by using a new, clean razor blade. Cut the tissue on a clean glass plate in

Protocol 1 continued

    pieces as small as possible. Mount tissue pieces in a few drops of fixative while trimming them to avoid drying of the samples.

4   Accumulate 4–6 tissue samples in a falcon tube containing 30 ml of fresh fixative.

5   Apply vacuum to improve penetration of the fixative. A wire net can be used to keep the tissue in the fixative in order to avoid damage of tissue not maintained in the fixative. After slow release of the vacuum the tissue should sink. If tissue does not sink, repeat this step.[b]

6   Fix 16 h at 4 °C.[c]

7   Wash twice with 0.1 M sodium phosphate buffer on ice.

8   Partly dehydrate in 30 and 50% ethanol for 30 min each on ice, and leave overnight at 4 °C in 70% ethanol. This step can be prolonged until the tissue is cleared and almost all chlorophyll is removed.

[a]   Stock solutions can also be purchased from microscopy suppliers in sealed glass vials.

[b]   Vacuum application can disrupt morphology of young and soft tissues. Addition of detergent, like Triton-X, to fixative (0.1%) helps to circumvent or reduce the strength of vacuum needed for infiltration of some tissues.

[c]   Fixation times vary for different tissues and need to be determined empirically (see text).

Protocol 2

FAE fixation for tissue sections

Reagents

- Concentrated formalin solution (37% formaldehyde)
- Acetic acid
- Ethanol
- FAE fixative: 2% formaldehyde (prepared from concentrated formalin solution), 5% acetic acid, 60% ethanol solution

Method

1   Prepare fresh FAE fixative.

2   Harvest and prepare the tissue as described in *Protocol 1* (steps 3–5) and fix 4–6 samples in 30 ml FAE for 24–48 h at 4 °C.

3   Wash twice for 5 min in 70% ethanol on ice and leave overnight in 70% ethanol at 4 °C.

Protocol 3

Fixation for whole mount *in situ*[a]

Reagents

- Paraformaldehyde (Sigma or Merck)
- Ethanol
- Heptane
- Dimethylsulphoxide (DMSO)
- $10\times$ Phosphate-buffered saline (PBS): 1.3 M NaCl, 2.7 mM KCl, 70 mM Na_2HPO_4, 30 mM NaH_2PO_4, pH 7.4
- Tween 20 (Sigma)
- 0.5 M EGTA
- Fixative solution: $1\times$ PBS solution, pH 7.4, containing 5% paraformaldehyde (for preparation see *Protocol 1*, step 1), 0.1% Tween 20, 0.08 M EGTA and 10% DMSO
- Methanol

Protocol 3 continued

Method

1 Dilute fixative solution 1:1 with heptane. This step results in an improved fixative penetration when samples are lipophilic.

2 Quickly harvest and prepare samples using new razor blades and clean forceps. If necessary, monitor the removal from tissue that might cover structures that are to be analysed under the stereomicroscope. However, large tissue wounding is later a source for background staining. Avoid drying of the whole mount tissue, after each single preparation immerse the tissue immediately into a falcon tube with 30 ml of the fixative:heptane solution. After accumulation of 4–6 tissue samples per tube fix at room temperature for 30 min up to 2 h, depending on the type and size of plant material.

3 Dehydrate twice with 100% methanol for 5 min and four times with 100% ethanol for 5 min.

4 Store samples in 100% ethanol at −20 °C until hybridization.

[a] The whole mount fixation and hybridization procedure is modified with permission from ref. 11.

3 Embedding and sectioning of tissue

3.1 Sample embedding

Following fixation the tissue must be embedded in paraffin for sectioning. For tissue embedding an organic solvent gradually replaces the ethanol present from the final stage of fixation. For many years xylene was used, which has now been replaced by non-toxic histo-clear. It is important not to remove solutions completely from the tissue, as they might become dry. Histo-clear is then replaced by paraffin by incubating the tissues for up to 3 days in paraffin. The embedded tissue can be stored at room temperature or 4 °C up to several months. *Protocol 4* usually gives good results for soft tissues like young inflorescences, young stems and young leaves. For other samples, like kernels, that are not easily penetrated by the wax it is useful to prolong all infiltration steps.

Protocol 4

Embedding of tissue for sectioning

Reagents

- Histo-clear (National Diagnostics)
- Paraffin (Paraplast Plus, containing DMSO, Sherwood)
- Ethanol
- Petri dishes

Method

1 Completely dehydrate the tissue at room temperature by sequential incubation in 30 ml of 85, 95, and 100% ethanol for 1 h each. If necessary de-gas the tissue again (see *Protocol 1*, step 5).

2 Transfer the tissue progressively to 30 ml mixtures of ethanol:histo-clear (2:1), (1:1), and (1:2) for 30 min each.

3 Soak the tissue in 100% histo-clear for 30 min, then incubate in fresh histo-clear and leave for about 4 h until the tissue is translucent.

4 Replace the solution with fresh histo-clear and pour the tissue on top of solid paraffin in a small beaker glass. Excess histo-clear is removed, such that it still covers all tissue and the histo-clear:paraffin ratio is about 1:5.

Protocol 4 continued

5 Incubate overnight in an oven set at 60 °C. Temperatures above 62 °C destroy the plastic polymers in Paraplast Plus.

6 Replace the paraffin with fresh melted paraffin twice a day for 2–3 days.

7 Transfer the tissue to a mold of a suitable size, for example, a Petri dish. If possible de-gas in a vacuum oven at 60 °C.

8 Orient the samples as required for sectioning with pre-warmed forceps and let the paraffin quickly solidify by placing the mold on ice water.

9 The tissue can be stored in paraffin at room temperature or 4 °C for several months.

3.2 Sectioning of tissue

Until a few years ago, adhesive slides had to be prepared by the researcher using a variety of adhesives like silane, poly-L-lysine or gelatin (17). Nowadays, adhesive slides of good and reliable quality can be purchased from many manufacturers (see *Protocol 5*). Conducting a pilot experiment to test the stickiness of purchased slides is recommended. If necessary, microtome blades are cleaned with histo-clear prior to sectioning.

Protocol 5

Sectioning of embedded tissue

Equipment
- Microtome
- Slide warmer
- Adhesion slides (e.g. Menzel–Gläser)
- Water bath

Method

1 Cut the paraffin block with a sharp, single-edge razor blade, leaving about 3–5 mm of paraffin around the embedded samples. Mount the block by melting some paraffin on the object holder.

2 Trim the block further to a rectangular or trapezoid shape, leaving about 2 mm of paraffin around the tissue. It is important that top and bottom edges are parallel. Mounting should be conducted such that the microtome knife will strike the longest side (parallel to the blade) first.

3 Using a microtome cut a ribbon with 7–8 µm thick sections.

4 For relaxation of the sections transfer the sections with forceps or a small paintbrush onto distilled water heated to 42 °C.

5 After relaxation pick up sections with the coated side of an adhesive slide. Drain off any excess water and place the slide on a heating plate set at 42 °C.

6 Let sections dry on the plate for 24–48 h at 42 °C. Cover with a lid to prevent dust falling on the slides. Sections can be stored at room temperature or 4 °C for several months.

4 Probe preparation

4.1 Probe labelling

Initially, the *in situ* technique was established using radioactive labels (*Protocol 10*), which were more sensitive than non-radioactive labelling methods. The improvement of the indirect labelling methodology (*Protocol 8*) has led to an equivalent sensitivity in signal detection.

DIG is most commonly used as an indirect labelling hapten coupled to UTP. Immunochemical detection is conducted by an antibody directed against the DIG molecule, which is linked to an enzyme, for example, alkaline phosphatase or horseradish peroxidase. Digoxigenin is a steroid that occurs naturally only in leaves and flowers of digitalis plants. Therefore, background staining due to cross-binding of the antibody in other plant tissues is usually not observed.

Non-radioactive detection techniques require a shorter processing time circumventing the extensive time required for autoradiography and allow easier handling of the probes, which have an extended shelf live. Probes that have been used successfully can be stored for several months for reuse. Furthermore, as there is no scattering of the detected signal as with the S^{35}-labelled radioactive probes, the spatial resolution of the signal is increased.

Another non-radioactive labelling technique makes use of the hapten biotin (vitamin H), which can be detected by avidin, a glycoprotein from egg white that has a very high binding affinity for biotin. However, in our hands the sensitivity of biotin-labelled probes seems to be lower than DIG-labelled probes.

4.2 Template preparation for probe production

Riboprobes can be made by run-off transcription from either a linearized plasmid vector template (*Protocol 6*) or a PCR template (*Protocol 7*). Use of a linearized vector plasmid requires cloning of the cDNA into a suitable vector, plasmid preparation, and linearization of the plasmid before preparation of the probe. A PCR-generated template is produced with primers containing sequences from the 5′ and 3′ end of the cDNA under investigation and, in addition, the antisense primer contains sequences for a RNA polymerase binding site (18). A PCR-generated template, produced with such a primer pair, contains an incorporated RNA polymerase binding site and can directly be used for synthesis of an antisense probe. Thereby, the above-mentioned time-consuming steps to produce a linearized plasmid vector template can be circumvented. Incorporation of the RNA polymerase binding site into the sense primer allows the production of a PCR-template to prepare a sense probe that can be used as a negative control. In our hands, transcriptions with T7 and T3 polymerases give more reliable results than using SP6 polymerase. Therefore, we design primers with incorporated T7 or T3 polymerase binding sites.

4.3 Quantification of DIG-labelled probe

Estimation of the amount of DIG labelling is important to optimize *in situ* hybridization. A highly concentrated probe may cause background staining, whereas a dilute probe will result in weak staining. Yield estimation can be achieved by spotting a dilution series of the labelled probe side by side with a dilution series of an appropriate standard (*Protocol 9*). The standard can be a commercially labelled RNA or a probe that has been used successfully in previous experiments. The optimal concentration of probes derived from different cDNAs may vary and has to be determined empirically.

4.4 Precautions working with RNA

As contamination with RNases results in degradation of the RNA probe, it is important to protect the RNA. Gloves should always be worn and solutions prepared with diethyl pyro-carbonate (DEPC) treated water. Stir H_2O overnight with 0.1% DEPC and autoclave before use. We also recommend the use of baked glassware (see also *Chapter 1*).

Protocol 6

Preparation of linearized plasmid vector DNA template for transcription

Reagents

- Purified plasmid vector, containing a RNA polymerase binding site
- Appropriate restriction enzyme
- Phenol
- Chloroform
- Isoamylalcohol
- 3 M sodium acetate, pH 6.0
- 10 mM Tris–HCl, pH 8.5

Method

1 Cut 10–20 μg of the plasmid with 5–10-fold excess of an appropriate restriction enzyme for 4 h. The enzyme should preferably create a 5′ overhang. Enzymes that produce 3′ overhangs or blunt ends might lead to RNA synthesis artifacts.

2 Purify the linearized plasmid with phenol/chloroform extraction by adding an equal amount of a 1:1 phenol/chloroform mixture to the plasmid solution. Vortex thoroughly and centrifuge in a microfuge for 5 min at 13 000 g. Transfer the aqueous phase into a new tube. Repeat this step by adding the same volume of a 24:1 chloroform:isoamylalcohol mixture.

3 Precipitate the DNA by adding 1 volume of isopropanol and 1/10 volume of 3 M NaOAc, pH 6.0 to the aqueous phase and mix. After incubation on ice for 30 min centrifuge at 10 °C and 13 000 g. Wash the pellet with 80 % ethanol and resuspend the DNA in 10 mM Tris–HCl, pH 8.5.

4 Check for completeness of digestion on a 1% agarose minigel.

Protocol 7

Preparation of PCR template

Reagents

- DNA (50 ng/μl)
- Primers (10 μM) containing 13–17 bases corresponding to the specific cDNA sequence with a consensus for either the
  - T7 polymerase binding site: 5′ TAATACGACTCACTATAGGG3′ or
  - T3 polymerase binding site: 5′ ATTAACCCTCACTAAAGGGA 3′
  at their 5′ ends.
- PCR reagents:
  - dNTPs, 10 mM each (Roche/Boehringer Mannheim)
  - 10× PCR buffer (Roche/Boehringer Mannheim)
  - Taq polymerase (Roche/Boehringer Mannheim)
- PCR purification spin columns (QIAquick, Qiagen)

Method

1 For a 50 μl PCR reaction add 5 μl 10× buffer, 1 μl dNTP mix, 2 μl from each primer, 1 μl DNA, and 0.5 μl Taq polymerase together in a PCR tube, made up to 50 μl with H_2O. After denaturation for 5 min at 95 °C conduct 33 amplification cycles: 1 min at 94 °C, 1 min at 57 °C, and 1–2 min at 72 °C, depending on the length of the DNA fragment. Terminate the reaction with a final elongation step at 72 °C for 5 min.

2 Purify the PCR product by phenol/chloroform extraction and ethanol precipitation (see *Protocol 6*) or by using a PCR purification spin column and adjust the DNA concentration to 200 ng/μl.

Protocol 8

Transcription reaction to produce a DIG-labelled RNA probe

This protocol is adapted from the manual of the 10× DIG RNA labelling mix purchased from Roche/Boehringer Mannheim. The RNA is labelled to a density of 1 digoxigenin for every 20–25 nucleotides.

Reagents

- 10× transcription buffer (Roche/Boehringer Mannheim)
- 10× DIG RNA labelling mix with digoxigenin-UTP (Roche/Boehringer Mannheim)
- DNA (linearized plasmid or PCR product)
- RNA polymerase 20 U/μl (Roche/Boehringer Mannheim)
- RNase inhibitor, 50 U/μl (Roche/Boehringer Mannheim)

- 7.8 M NH_4AcOH, pH 6.0
- tRNA, 20 mg/ml (Roche/Boehringer Mannheim)
- DNase I, 10 U/μl (Roche/Boehringer Mannheim)
- 0.2 M EDTA, pH 8.0
- Ethanol

Method

1 Prepare the transcription mix by combining, in a 1.5 ml microfuge tube, 1 μg linearized plasmid DNA or 200 ng PCR-template DNA with 2 μl 10× transcription buffer, 3 μl 10× DIG-NTP labelling mix, 2 μl RNA polymerase, 0.5 μl RNase inhibitor, and add H_2O to a total volume of 20 μl.

2 Mix the components, centrifuge briefly in a microfuge and incubate at 37 °C for 2 h.[a]

3 Load 1 μl of the transcription reaction on a non-denaturing 1% agarose gel to check RNA synthesis.

4 Add 2 μl of carrier tRNA (20 mg/ml) and 2 μl DNase I (10 U/μl).

5 Mix carefully, centrifuge briefly, and incubate at 37 °C for 15 min.

6 Take a 1 μl sample to test on a 1% agarose gel if the DNA digestion is complete.

7 Stop the reaction with 2 μl of 0.2 M EDTA and precipitate with 0.5 vol of 7.8 ammonium acetate, pH 6.0 and 2 vols of ethanol for at least 2 h at −20 °C.

8 Centrifuge at 13 000 g in a microfuge for 30 min.

9 Wash with 80% ethanol, dry the pellet and dissolve in 100 μl H_2O.

10 The yield of transcription can be estimated by running an aliquot of the probe on an agarose gel next to an RNA standard of known concentration. Exact size determination can be assessed by conducting a northern transfer (17) (see *Chapter 1*).

[a] A higher transcript yield is not achieved by longer incubation times. Instead, reaction components should be scaled up.

Protocol 9

Quantification of the DIG-labelled probe by dot blot analysis

All steps are conducted at room temperature.

Reagents

- Control RNA (Roche/Boehringer Mannheim)
- 2× SSPE: 0.3 M NaCl, 2 mM EDTA, 20 mM $NaH_2PO_4 \cdot H_2O$, pH 7.4

- 0.2 M NaOH
- Blocking reagent (Roche/Boehringer Mannheim)

Protocol 9 continued

- Buffer 1: 100 mM Tris–HCl, pH 7.5, 150 mM NaCl
- Buffer 2: 100 mM Tris–HCl, pH 9.5, 100 mM NaCl, 50 mM MgCl$_2$
- Tween 20
- Sheep anti-DIG-AP conjugated antibody (Fab fragment, Roche/Boehringer Mannheim)
- Nitro blue tetrazolium chloride (NBT, Roche/Boehringer Mannheim): 100 mg/ml stock solution in 70% dimethylformamide (DMF)
- 5-Bromo-4-chloro-3-indolyl-phosphate (BCIP, Roche/Boehringer Mannheim): 50 mg/ml stock solution

Method

1 Make a 1:10, 1:100, 1:1000, and 1:10 000 dilution of the probe and the control RNA in sterile distilled H$_2$O and spot 1 μl of each on a small piece of nylon membrane.

2 Denature RNA by floating the membrane on 0.2 M NaOH for 5 min.

3 Neutralize by floating the membrane for 5 min on 2× SSPE. All following steps are conducted by immersing the membrane into the solutions.

4 Equilibrate the filter in buffer 1 containing 0.1% Tween 20 and equilibrate for 5 min.

5 Block the membrane in buffer 1 containing 0.1% Tween 20 and 0.5% blocking reagent for 20 min.

6 Incubate with a 1:3000 dilution of the anti-DIG-AP Fab fragment in buffer 1 for 1 h at room temperature.

7 Wash three times in buffer 1 containing 0.1% Tween 20 for 5, 10, and 15 min.

8 Equilibrate the filter for staining in buffer 2 for 5 min.

9 Incubate the membrane in buffer 2 containing 1.5 μl NBT/ml and 1.5 μl BCIP/ml. The signal will become visible after 5–10 min.

10 Estimate the amount of probe by comparison with the control RNA.

Protocol 10

Transcription reaction to produce a S^{35}-labelled RNA probe

Reagents

- DNA (linearized plasmid DNA or PCR product, see *Protocols 6 and 7*)
- 10× transcription buffer (Roche/Boehringer Mannheim)
- RNA polymerase 20 U/μl (Roche/Boehringer Mannheim)
- RNase inhibitor, 50 U/μl (Roche/Boehringer Mannheim)
- tRNA, 20 mg/ml (Roche/Boehringer Mannheim)
- DNase I, 10 U/μl (Roche/Boehringer Mannheim)
- 0.2 M EDTA, pH 8.0
- 0.75 M dithiothreitol (DTT)
- ATP, CTP, GTP (each 10 mM)
- αS^{35}-UTP (10 mCi/ml), Amersham
- Sephadex G-50 column (Pharmacia)
- TE, pH 7.5: 10 mM Tris–HCl pH 7.5, 1 mM EDTA, pH 8.0
- 3 M Sodium acetate, pH 5.0
- Ready Cap, XtalScint (Beckman Instruments)

Method

1 Prepare the transcription mix by combining, in a 1.5-ml microfuge tube, 1 μg linearized plasmid or 200 ng PCR-template DNA, 2.5 μl 10× transcription buffer, 1 μl 0.75 M DTT, 2 μl RNA polymerase, 0.5 μl RNase inhibitor, 1 μl ATP, CTP, and GTP, respectively, 5 μl αS^{35}-UTP (10 mCi/ml) and add H$_2$O to the final volume of 25 μl.

Protocol 10 continued

2 Mix the components, centrifuge briefly in a microfuge and incubate at 37 °C for 1 h.

3 Add 2 µl tRNA and 2 µl DNase I to the reaction mix and incubate at 37 °C for 15 min.

4 Transfer the tube on ice and add 2.5 µl of 0.2 M EDTA.

5 Remove non-incorporated ribonucleotides by gel filtration through a Sephadex G-50 column. Elute with TE, pH 7.5. Collect about seven fractions each with 150 µl volume.

6 Measure cpm of 1-µl aliquots in a Ready Cap, a convenient small container that is coated on the bottom with solid scintillator. The labelled RNA is expected in fractions 4-6.

7 Pool the peak fractions and precipitate the RNA with 1/10 volume 3 M sodium acetate, pH 5.0 and 1.1 volumes isopropanol at −20 °C for at least 1 h.

8 Pellet the RNA by centrifugation in a microfuge for 30 min at 13 000 g at 4 °C.

9 Wash the pellet with 80% ethanol and dissolve the RNA after drying in 100 µl sterile distilled H_2O.

10 Determine activity by measuring 1 µl in a Ready Cap. The read out should be between $0.5–1 \times 10^6$ cpm/µl.

Note: Suitable safety precautions must be taken when using radiolabelled nucleotides.

5 Hydrolysis of RNA probe

The length of the probe used for hybridization is important. Probes need to be small enough to easily penetrate the tissue. Long probes tend to stick randomly to the sample and give unspecific background. An optimal length of the synthesized RNA probe for hybridization is 150–200 bp. Therefore, a limited hydrolysis is conducted with radioactive and non-radioactive labelled probes, producing short fragments that result in stronger *in situ* signals with less background. However, in our hands, probes in the size range of 200–400 bp length do not need to be hydrolysed because loss of probe during the hydrolysis procedure outweighs the positive penetration effect.

The appropriate hydrolysis time is calculated from the following equation:

$$t = (L_i − L_f)/(k \times L_i \times L_f)$$

where t is the hydrolysis time in min, L_i the initial length of the RNA probe in kb, L_f is the final length of RNA in kb, and k is the rate constant with 0.11 kb/min.

Protocol 11

Hydrolysis of RNA probe

Reagents

- 2× hydrolysis buffer: 80 mM $NaHCO_3$, 120 mM Na_2CO_3, pH 10.2

- For 1 ml buffer of a 2× stock solution combine 160 µl 0.5 M $NaHCO_3$ with 240 µl 0.5 M Na_2CO_3 and add 600 µl H_2O. Without adjustment solution should be pH 10.2, store at −20 °C

- Acetic acid
- 3 M sodium acetate, pH 6.0
- tRNA, 20 mg/ml (Roche/Boehringer Mannheim)
- Poly(A) RNA, 10 mg/ml (Roche/Boehringer Mannheim)
- Isopropanol

Method

1 Hydrolyse RNA by adding 100 µl 2× hydrolysis buffer to 100 µl probe and incubate at 60 °C for the calculated time to get fragments of 150–200 bp length.

2 Stop reaction immediately by placing tubes on ice and adding 1 µl acetic acid, 7 µl 3 M sodium acetate, 1 µl tRNA, 1 µl poly(A) RNA, and 210 µl isopropanol.

3 Mix solutions and incubate at −20 °C for at least 2 h.

4 Centrifuge in a microcentrifuge at 13 000 g and 4 °C for 30 min.

5 Wash the pellet twice with 80% ethanol, dry for 10 min at room temperature and resuspend the pellet in 100 µl DEPC-treated H_2O.

6 Pre-hybridization treatment

Incubation of the tissue sample with a protease is conducted as a pretreatment to permeabilize the tissue and thereby facilitate penetration of the probe. The concentration of the protease is critical and has to be determined empirically for each tissue. Proteinase K is a strong and effective protease, alternatively, the milder Pronase can be used. Over extensive protein degradation results in a loss of tissue morphology, whereas under digestion hinders probe penetration causing a weaker hybridization signal.

A weak diluted acid hydrolysis step of the tissue enhances permeabilization and thereby increases the signal, probably by partially solubilizing highly cross-linked basic nuclear proteins. Acetylation of tissue sections is an efficient way to reduce non-specific binding of negatively charged probes due to electrostatic forces (19). The benefit derived from this treatment is possibly exerted via blocking of cytoplasmic amine groups.

The pre-hybridization Protocol 12 is used for tissue sections labelled with non-radioactive and radioactive probe. Protocol 13, used for whole mount tissue includes an additional fixation step after the protease digestion to conserve the three-dimensional integrity of the sample.

Pre-hybridization and hybridization steps are performed under RNase-free conditions (see Section 4.4).

Protocol 12

Pre-hybridization treatment of paraffin embedded tissue sections

All following steps, unless otherwise indicated, are conducted at room temperature by putting the slides into a metal rack fitting into a glass container holding 200 ml solution. Pre-warm solutions to the indicated temperature before immersing slides.

Reagents

- Histo-clear (National Diagnostics)
- Ethanol
- 0.2 M HCl
- 2× SSPE: 0.3 M NaCl, 2 mM EDTA, 20 mM $NaH_2PO_4 \cdot H_2O$, pH 7.4
- Proteinase K (Roche/Boehringer Mannheim)
- Proteinase K buffer: 20 mM Tris–HCl, pH 7.0, 2 mM $CaCl_2$
- 100 mM triethanolamine
- Acetic anhydrate

Method

1 De-wax the tissue sections by transferring the slides with the attached sections through a series of histo-clear and ethanol solutions and, thereby, gradually replacing the wax by ethanol. Slides

Protocol 12 continued

are immersed in solutions of 100% histo-clear three times, 1:1 histo-clear:ethanol once, and 100% ethanol twice with 5 min incubation time for each step.

2 Hydrate sections by dipping them sequentially into a graded series of ethanol dilutions of 95, 85, 70, 50, and 30% made in 0.85% NaCl for 2 min each step.

3 Immerse the slides in 0.2 M HCl for 20 min.

4 Wash twice for 10 s in H_2O.

5 Soak in 2× SSPE at 70 °C for 20 min.

6 Incubate with Proteinase K in Proteinase K buffer.[a]

7 Stop digestion by washing the slides in 2× SSPE for 5 min.

8 Incubate sections in 0.5% acetic anhydrate (added just before use) in 100 mM triethanolamine,[b] pH 8.0 for 10 min on a magnetic stirrer. Wash twice with 2× SSPE.

9 Wash with 0.85% NaCl for 5 min and dehydrate sections by passing them through 70, 90, and 100% ethanol for 5 min each.

10 Let slides air dry.

[a] Optimal conditions have to be tested for each tissue. Try 1, 10, and 100 µg/ml Proteinase K for 20 min at 37 °C.

[b] This step should be conducted with radioactive-labelled probes to block non-specific probe interactions. Acetylation is optional for non-radioactive probes.

Protocol 13

Pre-hybridization treatment of whole mount tissues

Reagents

- 1× PBS (see *Protocol 3*)
- Ethanol
- Methanol
- PBT:PBS containing 0.1% Tween 20
- Proteinase K (Roche/Boehringer Mannheim)

- 5% Formaldehyde
- HS (hybridization solution): 50% formamide, 5× SSC: 750 mM NaCl, 75 mM sodium citrate, pH 7.0, 50 µg/ml heparin

Method

1 Leave the fixed whole mount pieces from *Protocol 3* for the following steps in batches of 4–6 samples/tube and conduct all steps with gentle agitation of the samples.

2 Wash fixed whole mount tissue pieces twice in 100% ethanol for 2 min and then for 30 min in a 1:1 ethanol:histo-clear mixture.

3 Wash the tissue twice in ethanol and twice in methanol, each step for 5 min.

4 Soak the tissue in a 1:1 methanol:PBT mixture for 10 min.

5 Wash three times in PBT for 2 min each step.

6 Digest samples with 40 µg/ml Proteinase K in PBT for 20 min.[a]

7 Wash twice with PBT for 5 min.

8 In order to preserve the morphology after the Proteinase K digestion, postfix samples by incubating them for 30 min in PBT containing 5% formaldehyde.

SABINE ZACHGO

Protocol 13 continued

  9 Equilibrate the tissue in a 1:1 PBT:HS mixture for 5 min.

  10 Rinse twice in HS and pre-hybridize in fresh HS at 62°C for 2 h.

 ª The conditions have to be optimized for each tissue, see *Protocol 12*, step 6.

7 Hybridization

7.1 Hybridization parameters

Different parameters affect the ability to form duplexes during hybridization and have to be considered to determine the optimal hybridization conditions:

7.1.1 Temperature

During hybridization the denatured RNA probe reanneals with complementary RNA molecules under conditions just below their melting point (T_m). T_m is the temperature at which 50% of the probe is dissociated. However, tissue sections do not follow classical solution hybridization kinetics and the T_m cannot be calculated but has to be determined empirically.

7.1.2 Formamide

Hybridization is carried out in a formamide solution. This is the organic solvent of choice to reduce the melting temperature of RNA-RNA duplexes and conduct the hybridization at a lower temperature. It acts by destabilizing hydrogen bonding between probe and target sequence.

7.1.3 Monovalent cations

In addition to temperature and formamide concentration, the hybridization stringency is determined by the concentration of monovalent cations (Na^+). Under high stringency conditions, with high Na^+ concentrations, only sequences with a high degree of homology form stable duplexes, whereas a low Na^+ concentration results in low hybridization stringency, allowing the formation of hybrids of lower sequence homology. As a basic guideline, it is recommended that hybridization is started with a 50% formamide solution at 50°C at 300 mM (Na^+) and then optimize to specific requirements.

7.1.4 Probe concentration

An increase in probe concentration leads to an increased signal until the saturation point is reached. Further increase results in non-specific background binding that masks the true signal. We use 0.1–0.3 ng/µl/kb of labelled probe to achieve a good target RNA sequence saturation. For estimation of the yield of the DIG-labelled probe see *Protocol 9*.

7.1.5 Dextran sulphate

Dextran sulphate is an inert polymer that is strongly hydrated in aqueous solutions. Thereby, macromolecules have no access to the hydrating water, which results in an apparent increase in probe concentration, which causes higher hybridization rates.

7.1.6 Blocking reagents

Yeast tRNA and poly(A) RNA block non-specific binding of RNA probes. Denhardt's solution, containing Ficoll, polyvinylpyrrolidone and bovine serum albumin (BSA) further reduces non-specific binding of the probe.

7.1.7 pH

Hybridization is carried out at a neutral pH (6.5–7.5) provided by a Tris buffer.

7.1.8 Coverslips

Commonly, hybridization is performed with coverslips on the hybridization mix. Incubation can be conducted without coverslips in airtight, humidified chambers. It is reported that this treatment results in less background for DIG-labelled probes (20).

Protocol 14

In situ hybridization with DIG- and S^{35}- labelled probes on tissue sections

Reagents

- Deionized formamide
- 5 M NaCl
- 50× TE, pH 7.0: 0.5 M Tris–HCl, pH 7.0, 50 mM EDTA, pH 8.0
- 20 mg/ml tRNA (Roche/Boehringer Mannheim)
- 50× Denhardts mix: 5 g Ficoll (Type 400, Pharmacia), 5 g polyvinylpyrrolidone (Serva), 5 g bovine serum albumin (Fraction V, Sigma)

- Poly(A) RNA (Roche/Boehringer Mannheim)
- 50% Dextran sulphate (Sigma)
- Probe, labelled with DIG (0.1–0.3 ng/µl/kb) or S^{35} (3 × 10^4 cpm/µl)
- 4× SSPE: 600 mM NaCl, 4 mM EDTA, 40 mM NaH$_2$PO$_4$·H$_2$O, pH 7.4
- Herring sperm DNA, 5 mg/ml

Method

1 Use 20–100 µl probe for hybridization on a 25 × 75 × 1.0 mm slide (hybridization volume depends on the number of sections per slide).

2 For 100 µl hybridization combine 6 µl 5 M NaCl, 4 µl 50× TE, 2.5 µl tRNA, 1 µl poly(A) RNA, 2 µl 50× Denhardt mix, 50 µl deionized formamide, and 20 µl 50% dextran sulphate in a micro-centrifuge tube and mix.

3 Take 1–5 µl of labelled and hydrolysed RNA probe, and add H$_2$O up to 100 µl total hybridization mix/slide. Heat mixture at 85 °C for 2 min.

4 Cool RNA probe immediately and add to the hybridization mix.

5 Apply hybridization mix with the RNA probe to the slides with the pre-hybridized tissue sections (see *Protocol 12*).

6 Overlay tissue sections with a coverslip avoiding air bubbles.[a]

7 Incubate slides overnight in a humid chamber (filter paper moistened with 4× SSPE) and incubate at 50 °C.

[a] Some brands of coverslips are difficult to remove after hybridization and, therefore, they should be treated with silane before use. Soak clean coverslips in 2% dichlorodimethylsilane in 1,1,1-trichloroethane (Merck) for 1 min, wash with several changes in ethanol, and dry. If incubation is carried out without coverslip (see text) avoid displacement of the probe by encircling the sections with a water-impervious substance, like a rapid mountain media (for example, Entellan, Merck).

Protocol 15

Hybridization with DIG-labelled probes on whole mount tissue

Reagents

- HS (see *Protocol 13*)
- DIG-labelled probe, 0.1–0.3 ng/µl/kb

- Herring sperm DNA, 5 mg/ml

Protocol 15 continued

Method

1 Incubation steps are done on a shaker/roller to ensure that samples are moved slowly in the hybridization mix to allow penetration from all sides.

2 After pre-hybridization (*Protocol 13*, step 9) incubate tissue[a] in a microcentrifuge tube with 400 µl of fresh HS containing 3–10 µl DIG-labelled probe (0.1–0.3 ng/µl/kb) and 8 µl Herring sperm DNA. Denature probe and Herring sperm before incubation for 5 min at 85 °C.

3 Incubate for 16 h at 60 °C.

[a] Occasionally, whole mount tissue is not equally penetrated because samples sometimes tend to stick together due to their architecture during incubation with one of the substances. As a result a staining gradient along the sample can be observed. Hybridization of several discrete samples (4–6) in one hybridization reaction allows the distinction to be made between staining artifacts, staining gradients, and genuine staining signals.

8 Post-hybridization washes

Background due to non-specific hybridization is removed by post-hybridization washes. Samples are soaked through a series of washes with increasing stringency. Starting with high salt concentration buffers, unbound probe is removed. Final washes at a lower ionic strength carried out at an elevated temperature remove mis-matched hybrids. Non-specific binding, due to the stickiness of RNA probes, is removed by an RNase digestion step. This step also degrades all left over non-hybridized labelled ssRNA molecules.

Protocol 16

Treatment of slides after hybridization with DIG- and ^{35}S-labelled probes

All post-hybridization washing steps are carried out with gentle agitation in a glass container holding up to 16 slides in a metal rack immersed into 200 ml volume of washing solution. Solutions above room temperature are pre-warmed and should be de-gassed.

Reagents

- 20× SSPE stock solution: 3 M NaCl, 20 mM EDTA, 0.2 M $NaH_2PO_4 \cdot H_2O$, pH 7.4
- RNase A (Roche/Boehringer Mannheim)
- NTE: 0.5 M NaCl, 10 mM Tris–HCl, pH 7.5, 1 mM EDTA, pH 8.0

Method

1 Gently rinse the coverslip off from the tissue sections in 3× SSPE.

2 Wash the samples three times for 30 min each in 3× SSPE at 45°C.

3 Incubate the samples for 20 min in NTE, pH 8.0 at 37 °C.

4 Treat the sections with 20 µg/ml RNase A in NTE for 30 min at 37°C.

5 Stop RNase digestion by washing twice for 5 min in NTE at 37°C.

6 Rinse the sections sequentially in 1.5× SSPE, 1.0× SSPE, and 0.5× SSPE each for 30 min at 52°C.

Protocol 17

Treatment of whole mount tissues after hybridization

Reagents

- 20× SSPE stock solution: 3 M NaCl, 20 mM EDTA, 0.2 M $NaH_2PO_4 \cdot H_2O$, pH 7.4
- NTE: 0.5 M NaCl, 10 mM Tris–HCl, pH 7.5, 1 mM EDTA, pH 8.0
- RNase A (Roche/Boehringer Mannheim)
- HS (see *Protocol 13*)

Method

Due to the preceding treatments, the whole mount tissues are now soft and their three-dimensional morphology is easily damaged. Change solutions carefully by removing old solutions with a pipette without touching the samples and slowly add fresh solution.

1 Remove hybridization solution and wash twice in fresh HS for 30 min at 60 °C.

2 Wash tissue for 60 min in 1:1 HS:NTE at room temperature and repeat with NTE.

3 Incubate the tissue pieces in NTE containing 40 µg/ml RNase A for 45 min at 37 °C.

4 Stop the RNase digestion by washing with NTE at 37 °C for 15 min.

5 Wash the tissue sequentially with 1.5× SSPE, 1× SSPE, and 0.5× SSPE, each for 45 min at 50 °C.

9 Signal detection

Signal detection can be carried out by a number of substrates and chromogens. Currently, we obtain best results using an alkaline phosphatase (AP) reaction with nitroblue tetrazolium (NBT) and 5-bromo-4-chloro-3-indolyl phosphate (BCIP) substrate. The reaction product formazan is a stable precipitate of a blue/purple colour that has bright reflective properties. Formazan is soluble in organic solvents and the tissue is therefore embedded in aqueous mounting media after staining. Other substrates are less suited for overnight detection times, as they tend to precipitate non-specifically after long incubation times. Endogenous AP activity is normally not a problem in plant tissues. If necessary, it can be blocked by adding levamisole in the detection solution. Addition of high molecular weight polyvinyl alcohol (PVA) in the detection reaction has been shown to strongly increase the sensitivity of the AP detection. PVA works by enhancing the AP reaction and prevents diffusion of formazan, thereby also allowing a more precise localization of the signal (21). Application of PVA can decrease signal detection times from overnight down to four hours incubation.

Another appropriate detection system is the enzyme horseradish peroxidase (HRP) used with the chromogen 3,3'-diaminobenzidine (DAB). It has a rapid development time, resulting in the formation of a stable insoluble end-product with a brown colour. As it contrasts the blue of the formazan AP and HRP can be used in double labelling experiments. Both colorimetric detection systems reveal good localization properties. However, AP is considered to be the more sensitive of the two. The staining colour of DAB can be modified with heavy metals like nickel and cobalt (22). Furthermore, fluorescent techniques are becoming available for plant *in situ* hybridization experiments using direct and indirect labelling techniques (11). A variety of protocols exist using different labelling, enzyme and substrate systems.

The following protocols are optimized for plant *in situ* experiments, and most commonly used for tissue sections and whole mount tissues. They are also intended to be used as basic guidelines to adapt protocols established for cell lines or tissues from other species described in refs 15 and 16.

Protocol 18

Immunological detection of DIG-labelled probes on tissue sections using alkaline phosphatase (AP)

Reagents

- Tween 20
- Blocking reagent (Roche/Boehringer Mannheim)
- Buffer 1: 100 mM Tris–HCl, pH 7.5; 150 mM NaCl
- Buffer 2: 100 mM Tris–HCl, pH 9.5; 100 mM NaCl, 50 mM $MgCl_2$
- Sheep anti-DIG AP-conjugated antibody (Fab fragment, Roche/Boehringer Mannheim)
- Nitro blue tetrazolium chloride (NBT): 100 mg/ml stock solution (Roche/Boehringer Mannheim)
- 5-Bromo-4-chloro-3-indolyl-phosphate (BCIP): 50 mg/ml stock solution (Roche/Boehringer Mannheim)
- 10% polyvinyl alcohol (PVA), 70–100 kD, (Sigma)
- Bovine serum albumin (BSA), fraction V (Sigma)
- Fluorescent brightener 28, Calcofluor (Sigma)
- Entellan (Merck)

Method

1 All washing and blocking steps are carried out at room temperature using glass containers holding 200 ml washing buffer.

2 Equilibrate tissue sections on slides after the post-hybridization washes (*Protocol 16*, step 6) in buffer 1 for 5 min.

3 Block the samples for 30 min in buffer 1 containing 0.5% blocking reagent, and transfer for 30 min in buffer 1 containing 1% BSA and 0.1% Tween 20.

4 Incubate sections with an anti-DIG AP-conjugated Fab fragment antibody diluted 1:3000 in buffer 1 containing 1% BSA for 1 h in a moisture chamber. In order to minimize waste of the antibody apply the buffer directly on the slides (500–1000 μl/slide, depending on the number of sections/slide).

5 Wash the slides in buffer 1 with 0.1% Tween 20 four times for 20 min.

6 Equilibrate the sections in buffer 2 with 0.1% Tween 20 for 5 min.

7 Prepare 30 ml NBT/BCIP detection solution by adding 1.5 μl NBT/ml and 1.5 μl BCIP/ml dropwise under stirring to buffer 2 containing 0.1% Tween 20 to avoid formation of precipitates. If necessary filter solution through filter paper.

8 Develop for 4–16 h in a vertical glass staining dish (Coplin jar) suited for eight slides in the dark and monitor colour formation visually.[a]

9 Stop the colour reaction by washing twice in H_2O for 5 min.

10 Wash off background by passing slides sequentially through an ethanol dilution series of 50, 70, 95, and 100% for 30 s each.

11 Cell walls can be stained for better visualization of the tissue by incubating the stained sections for 15 min with 0.01% fluorescent brightener 28. Examine the tissue using UV fluorescence.

12 Dry slides and add 3–5 drops of a rapid mounting media such as Entellan and cover with a coverslip.

[a] If colour reaction takes longer than 16 h and/or the signal detection has to be increased add polyvinyl alcohol (PVA) in the detection solution (see text). Dissolve 10% PVA (70–100 kD) in buffer 2 without $MgCl_2$ at 90 °C under stirring. Cool to room temperature and filter if aggregates have formed. Add Mg_2Cl (50 mM) and NBT/BCIP (see above).

Protocol 19

Detection of DIG-labelled probes on tissue sections using horseradish peroxidase(HRP)

Reagents

- PBS (see *Protocol 3*)
- Blocking reagent (Roche/Boehringer Mannheim)
- Bovine serum albumin (BSA), fraction V (Sigma)
- Anti-digoxigenin-HRP, Fab fragments from sheep (Roche/Boehringer Mannheim)

- SIGMA FAST™ 3,3′-diaminobenzidine tablet (DAB peroxidase substrate)
- TE buffer (see *Protocol 10*)
- Entellan (Merck)

Method

1 All washing and blocking steps are conducted in glass containers holding 200 ml washing buffer unless otherwise indicated. All steps are carried out at room temperature.

2 Equilibrate the tissue sections after the post-hybridization washes (*Protocol 16*, step 6) in PBS for 5 min.

3 Block the samples in PBS containing 0.5% blocking reagent and in buffer 1 containing 1% BSA, 0.1% Tween 20, each for 30 min.

4 Incubate the tissue sections with an anti-DIG HRP-conjugated Fab fragment antibody diluted 1:3000 in PBS containing 1% BSA for 1 h in a moisture chamber. Depending on the number of the sections apply between 500 and 1000 μl of the antibody solution per slide.

5 Wash the samples four times for 10 min in PBS.

6 Drain off excess washing buffer and apply about 500 μl DAB working solution onto the sections. As colour detection is quick, monitor its development while incubating and stop the reaction by rinsing the slides in TE buffer. Detection times over 30 min usually result only in increase of background.

7 Dry slides and mount with a rapid mountant, such as Entellan and a coverslip.

Protocol 20

Detection of DIG-labelled probes on whole mount tissue using an alkaline phosphatase detection system

Reagents

PBT: PBS (*Protocol 3*) containing 0.1% Tween 20

Buffer 2: 100 mM Tris–HCl, pH 9.5; 100 mM NaCl, 50 mM $MgCl_2$

Nitro blue tetrazolium chloride (NBT): 100 mg/ml stock solution (Roche/Boehringer Mannheim)

5-Bromo-4-chloro-3-indolyl-phosphate (BCIP): 50 mg/ml stock solution (Roche/Boehringer Mannheim)

20 mM EDTA, pH 8.0

Method

1 All steps are carried out with gentle agitation of the whole mount samples.

2 Equilibrate the whole mount tissue pieces after post-hybridization washes (*Protocol 17*, step 5) in PBT for 5 min.

SABINE ZACHGO

Protocol 20 continued

3 Block tissue for 1 h in PBT containing 2% BSA.

4 Incubate the tissue for 16 h in PBT containing the anti-DIG AP-conjugated Fab fragment antibody diluted 1:1500 at 4 °C while gently shaking.

5 Wash the samples four times with PBT, each time for 60 min.

6 Equilibrate tissue in buffer 2 with 0.1% Tween 20 for 5 min.

7 Conduct chromogenic reaction as described in *Protocol 18*, step 6.

8 Monitor the rapid colour formation under a stereo microscope. The time for the colour reaction is about 3–30 min. Prolonged detection time normally results only in increased background staining.

9 Stop the reaction by washing in PBT with 20 mM EDTA, pH 8.0.

10 Tissue pieces can be stored for several months in H_2O at 4 °C.

Protocol 21

Autogradiography

Reagents

- Autoradiography emulsion: Kodak NTB-2 nuclear track emulsion
- X-ray film developer: Kodak D-19b developer
- Fluorescent Brightener 28, Calcofluor (Sigma)
- Standard photographic fixer like Agefix (Agfa), diluted 1:5
- Entellan (Merck)

Method

1 All steps with the unfixed autoradiography emulsion are carried out in the dark or under a dark-red safelight (Kodak Wratten filter series II).

2 After the post-hybridization washes (*Protocol 16*, step 6) dehydrate the slides in 80 and 95% ethanol for 5 min each and let them air dry.

3 Melt Kodak NTB-2 nuclear track emulsion at 45 °C for 1 h in water bath. Mix with an equal volume of distilled water pre-equilibrated to 45 °C. Aliquots of this mix can be stored at 4 °C.

4 Dip slides slowly in diluted solution stored in a beaker kept at 45 °C. Avoid bubbles and unnecessary agitation that result in increased background.

5 Put the slides in a rack and let dry for at least 1 h.

6 Place the dried slides in a light-tight dark box and expose them at 4 °C for an appropriate length of time. Exposure time needs to be determined empirically. Make extra slides to determine the correct exposure time and develop them after, for example, 2 days, 6 days, and 14 days.

7 To develop the slides, warm them up to room temperature. Place slides in fresh X-ray film developer at 16 °C for 2 min.[a]

8 Stop development by rinsing for 30 s in H_2O at 16 °C.

9 Fix the slides for 5 min in fixer at 16 °C.

10 Wash the slides in cold running H_2O for 20 min.

11 Stain cell walls of the sectioned tissue with 0.01% fluorescent brightener 28 (Calcofluor) for 15 min.

12 Wash the sections three times in distilled H_2O for 5 min each and let air dry. Mount the slides with a rapid mountant, such as Entellan and a coverslip.

13 Dark field microscopy is used to detect the silver grains, indicating the presence of bound probe. Underlying cell walls from the tissue can be visualized using UV-epifluorescence.

[a] Longer development of the emulsion might increase background.

10 Confirmation of signals and trouble shooting

10.1 Controls

Several different control steps can be conducted to confirm whether the observed signal is due to a real hybridization complex or if it is an artifact produced by non-specific interactions. Furthermore, one should be aware that a true signal in older differentiated cells like leaf mesophyll cells can be easily overlooked and interpreted as a negative result. Due to the presence of large vacuoles, only a small cytoplasmic area is left close to the cell wall and, therefore, the stained area is relatively small compared to the whole cell size.

In order to control the specificity and the sensitivity of the probe a control reaction must be included when a probe for a new gene with an unknown expression pattern is used, or when a new technique is established:

10.1.1 Positive controls

1 Use tissue known to contain the mRNA of interest.

2 Check the fixation and mRNA retention in the sample by using a known RNA, for example, an abundant housekeeping gene. This allows further optimization of the technique.

10.1.2 Negative controls

1 Use tissue known not to express the target sequence.

2 Control if staining artifacts are observed due the unspecific hybridization by using a sense probe. However, there are a number of reports of problems with sense probes, obtaining the same result as with an antisense probe, which is probably due to artifact RNA synthesis. In these cases, any non-specific probe (like plasmid DNA) can be used.

3 Control for detection by hybridizing without probe and by omitting the anti-DIG antibody.

10.2 Trouble shooting

Table 2 (adapted from ref. 16) describes common problems occurring during *in situ* hybridization experiments and their possible causes. Several suggestions are given that can help to correct the problems.

Table 2. Trouble shooting guide[a]

Problem	Possible cause(s)	Possible solution
Poor morphology	Fixation time too short	Try longer fixation times
	Inappropriate fixative	Check other fixatives
	Over digestion with protease	Decrease incubation time. Try pronase instead of proteinase K
	Vacuum too strong	Try shorter vacuum infiltration time. Add detergent (Tween-20 or Triton-X) to improve infiltration of fixative
No or weak signal	Inadequate fixation of samples	Check tissue with a positive control (nuclear or mitochondrial mRNA) to assess nucleic acid preservation, try weaker fixation reagent
	Over-fixation	Shorter fixation time
	Insufficient protein digestion preventing protein penetration	Increase protease digestion
	Poor probe labeling	Estimate incorporated yield of labelling (see *Protocol 9*). Increase DIG-UTP concentration in the labelling reaction by preparing own labelling mix from separate NTPs with elevated amount of DIG-UTP
	Probe concentration too low	Increase probe concentration
	Hybridization conditions too stringent	Decrease stringency
	Nuclease contamination	Increase working precaution. Make fresh RNase-free solutions
	Weak detection system	Check enzyme and substrate by incubating them and monitor color change of solution. Check antibody by dot blot analysis with an old probe, known to function (see *Protocol 9*). If potency is lost over time use higher concentration. Include Tween 20 or increase concentration in detection system
High background	Probe concentration too high	Titrate probe
	Antibody concentration too high	Titrate antibody
	Stringency of post-hybridization washes too low	Optimize washes with salt concentration and temperature
	Non-specific binding of probe to tissue components	Acetylate tissues (see *Protocol 12*, step 8)
	Retained unbound probe	Check and optimize RNase digestion
	Non-specific binding of detection system reagent	Optimize pre-incubation with blocking reagents. Use an immunohistochemical grade of BSA, certain grades of BSA might contain contaminants that contribute to background staining
	Endogenous enzyme develops substrate	Lock endogenous enzyme

[a] Adapted from ref. 16.

References

1. John, H., Birnstiel, M. and Jones, K. (1969). *Nature*, **223**, 582.
2. Pardue, M. L. and Gall, J. G. (1969). *Proc. Natl. Acad. Sci. USA*, **83**, 600.
3. McDougall, J. K., Myerson, D. and Beckmann, A. M. (1986). *J. Histochem. Cytochem.*, **34**, 33.
4. Motte, P., Saedler, H. and Schwarz-Sommer, Z. (1998). *Development*, **125**, 71.
5. Fobert, R. P., Coen, E. S., Murphy, G. J. P. and Doonan, J. H. (1994). *EMBO J.*, **13**, 616.
6. Raikhel, N. V., Bednarek, S. Y. and Lerner, D. R. (1989). In *Plant molecular biology manual*, vol. B9, pp. 1–32, editors Gelvin, S. B.; Schilperoort, R. A.; Verma, D. P. S. Kluwer Academic Publisher, Dordrecht.
7. Trask, B. (1991) *Trends in Genetics*, **7**(5), 149.
8. Ambros, P. F., Matzke, M. A. and Matzke, A. J. M. (1986). *Chromosoma*, **94**, 11.
9. Bauwens, S., Katsanis, K., Van Montagu, M., Van Oostveldt, P. and Engler, G. (1994). *Plant J.*, **6**(1), 123.
10. Rosen, B. and Beddington, R. S. P. (1993). *Trends in Genetics*, **9**, 162.
11. de Almeida Engler, J., Van Montagu, M. and Engler, G. (1994). *Plant Mol. Biol. Reporter*, **12** (4), 321.
12. Jackson, D. and Hake, S. (1999). *Development* **126**, 315.
13. Zachgo, S., Perbal, M-C., Saedler, H. and Schwarz-Sommer, Z. (2000). *Plant J.* **23** (5) 697–702.
14. Perbal, M-C., Haughn, G., Saedler, H. and Schwarz-Sommer, Z. (1996). *Development*, **122**, 3433.
15. Williamson, D. G. (ed.) (1996). In situ *hybridization: a practical approach*, 2nd edn. IRL Press, Oxford.
16. Herrington, C. S. and Levy, E. R. (eds) (1995). *Nonisotopic methods in molecular biology: a practical approach*. IRL Press, Oxford.
17. Spector, D. L., Goldman, R. D. and Leinwand, L. A. (eds) (1998). In *Cells: a laboratory manual*, Vol. 2, 98.4. Cold Spring Harbor Laboratory Press, Ann Arbor.
18. Gandrillon, O., Solare, F., Legrand, C., Jurdic, P. and Samarut, J. (1996). *Mol. Cellular Probes*, **10**, 51.
19. Hayashi. S., Gillham, I. C., Delaney, S. D. and Tener, G. M. (1978) *J. Histochem. Cytochem.*, **26**, 677.
20. Borneman, J. and Altschuler, M. (1995). *BioTechniques*, **18** (3), 406.
21. DeBlock, M. and Debrouwer, D. (1993). *Analyt. Biochem.*, **216**, 88.
22. Hsu, S. and Soban, E. (1982). *J. Histochem. Cytochem.*, **30**, 1079.

Chapter 3
DNA micro-arrays for gene expression analysis

Shanna Moore, Paxton Payton and Jim Giovannoni
Boyce Thompson Institute and USDA/ARS Plant, Soil, and Nutrition Laboratory, Cornell University Campus, Tower Road, Ithaca, New York, 14853

1 Introduction

Micro-arrays allow analysis of expression patterns of thousands of genes within the confines of one experiment. Arrays are direct descendents of DNA gel-blot (Southern) based assays that exploit interactions between complimentary strands of DNA (1). The addition of a solid glass substrate, precision robotics, and the use of fluorescence provide expression arrays with increased precision, speed, and scale over their filter- and radioactivity-based cousins. These qualities make micro-arrays a particularly attractive partner for the huge amounts of gene sequence information generated in the current genomics era. Although transcript monitoring is currently the most popular use for arrays, they have been successfully utilized in fields ranging from mutation detection to evolutionary sequence analysis (2–6). Micro-array publications seem to appear with increasing fervour; however, very few concern work in the plant sector (7–12). *Figure 1* provides a basic outline of micro-array construction and use as we have optimized for our work with plant species (in particular tomato), with focus on sample preparation, probe synthesis, and data generation.

2 Types of arrays

Micro-arrays can be constructed using either PCR-amplified cDNAs or oligonucleotides. There are three techniques commonly utilized to generate micro-arrays: photolithography, ink-jet micro-spotting, and physical micro-spotting (13, 14). High quality arrays can be successfully constructed by any of these techniques. Different techniques lend themselves to different research platforms, and these must be considered before undertaking array construction. Additionally, construction of arrays and probes, as well as hybridization and data generation, are available from a number of commercial sources.

Photolithography involves a novel application of microelectronics technology, utilizing light to direct the synthesis of oligonucleotides onto a solid support. This allows for high precision, high density chips (over 200,000 spots/cm). Because the target DNA is synthesized on the chip, knowledge of the sequence of the gene or chromosomal region of interest is requisite. Arrays created by photolithography have been utilized for a multitude of analyses (15–17).

Ink-jet technology employs propulsion of cDNAs or synthesized oligonucleotides, through small jets onto a solid support without any direct contact between the print head and the substrate. Array density can approach 10,000 spot/cm, and arrays can be used for gene expression studies and mutation detection (18, 19).

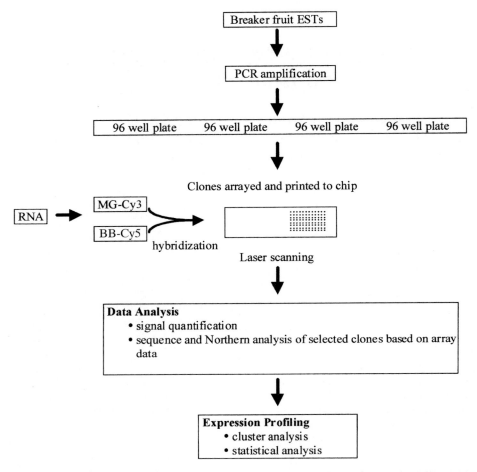

Figure 1. Construction of expression arrays. ESTs were sequenced from a Breaker fruit library. Sequences were aligned to determine redundancy using TIGR Assembler software (TIGR) and a non-redundant tomato EST database. The set of non-redundant cDNA clones were re-arrayed into 96-well plates, amplified by PCR, and arrayed onto poly-L-lysine-coated glass microscope slides. Fluorescently-labelled cDNA probes prepared from total RNA isolated from tissues of interest (MG, mature green; BB, breaker) are labelled with either Cy3 or Cy5, and simultaneously hybridized at high stringency to arrayed EST sequences. Hybridization intensities are measured and quantified to generate gene expression profiles.

Mechanical microspotting is accomplished by direct contact of a printing tool (e.g. a pin) with the printing surface. Spot sizes typically average 50–200 μ, and 96- and 384-well formats are commercially available. The popularity of microspotting is quickly growing as commercial arrayers become more available (*Table 1*) and more affordable, and as this technology readily lends itself to traditional means of DNA preparation and handling, once the primary array and scanning infrastructure is in place. Increased availability of cDNA clones and libraries is also a factor. The ease and low cost of printing via microspotting are contributing to its increasing utilization and popularity for transcript analysis and will be the focus for the duration of this chapter. (2, 6, 20).

Table 1 Commercial array sources

Vendor	Contact	Arrayer
Affymetrix	www.affymetrix.com	Oligo arrays; microspotters
BioRobotics	www.biorobotics.co.uk	Microspotters; ink-jet printers
Cartesian Technologies	www.cartesiantech.com	Microspotters; ink-jet printers
Genemachines	www.genemachines.com	Microspotters; ink-jet printers
Genetix	www.genetix.co.uk	Q-Bot micro-arrayer
Labman Automation	www.labman.co.uk	Microspotter
MicroFab Technologies	www.microfab.com	Ink-jet arrayer
GeSiM	www.gesim.de	Ink-jet arrayer

3 Target preparation

Micro-arrays based on amplified expressed sequence tags (ESTs) are the most popular candidates for microspotting. ESTs are usually generated by single-pass sequencing 300–900 bases from the 5′ end of cDNA clones, which have been directionally ligated into a plasmid or excised phagemid vector. EST sequence and homology information provide a distinct and obvious advantage in expression studies compared to the use of anonymous clones as functional implications can often be made based on sequence homologies. In addition to *Arabidopsis*, a number of EST projects are underway in agronomically important crops including tomato, potato, cotton, rice, maize, sorghum, woody tree species, and soybean. These projects will facilitate interest and development of additional micro-array applications in plants (21). Expression studies using anonymous clones, however, are also feasible in certain applications. For example, non-sequenced and arrayed target genes that exhibit interesting patterns of expression can be sequenced following the results of micro-array experiments as in traditional differential screening approaches.

3.1 PCR

Constructing EST arrays lends itself easily to high throughput methods and automation. By utilizing a 96-well format for clone amplification, purification, and processing, one increases speed of production, and reduces sample handling and necessary storage.

Protocol 1

PCR amplification of cDNA clones in 96 – well plate format

Equipment and reagents

- MJ Research PTC-225 thermal cycler with heated lid
- Rubber sealing mat (Perkin Elmer)
- 96-well PCR plates (Phenix)
- 10× PCR buffer: 100 mM Tris pH 8.3, 500 mM KCL, 15 mM MgCl$_2$, 1% gelatin
- dNTPs : 2.5 mM final concentration dATP, dCTP, dGTP, dTTP
- 96-well replicator (Boekel)
- Primers: M13 Forward/Reverse, 10 μM each
- 96-well sterile U-bottom plates (Falcon)
- 50 ml polystyrene reagent reservoir (Corning)
- Media: 1× LB containing 100 μg/ml ampicillin

Protocol 1 continued

Method

1 Array clones of interest into 96-well plates.

2 For each plate to be amplified, inoculate 150 μl of fresh 1× LB + antibiotic utilizing the 96-well replicator. After each use rinse replicator in 50% bleach, ddH$_2$O, and 95% ethanol successively, followed by flame sterilization.

3 Incubate plates at 37 °C for 8–14 h until growth is confluent.

4 Prepare 10 ml of PCR 'master mix', per plate, in a sterile reservoir containing 1× buffer, 1 μM of each primer, 0.5 U/μl Taq, 250 μM dNTPs, 5% DMSO, aliquot 98 μl of master mix per well, and inoculate from overnight culture plates. (Note: In the absence of a pin tool, 2 μl of overnight culture may be added to each reaction.)

5 PCR conditions consist of an initial denaturation of 5 min at 94 °C, followed by 35–45 cycles of 94 °C for 1 min, 60°C for 1 min, and 72 °C for 1 min, with a final elongation step of 72 °C for 7 min.

6 PCR product presence and quality are visualized on a 1% agarose gel by staining with ethidium bromide. Electrophorese with a DNA mass marker to estimate DNA concentration. Clones that do not amplify or show multiple bands are re-amplified under modified conditions (adjusted annealing temperature, MgCl$_2$ concentration, specific primers, etc.).

7 Precipitate the amplified clones by adding 1/10 volume of 3 M sodium acetate followed by 2.5 volumes of 95% ethanol or isopropanol.[a]

8 Incubate at −20 °C for at least 1 h. Centrifuge for 1 h at 3500 rpm and 4 °C.[b]

9 Decant supernatant and wash the pellet with 70% ethanol or isopropanol. Centrifuge for 30 min at 3500 rpm and 4 °C.

10 Decant supernatant and dry the reactions in a concentrator for 10 min at 45 °C. Resuspend in 25–50 μl of preferred printing solution (we typically employ 3× SSC or 3× SSC with 0.2% sarcosyl).[c]

[a] We have found that samples may be printed from the PCR amplification mix without purification. However, for long-term storage of amplified clones, purification is recommended (see *Figure 1*).

[b] Samples may be precipitated overnight.

[c] Samples can be resuspended in 50μl of H$_2$O, aliquoted into five plates, dried, and stored at room temperature for extended periods.

4 Substrate/support

Micro-arrays are generally printed onto glass microscope slides. Glass provides an excellent platform with low inherent fluorescence, resulting in negligible intrinsic background levels, and a non-porous surface important for preventing diffusion of deposited samples and thus allowing utilization of minimum hybridization volumes (14, 22). Glass slides also allow for miniaturization and easy storage of arrays.

There are two commonly used binding chemistries: amine and aldehyde (23, 24). Amine slides are coated with amine-rich chemicals, such as poly-L-lysine, that allow ionic attachment of the DNA to the slide, which is, in turn, covalently bound by UV-cross-linking (25). Because amine chemistry relies simply on ionic interaction between the positively charged slide surface and the negatively charged DNA, the PCR product does not have to be modified in any way for attachment. The alternative method utilizes Schiff base aldehyde-amine chemistry.

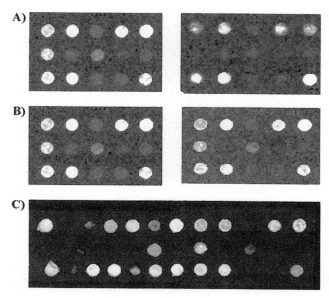

Figure 2. Example of DNA micro-array images. (A) Comparison of clones purified by ethanol precipitation (left) versus no purification (right). (B) Comparison of Cy3 (left) and Cy5 (right) labelled probes generated from the same RNA sample using the Genisphere 3DNA Expression Kit. C) False colour overlay image comparing expression levels in immature fruit (green) and ripe fruit (red) using Quantarray software. Yellow spots indicate similar expression levels (see plate 1).

The use of a synthetic primer containing an aliphatic-amine group on the 5′-end is necessary to generate a modified PCR product. The amines on the modified PCR product react with aldehydes, which are covalently attached to the slide surface and the DNA is bound via Schiff-base formation. Both slide types are available commercially from a number of vendors at varying costs. Poly-L-lysine coated slides can also be manufactured in the laboratory at a minimal cost (26).

Protocol 2

Poly-L-lysine coating glass slides

Equipment and reagents

- Poly-L-lysine solution, Sigma Cat. No. P8920
- Gold Seal micro-slides, Cat. No. 3010
- Glass slide washing tray
- Phosphate buffered saline (8.0 g NaCl, 0.2 g KCl, 1.44 g Na_2HPO_4, 0.24 g KH_2PO_4/l of ddH_2O)
- 45 °C oven
- 6.25 M NaOH
- 95% Ethanol
- Centrifuge
- Slide box

Method

1 Prepare wash solution by mixing 280 ml of 6.25 M NaOH with 420 ml of 95% ethanol.
2 Place slides in wash rack and wash 2 h on an orbital shaker.

Protocol 2 continued

3 Rinse slides by plunging racks up and down in ddH$_2$O. Wash five times using fresh H$_2$O with each wash.

4 Prepare poly-L-lysine solution by mixing 70 ml of poly-L-lysine with 70 ml of phosphate buffered saline in 560 ml ddH$_2$O.

5 Wash slides in poly-L-lysine solution for 1 h.

6 Rinse slides in ddH$_2$O by plunging rack up and down 10 times.

7 Centrifuge slides for 5 min at 500 rpm then dry 10 min at 45 °C.

8 Store in slide box at room temperature.

Once printed, slides may be processed and used for hybridization or stored in a slide box at room temperature. Processing/hydration evenly distributes the DNA in the spot, increases binding stability to the substrate and, in some cases, minimizes non-specific binding of the probe to the slide (22, 27).

Protocol 3

Slide processing

Equipment and reagents

- Heat block at 80–90 °C
- ddH$_2$O
- Pyrex dish containing boiling ddH$_2$O
- Ice-cold 95% ethanol
- Filter-sterile 0.2% SDS
- Slide washing/staining rack
- UV-cross-linker
- Centrifuge

Method

1 Amplify target cDNAs by PCR using flanking primers in the vector (100 μl reaction).

2 Verify the presence of the PCR product on an agarose gel.

3 Ethanol precipitate the DNA and resuspend in 30–40 μl of printing solution (e.g. 3× SSC and 0.2% sarcosyl).

4 Print target cDNAs on poly-L-lysine coated slides using a gridding robot (see *Table 1*).

5 'Snap-dry' slides DNA side up on a heat block or hot plate at 80–90 °C for 5–10 s.

6 Rehydrate the DNA by waving the slide over a 65 °C H$_2$O bath for 5 s or until spots glisten. The slide should fog over and the array will appear as a clear 'window' on the slide.

7 UV-crosslink DNA to the slides with 550 mJ energy (DNA side up).

8 Place slides in slide washing vessel containing 0.2% SDS and wash on an orbital shaker for 10 min at room temperature.

9 Briefly rinse by plunging the slide rack in ddH$_2$O.

10 Denature the DNA by placing the slides in 100 °C ddH$_2$O for 2 min.

11 Rinse slides in ice-cold 95% ethanol for 2 min.

12 Immediately dry slides by centrifugation at 500 rpm for 5 min.

13 Processed slides may be used immediately for hybridization or stored in slide boxes at room temperature.

5 Probe synthesis and hybridization

Micro-array probes, for transcript analysis, are constructed by incorporating fluorescent molecules into cDNAs created from a single round of reverse transcription. The most common labelling methods are direct incorporation of fluorescently-labelled dNTPs, incorporation of an amino-allyl modified dNTP, and attachment of a fluorescently-labelled oligonucleotide to 'primed' cDNAs (7, 26, 28, 29).

The direct incorporation method relies on the incorporation of Cy3 and Cy5-labelled nucleotides (dCTP or dUTP) during the reverse transcription of a mRNA template. While this is the most commonly used method to date, achieving balanced labelling between the Cy3 and Cy5 reactions is not always easily reproducible. The amino-allyl method utilizes a modified dUTP nucleotide that is incorporated into the cDNA during reverse transcription and is then conjugated to a monofunctional Cy3 or Cy5 molecule via CNS ester binding. The conjugation step bypasses the less consistent incorporation of the larger Cy5 molecule into the cDNA, but specific activity is still dependent upon base composition (sequence). Total RNA, in addition to polyA RNA, may be utilized as template.

The most reliable protocol, in our experience, is based on the hybridization of a labelled oligonucleotide dendrimer to a primed cDNA. This method, commercially available as a kit from Genisphere (Oakland, NJ), uses a polyT primer with a unique recognition sequence attached. After the cDNA is made, it is mixed with a fluorescently-labelled capture oligomer that is complementary to the recognition sequence on the polyT primer. The major advantage to this technology is that the signal generated is essentially independent of transcript length or composition, and labelling is based on the hybridization of a labelled oligomer to cDNAs synthesized from standard dNTPs. Currently, only Cy3 and Cy5 kits are available, but

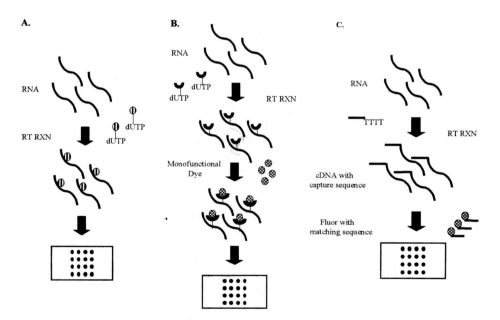

Figure 3. Construction of fluorescently-labelled cDNA probes by reverse transcription. (A) Direct incorporation of Cy3 and Cy5-labelled nucleotides (dCTP or dUTP). (B) Incorporation of an amino-allyl modified dUTP, and conjugation to a monofunctional Cy3 or Cy5 molecule. (C) Attachment of a fluorescently-labelled oligonucleotide to 'primed' cDNAs using the Genisphere 3DNA dendrimer labelling system.

71

work is being conducted on other fluors and additional products are expected in the near future. The dendrimer approach uses total RNA and with some variations calling for as little as 2.5 μg RNA template per reaction.

Typically, probes from two RNA pools (e.g. ripe and unripe fruit) and labelled with distinct fluors are hybridized to a chip. The two individual probes are combined, added to the micro-array, and incubated under a glass coverslip. Hybridization conditions vary depending on the hybridization buffer. Our probes are hybridized overnight at 50–60 °C in a humidified chamber. Hybridization chambers in a variety of forms, ranging from machined, watertight chambers to homemade racks using sealed plastic kitchenware are employed. One particularly useful hybridization device is a 50-ml conical polypropylene vial. With the use of conical vials and single-slide chambers, the probe renaturation and hybridization volume loss (evaporation) can be minimized. Additionally, conical vials are substantially cheaper than manufactured hybridization chambers.

Protocol 4

Hybridization

Equipment and reagents

- Incubator at 60 °C (or relevant hybridization temperature)
- ddH$_2$O
- 50-ml conical vial (Falcon or Corning)
- Test-tube rack
- Parafilm
- Plastic slide washing trays
- 2× SSC with 0.2% SDS, 2× SSC, 0.2× SSC
- Centrifuge
- Slide box

Method

1 Place each processed slide into a 50-ml conical vial containing 100 μl ddH$_2$O and a moistened strip of filter paper.
2 Incubate at 60 °C for 10–15 min to warm and moisten the slide.
3 Remove the slides one-at-a time, and pipette the probe mixture onto the micro-array, cover with a coverslip, and return to the conical vial.
4 Seal the lid of the conical vial with Parafilm and hybridize overnight at 60 °C.[a]
5 Remove slides, dip in 2× SSC with 0.2% SDS to remove coverslips, and wash 10 min at 55 °C in 2× SSC with 0.2% SDS.[b]
6 Wash for 10 min in 2× SSC at room temperature followed by 10 min on 0.2× SSC for 10 min at room temperature.
7 Immediately dry slides by centrifugation at 500 rpm for 5 min.
8 Place slides in a plastic slide box and scan immediately.

[a] Hybridization conditions vary depending on the hybridization buffer used. Lower hybridization temperatures reduce evaporation of the probe mixture, but increase background signal.
[b] Exposure to light must be kept to a minimum from this step forward.

6 Scanning

Visualizing hybridized arrays requires excitation of bound fluorochromes by a laser source, collection of the emitted fluorescence through a series of filters that block reflected and scattered excitation energy, and conversion of the focused energy to an electrical signal by a photomultiplier tube (30–32). This process is generally done through use of a commercially made scanner, but it is possible to construct one's own (see *Table 2* for contact). Most scanners are dual channel (two lasers), but the availability of multi-channel scanners is increasing. Multi-channel scanners allow for the use of multiple probes (currently up to four probes per chip) in a single hybridization. Scanner choices (or fluor choices) must be made with available capacity in mind (*Table 2*). It is advisable to choose fluors that are readily available, and have distinct and distant emission frequencies to avoid overlap of emission frequencies with probes to be compared simultaneously.

Table 2 Commercial scanners[a]

Vendor	Contact
Axon	www.axon.com
GMS	www.geneticmicro.com
GSI Lumonics	www.genscan.com
Applied Precision Inc	www.api.com

[a]Instructions for building scanners are available at: cmgm.stanford.edu/pbrown/sequence.aecom.yu.edu/bioinf/micro-array/reader.html

7 Data analysis/data management

Perhaps the most difficult and challenging aspect of micro-array experiments is data analysis (33–38). A single experiment may produce thousands of data points. Output from scanning a hybridized array is a simple TIFF or bitmap image. There are multiple software programs available and many under development (see *Table 3*). Most available programs are a variation of the same theme and do the following: locate spots, normalize signal, quantitate intensities, subtract out background, and generate a report.

Table 3 Array analysis software

Software package	Application	Vendor
ArrayScout	Spot finding; quantification; clustering	Lion Biosciences; www.lion-ag.de/ company
ArrayVision	Spot finding; quantification	Imaging Research, Inc.
ArrayDB	Data management	NHGRI;www.nhgri.nih.gov
Scanalyze, Cluster, and Tree View	Spot finding; clustering; cluster analysis	rana.stanford.edu/clustering
Clustering tools	Clustering	Pangea Systems, Inc.; www.panbio.com
GeneCluster	Clustering	www.genome.wi.mit.edu
GeneSight	Spot finding; clustering	Biodiscovery; www.biodiscovery.com
AutoGene	Image analysis; quantification	Biodiscovery
ImaGene	Spot finding	Biodiscovery
CloneTracker	Data management	Biodiscovery
GeneSpring	Spot finding; clustering; database links	Silicon Genetics; www.sigenetics.com
LifeExpress	Database management	Incyte Genomics; www.incyte.com
GeneChip Information Systems	Database management	Affymetrix; www.affymetrix.com

Locating spots generally consists of overlaying a defined grid onto the captured array image. Many software grids are flexible to account for any printing or substrate inconsistencies. Once the spots have been located, the fluorescent signal from each channel must be normalized. There are several methods for normalization including the use of median fluorescence of each channel or the use of total signal per channel. The preferred method, however, is the use of a subset (30–100) of 'housekeeping' genes that are expected to be expressed in all tissues or treatments. To quantitate the signal in each spot, the fluorescence signal is converted into an electronic signal and background fluorescence is removed by comparing fluorescence outside the spot, but within the array grid. Output may be in several forms, depending on the analysis software. All software packages report intensities in some form of tab delimited text file containing some or all of the following: intensities for each clone, relative percentage signals, relative ratios, standard deviations, mean, median, and mode. Software such as Quantarray (GSI Lumonics) also generate reports in the form of bar graphs, scatter plots, and pie charts. Genes can be grouped or clustered based on expression levels using programs like Cluster and Tree View; both developed by Michael Eisen at Stanford University (39). These programs use a phylogenetic approach to visualize expression patterns, and are useful for gaining insight into the possible function of unknown genes and the elucidation of biochemical pathways. Finally, a subset of data generated from micro-array experiments is best verified by a secondary method such as RNA gel blot (northern) or quantitative PCR.

Data management is a critical aspect of array analysis. Clone tracking, sequence information, expression data, and links to other databases are necessary to fully utilize array technology. In addition, as development and utilization of EST and micro-array databases involves numerous steps, there are many opportunities to introduce errors, which must be controlled for, e.g. through occasional re-sequencing of micro-arrayed ESTs to be certain of tracking integrity through the process. These aspects of information management are rapidly developing and becoming more automated, which will assist biologists in all fields.

8   Challenges facing the array community

Although arrays are becoming more commonplace, they are still in their fledgling state and face certain challenges. We have mentioned multiple methodologies for array construction, probe generation, and data analysis. As is typical of any new technology, the current status quo in the early stages of use is few widely accepted standards in the array community and some confusion for those initiating array experiments. The creation of forums, list servers, and the growing number of websites will certainly help drive and facilitate agreement upon many issues concerning array technology (see *Table 4*). The *Arabidopsis* consortium group, for

Table 4   Array resources

Site	Resource
Cmgm.stanford.edu/pbrown/	Protocols, software, arrayer construction instructions, array forum
Syntom.cit.cornell.edu/chips.html	Arabidopsis/tomato synteny database
plantarrays@genome.stanford.edu	Overview of genomics research; useful array links
www.bsi.vt.edu/ralscher/gridit/	Plant array list server—to subscribe send the word 'subscribe' to :
	plantarrays-request@genome.stanford.edu
www.gene-chips.com	Micro-array overview; useful links
afgc.stanford.edu	Arabidopsis Functional Genomics Consortium

instance, has worked extensively on generating a list of controls under many different conditions, in an effort to create a standard that may allow data from different sources and experiments to be compared.

Micro-arrays can also still be very costly between initial equipment requirements and continued demand for relatively expensive reagents. This may be an imposing obstacle to smaller groups or experiments. A growing number of collaborations and private facilities, however, are becoming available for the creation of small or specialized arrays.

Despite these and other challenges, the array community seems to have an unprecedented air of co-operation and information exchange that crosses the boundaries of institution, industry, and organism. Continued research into fluorescence and surface chemistry, improved methods of data analysis, and expansion of the public forum assure alleviation of current hindrances.

We have provided a brief overview of micro-array technology as adapted for our applications to plant species. For more in-depth coverage please consult *DNA Micro-arrays: A Practical Approach* (14).

Acknowledgements

This work was supported by the National Science Foundation Grant No. DBI-9872617 (Plant Genome Program).

References

1. Southern, E. M. (1975). *J. Mol. Biol.*, **98**, 503–517.
2. Schena, M. *et al.* (1996). *Proc. Natl Acad. Sci.*, **93**, 10614–10619.
3. Cheung, V. G. and Nelson, S. F. (1998). *Genomics*, **47**, 1–6.
4. Galitski, T., Saldanha, A. J., Styles, C. A., Lander, E. S. and Fink, G. R. (1999). *Science*, **285**, 251–254.
5. Hacia, J. G., *et al.* (1999). *Nat. Genet.*, **22**, 164–167.
6. Iyer, V. R., *et al.* (1999). *Science*, **283**, 83–87.
7. Schena, M., Shalon, D., Davis, R. W. and Brown, P. O. (1995). *Science*, **270**, 467–470.
8. Reymond, P., Weber, H., Damond, M. and Farmer, E. E. (2000). *Plant Cell*, **12**, 707–720.
9. Richmond, T. and Somerville, S. (2000). *Curr. Opin. Plant Biol.*, **3**, 108–116.
10. Aharoni, A. *et al.* (2000). *Plant Cell*, **12**, 647–662.
11. Ruan, Y., Gilmors, J. and Conner, T. (1998). *Plant J.*, **15**, 821–833.
12. Desprez, T., Amselem, J., Caboche, M. and Hofte, H. (1998). *Plant J.*, **14**, 643–652.
13. Lemieux, B., Aharoni, A. and Schena, M. (1998). *Mol. Breeding*, **4**, 277–289.
14. Schena, M. (Ed.) (1999). *DNA micro-arrays: a practical approach.* Oxford University Press, New York.
15. Cargill, M. *et al.* (1999). *Nat. Genet.*, **22**, 231–238.
16. Gentalen, E. and Chee, M. (1999). *Nucl. Acid Res.*, **27**, 1485–1491.
17. Hacia, J. G. *et al.* (1998). *Nat. Genet.*, **18**, 155–158.
18. Okamoto, T., Suzuki, T. and Yamamoto, N. (2000). *Nat. Biotechnol.*, **18**, 438–441.
19. Roda, A., Guardigli, M., Russo, C., Pasini, P. and Baraldini, M. (2000) *Biotechniques*, **28**, 492–496.
20. Khan, J. *et al.* (1998). *Cancer Res.*, **58**, 5009–5013.
21. http://fastlane.nsf.gov/cgi-bin/A6QueryList
22. Duggan, D. J., Bittner, M., Chen, Y., Meltzer, P. and Trent, J. (1999). *Nat. Genet.*, Suppl. 21, 10–14.
23. http://www.arrayit.com
24. Zammatteo, N. *et al.* (2000). *Anal. Biochem.*, **280**, 143–150.
25. Mazia, D., Schatten, G. and Sale, W. (1975). *J. Cell Biol.*, **66**, 198.
26. http://cmgm.stanford.edu/pbrown/
27. DeRisi, J. L., Iyer, V. R. and Brown, P. O. (1997). *Science*, **278**, 680–686.
28. DeRisi, J. *et al.* (1996). *Nat. Genet.*, **14**, 457–460.
29. http://www.genisphere.com/
30. http://www.gsilumonics.com

31. Montagu, J. and Weiner, N. (1999). *J. Ass. Lab. Automation*, **4**.
32. Brignac, S. J. Jr *et al.* (1999). *IEEE Engl. Med. Biol. Mag.*, **18**, 120–122.
33. Ermolaeva, O. *et al.* (1998). *Nat. Genet.*, **20**, 19–23.
34. Chen, Y., Dougherty, E. R. and Bittner, M. L. (1997). *J. Biomed. Optics*, **2**, 364–375.
35. Claverie, J. M. (1999). *Hum. Mol. Genet.*, **8**, 1821–1832.
36. Bard, J. B. (1999). *Int. J. Dev. Biol.*, **43**, 397–403.
37. Baldwin, D., Crane, V. and Rice, D. (1999). *Curr. Opin. Plant Biol.*, **2**, 96–103.
38. Bowtell, D. D. (1999). *Nat. Genet.*, **21**, 25–32.
39. Eisen, M. B., Spellman, P. T., Brown, P. O. and Botstein, D. (1998). *Proc. Natl Acad. Sci. USA*, **95**, 14863–14868.

Chapter 4
DNA–protein interactions

Paul J. Rushton and Bernd Weisshaar

Max-Planck-Institut für Züchtungsforschung, Abteilung Biochemie,
Carl-von-Linné-Weg 10, 50829 Köln, Germany

1 Introduction

The availability of techniques for studies of the regulation of promoter activity has helped increase our knowledge concerning the function and interplay of specific genes in diverse biological processes. Information on the regulation of gene activity and the factors involved is important for the understanding of regulatory networks and cascades of transcription factors. In this chapter, we summarize some of our experience in collecting data on DNA–protein interactions from plant systems, and show how such approaches can lead to the identification of both *cis*-acting elements and transcription factors involved in the expression of the gene under study.

The starting point in many studies is the isolation of the regulatory regions from a gene of interest. These genes are normally studied because of their inducibility (for example, by hormones, pathogens, stress), or their tissue-specific or developmentally regulated expression. The first step in these studies is usually to identify *cis*-acting elements in the promoter either by functional analysis (transient expression, transgenic plants) or via DNA–protein interaction studies (DNase I footprints, electrophoretic mobility shift assays). Once these *cis*-acting elements have been identified, further protein binding studies can be performed to define the core sequences and produce mutant versions that abolish not only function, but also binding of the cognate transcription factors. The availability of functional and mutant versions then facilitates the cloning of the corresponding transcription factors for example, by (South-western (see *Volume 1, Chapter 10*) or yeast one hybrid screening (see *Chapter 9*).

We have divided this chapter into four parts; the preparation of nuclear extracts, EMSA reactions, DNase I footprinting, and *in vivo* DNA footprinting. All of the protocols have worked well in our hands and we believe them to be suitable for many other plant systems.

2 Preparation of nuclear extracts

The quality of the nuclear proteins is one of the most critical factors when investigating DNA–protein interactions *in vitro*. Protein extracts should be of a sufficiently high concentration, such that EMSA and/or DNase I footprinting experiments are possible. Unfortunately, some starting materials from plants contain high levels of protease and/or nuclease activities making the preparation of good nuclear extracts more difficult. As a guide for optimization of experiments we present two different procedures (1, 2). The first has been successfully used with cereal aleurone cells (1). Aleurone cells contain large quantities of hydrolytic enzymes, including proteases and nucleases, and therefore represent a tissue that is not easily amenable to study. The procedure involves the preparation of protoplasts and their subsequent lysis in a large volume of nuclear lysis buffer. The second procedure is used routinely with cultured cells and does not include a protoplasting step (2). An additional important parameter for extract preparation is the phosphorylation state of the proteins extracted. The addition of phosphatase and/or kinase inhibitors during extract preparation may influence the results of later EMSA/footprinting experiments.

2.1 Preparation of nuclear extracts from protoplasts

Protocol 1 uses protoplasts as the starting material (1). Nuclei are released by lysing the protoplasts in a large volume of nuclei isolation buffer followed by passing the lysate gently through a 20 μm nylon sieve. The nuclei are washed and then lysed in a small volume using a high salt extraction buffer. To ensure high protein concentrations, the nuclear proteins are precipitated by ammonium sulphate prior to dialysis and storage.

Protocol 1

Preparation of nuclear proteins from protoplasts

Perform all steps at 0–4 °C in a cold room. All centrifuges, glassware and equipment should be pre-cooled. The amount of protoplasts used as starting material is dependent on the quantities available. We routinely use about 7×10^6 protoplasts per preparation as this requires 100 ml nuclei isolation buffer (NIB) during the lysis step and the lysed protoplasts fit into a 100 ml centrifuge tube. For larger amounts the volume of all buffers should be scaled up accordingly.

Reagents

- Nuclei isolation buffer (NIB): 10 mM Tris–HCl pH 7.5, 10 mM KCl, 10 mM $MgCl_2$, 20% glycerol. Dithiothreitol (DTT), phenylmethylsulfonylfluoride (PMSF), and Triton X-100 are added freshly as indicated
- 1 M DTT stock
- 0.1 M PMSF stock solution (17.4 mg in 1 ml isopropanol)

- Protease inhibitors: leupeptin (1 mg/ml in water), pepstatin A (1 mg/ml in ethanol), antipain (1 mg/ml in water), and bestatin (1 mg/ml in water)
- HEN-25 buffer: 20 mM HEPES pH 7.5, 1 mM EDTA, 50 mM NaCl, 25% glycerol
- Spermidine stock (50 mM)

Method

1 Pellet 7×10^6 protoplasts gently at approximately 50 g and then wash in an equal volume of fresh medium. Place washed protoplasts on ice.

2 To 150 ml NIB add 0.15 ml Triton, 1.5 ml PMSF stock, and 1.5 ml DTT stock. This is used for all steps up to and including step 6.

3 Lyse the pellet of washed protoplasts by the addition of 40 ml NIB from step 2 and transfer to a glass beaker.

4 Using a syringe, take up 15 ml of lysed protoplast suspension and filter through a 20 μm nylon sieve placed in a filter holder on the end of the syringe. Rinse the debris on the filter with 15 ml NIB and repeat until the entire sample has been passed through.

5 Pellet nuclei by centrifugation at 1000 g for 20 min at 4 °C.

6 Resuspend the pellet in 35 ml NIB and centrifuge at 500 g for 15 min.

7 Prepare 10 ml NIB without Triton, i.e. add 100 μl PMSF stock and 100 μl DTT stock.

8 Resuspend the pellet in 6.0 ml NIB without Triton, divide between four microcentrifuge tubes and centrifuge for 3 min at 500 g. Each tube should not contain more than about 1×10^6 nuclei to ensure a good high salt extraction of proteins.

Protocol 1 continued

9 Remove as much NIB as possible and place the nuclei on ice.

10 Prepare 10 ml HEN-25 buffer by adding 50 µl DTT stock, 100 µl PMSF stock, 50 µl leupeptin stock, 250 µl antipain stock, 35 µl bestatin stock, and 40 µl pepstatin stock.

11 Resuspend each pellet in 0.5 ml of the HEN-25 buffer.

12 Transfer to a larger tube and add 1.2 ml HEN-25 buffer, 200 µl 5 M NaCl, and 200 µl spermidine stock. Mix gently.

13 Extract the nuclear proteins by gentle agitation for 1–2 h.

14 Centrifuge at 88 000 g for 15 min.

15 Remove the supernatant to a suitable tube and add solid ammonium sulphate to near saturation (this is approximately 0.475 g per ml).

16 Precipitate the proteins by gentle agitation for 1 h.

17 Transfer the samples to appropriate centrifuge tubes and centrifuge at 50 000 g for 30 min.

18 Discard the supernatant and add 100 µl HEN-25 buffer containing protease inhibitors to gently dissolve the precipitated proteins. The samples can be pooled at step 15 for ease of processing and it may then be necessary to add more HEN-25 buffer.

19 Dialyse the samples against HEN-25 buffer containing 5 µl DTT stock and 10 µl PMSF stock per ml. Because of the small volume we find a micro-dialysis apparatus most convenient. Remove insoluble material with a 5 min centrifugation at full speed in a microcentrifuge at 4 °C.

20 Store aliquots of the extract at −80 °C after freezing in liquid nitrogen.

2.2 Preparation of nuclear extracts from cultured cells

Protocol 2 uses cultured cells as the starting material. Nuclei are released by grinding the cells (2). The protocol is aimed at producing high nuclei concentrations prior to high-salt extraction to avoid the labour-intensive (and potentially damaging) ammonium sulphate precipitation necessary to reach sufficiently high protein concentrations.

Protocol 2

Preparation of nuclear protein from cultured cells

Reagents

- Nuclei isolation buffer B (NIB-B): 70% glycerol (v/v), 20 mM Tris–HCl pH 7.8, 5 mM $MgCl_2$, 5 mM KCl, 250 mM sucrose, 0.1% β-mercaptoethanol, 0.1 µM E-64, 200 µM PMSF. The proteinase inhibitors (Sigma) are added freshly prior to use.

- High-salt extraction buffer (HSE): 20 mM Tris–HCl, pH 7.8, 5 mM $MgCl_2$, 0.5 M NaCl,

- 0.1% β-mercaptoethanol, 0.1 µM E-64, 200 µM PMSF

- Dialysis buffer: 20% glycerol (v/v), 0.1 mM EDTA, 50 mM KCl, 25 mM Hepes-KOH, pH 7.8, 0.1% β-mercaptoethanol, 0.1 µM E-64, 200 µM PMSF

Method

Perform all steps at 0–4 °C in a cold room. We have used this protocol to prepare extracts from *Petroselinum crispum* and *Arabidopsis thaliana* cultured cells, but it may need to be optimized for other cell cultures.

1 Grind 50 g of cells to a fine powder in liquid nitrogen.

Protocol 2 continued

2 Suspend the powder in 2.2 volumes (v/w) of NIB-B.

3 Strain the homogenate through nylon meshes of first 80 μm and then 20 μm.

4 Centrifuge the filtrate at 3500 g for 1 h at 4 °C and discard the supernatant.

5 Resuspend the pellet in NIB-B [1.5 vols (v/w) of initial cells] and centrifuge as above for 40 min. Remove the supernatant.

6 Resuspend the pellet in 10 ml of HSE, incubate on ice for 15 min with slow stirring and centrifuge at 20 000 g for 20 min at 4 °C.

7 Dialyse the final supernatant against 2 l of dialysis buffer at 4 °C for 2 h. Try to avoid an increase in volume by tightly closing the dialysis bag.

8 Aliquot dialysed extracts and freeze in liquid nitrogen and store at −80 °C. Prepare small aliquots and use them only once.

3 Electrophoretic mobility shift assay (EMSA)

The electrophoretic mobility shift assay (EMSA) is a method to determine if a given DNA sequence is recognized as a binding site by DNA binding proteins (3–5). The main advantage of the method is that competition experiments can be performed. These experiments allow the determination of relative binding affinities of variant (mutant) elements and provide the opportunity to correlate protein binding with the effect, which the given mutation may have in functional assays.

When working with nuclear extracts, it is important to consider two types of competitor DNAs. First, 'non-specific' competitor may be required to allow detection of the DNA–protein interaction of interest. This 'non-specific' competitor is included so that it can be bound by non-sequence-specific DNA-binding proteins, which would otherwise recognize the low amount of labelled DNA probe and cause the probe to remain in the gel slot. Poly(dI-dC) and poly(dA-dT) work well as non-specific competitors (6, 7). The optimal concentration in the binding buffer needs to be determined experimentally. As a rule of thumb, it can be noted that poly(dI-dC) is better for the analysis of A/T-rich binding sites and poly(dA-dT) is better for G/C-rich binding sites. Usually, the non-specific competitor is not required when purified recombinant proteins are studied.

The second type of competitor is the specific competitor, which is used to demonstrate the specificity of the interaction. It is most important to show that a moderate excess of cold probe causes the slow-moving band representing the DNA–protein complex to disappear, but that it remains if a mutant probe is used. In the ideal case, a small mutation in the original cis-acting element, which is known to knock out function, should be used as a mutant competitor.

The conditions for competition experiments must be set up in such a way that the probe is in excess over the DNA binding protein at the standard conditions. Some free probe must be left all the time. A good initial test is to determine if an increase in the amount of extract used causes a similar increase in the intensity of the detected band (while keeping the amount of probe constant). Further assays should then be performed using an amount of extract which lies at the beginning of the area of linear correlation between amount of extract and band intensity. It should be emphasized again that meaningful results with EMSAs can only be produced if the test conditions are carefully optimized and adapted to the interaction under study. Among the conditions to be optimized are the buffer system to be used in the native gel, the composition of the binding buffer, such as the presence or absence of specific salts, e.g. zinc, the loading buffer and the temperature at which the binding reaction is performed.

3.1 DNA probes for use in EMSAs

In our hands, the best EMSA probes are short dsDNA oligonucleotides of about 30 bp in length. The design of 5′ overhangs, formed after hybridization of the single stranded oligos, allows easy labelling with DNA polymerase I Klenow fragment (*Protocol 3*). Larger DNA fragments prepared from plasmid DNA can also be used, but the results in most cases are not as good. The main problem is to define the promoter region containing the protein binding site or *cis*-acting element. This information may be derived from sequence comparisons, from *in vivo* or *in vitro* footprinting, or from functional promoter analyses. Larger areas detected by deletion analyses can be dissected by designing overlapping oligos (8).

Special care must be taken to ensure correct relative concentrations between probe and competitor DNA. The specific competitor oligonucleotides, which usually contain designed mutations or represent binding site variants of the *cis*-acting element under study, should also be filled in according to the protocol given below so that all DNAs are identical, i.e. blunt ended.

Protocol 3

Probe preparation for EMSA

Reagents

- Gel- or HPLC-purified oligonucleotides
- Klenow buffer: 50 mM Tris (pH 7.5), 10 mM MgCl$_2$, 1 mM DTT
- Klenow enzyme
- Four separate dNTP solutions: 10 mM dATP, 10 mM dCTP, 10 mM dGTP, and 10 mM TTP.
- Sephadex G-25 column with 1 ml bed volume
- TE: 10 mM Tris–HCl pH 7.5, 1 mM EDTA

Method

1. To prepare double stranded oligonucleotides, equimolar amounts of both single stranded oligos are combined in Klenow buffer, heated for a few min to 65 °C in a water bath and allowed to cool. Practically, this can be done by placing the small tube in a 50 ml tube containing the hot water and letting it sit at room temperature (RT) for about 30 min.

2. Add the following components to a 1.5 ml microcentrifuge tube: 100 ng ds oligonucleotide, 2 μl 10× Klenow buffer, 1 μl each of three dNTPs (the three unlabelled dNTPs), 25 μCi dXTP,[a] 2 U Klenow Fragment of DNA polymerase I, and water to a final volume of 20 μl. Mix and incubate at RT.

3. After 10 min, add 4 μl of a mixture of each dNTP (all four unlabelled), mix, and incubate for 30 min at RT.

4. Separate the labelled oligonucleotide from unincorporated nucleotides by centrifuging the probe for 5 min through a Sephadex G-25 spin column, which has been equilibrated with TE.

5. Bring the eluate of the spin column to a volume of 500 μl with TE and determine the incorporated radioactivity. Expect about 20 000 to 30 000 cpm/μl (Czerenkov counts).

6. The probe can be used for about 10 days. Store at −20 °C or 4 °C depending on how often you need to use it. Avoid too many freeze-thaw cycles.

[a] dXTP is the labelled, fourth, dNTP. The overhang should be checked to ensure this dNTP will be incorporated.

3.2 EMSA reactions

These experiments are performed in a cold room. In our hands, EMSAs performed at 4 °C show much better resolved (sharper) bands (*Protocol 4*).

Protocol 4

Electrophoretic mobility shift assay (EMSA)

Reagents

- Running buffer for PAGE (TBE): 45 mM Tris–HCl, pH 7.5, 45 mM boric acid, 1 mM EDTA
- Binding buffer (1×): 25 mM HEPES-KOH pH 7.4, 4 mM KCl, 5 mM MgCl$_2$, 1 mM EDTA, 7% glycerol (9)

- Solutions to set up a 5% polyacrylamide gel: 30% acrylamide with 1/40 bisacrylamide; 10% ammonium persulphate (APS), N,N,N′,N′-tetramethylethylenediamine (TEMED)
- Whatman DE81 paper

Method

1 Prepare a native 5% polyacrylamide (40:1) gel in TBE. Set up the gel with TBE as running buffer in the cold room (4 °C) and pre-run it prior to loading for 30 min at 250 V. Our gels are 29 × 17 × 0.1 cm (including spacers) with 20 slots.

2 Combine in a 1.5 ml microcentrifuge tube (placed on ice): 4 μl 5× binding buffer, 2 μg nuclear proteins,[a] 400 ng unspecific competitor,[b] specific competitor,[c] labelled probe (200 pg), and water to a final volume of 20 μl. Mix and incubate at room temperature for 10 min.

3 Load samples onto the gel with the power on (we use 100 V—take extreme care!). Once conditions have been optimized, it may be possible to load gels without the power on, but this must be determined for the interactions under investigation. The glycerol in the binding buffer is sufficient to keep it in the slot. No 'stop mix' or 'loading buffer' need be added. Bromophenol blue-containing loading buffer is added to the sample that contains the probe only, in order to monitor migration.

4 Run the gel at 250 V, 30 mA. We run our gels for 135 min, but the optimum time for your equipment and samples will have to be determined empirically.

5 Remove the gel and dry onto Whatman DE81 paper. This paper binds (radioactive) DNA, and therefore avoids loss of label and contamination of the gel dryer.

6 Expose for between 12 and 24 h.

[a] Or different amount of DNA-binding protein from other sources: e.g. recombinant protein from *E. coli*, *in vitro* translated protein.

[b] The actual amount should be determined empirically for each extract preparation.

[c] If desired; 5-, 25-, or 125-fold excess. The specific competitor is mixed with the other components and the labelled probe.

---

4 DNase I footprinting

DNase I footprinting can be a very powerful tool for establishing multiple areas of DNA–protein interaction within a promoter fragment of several hundred base pairs. These experiments can be very useful starting points for promoter analysis because the positions of many putative *cis*-acting elements can be revealed (*Figure 1*). This information can be then used to

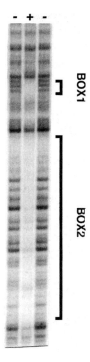

Figure 1 DNase I footprinting. A DNA fragment from the promoter region of the parsley *BPF1* gene was 3´ end-labelled with T4 DNA polymerase and DNase I footprinting reactions were performed using 50 µg parsley nuclear protein. –, the pattern of DNA cleavage in the absence of added nuclear protein; +, the pattern obtained after the addition of nuclear proteins. Two footprint regions are indicated by brackets. Box 2 is the result of a number of DNA–protein interactions.

design functional experiments and further binding analyses. DNase I footprinting is often more powerful than mobility shift analysis because the experiments yield more detailed information about the number and position of DNA–protein interactions. The major disadvantage of the technique is that the majority of potential binding sites must be occupied by a DNA-binding protein for a footprint to be visible. Note that for reliable data the analysis of both DNA strands is required.

The most important requirement for DNase I footprint experiments is a nuclear protein preparation of sufficiently high concentration and which is free from nuclease and protease activity. The best method for extract preparation may vary with the tissue used and must be determined experimentally. In this chapter we have presented two methods, one using whole cells and the other using protoplasts as starting material (*Protocols 1 and 2*).

4.1 DNA probes for DNase I footprinting

DNA probes are prepared by radiolabelling at the 3′ ends, followed by cutting with a restriction enzyme to create a double-stranded DNA fragment with an unlabelled 5′ end. The probes can be labelled using Klenow fragment, but our standard protocol uses T4 DNA polymerase to recess the 3′ ends using its 3′ to 5′ exonuclease activity, which occurs in the absence of dNTPs, followed by a fill-in reaction using the 5′ to 3′ polymerase activity, which occurs after the addition of dNTPs. In this way, probes of very high specific activity can be generated. Plasmid DNA containing the promoter to be studied is digested with a restriction enzyme that cuts once close to the 3′ end of the promoter. Enzymes should leave 5′ overhangs as the reaction works best from a recessed 3′end. The method has been used successfully (1) and is presented in *Protocol 5* with minor modifications.

Protocol 5

3′-end-labelling using T4 DNA polymerase

Reagents

- T4 DNA polymerase (1 U/μl)
- 10× T4 buffer: 0.33 M Tris-acetate pH 8.0, 0.66 M potassium acetate, 0.1 M magnesium acetate, 5 mM DTT, 1 mg/ml BSA
- dNTP stock 1 (2 mM each of dATP, dCTP, and TTP)
- dNTP stock 2 (2 mM each of dGTP, dATP, dCTP, and TTP)
- [α-^{32}P]dGTP
- Klenow fragment (5 U/μl)
- 1 M DTT

- 0.5 M EDTA pH 8.0
- Phenol:chloroform:isoamyl alcohol (25:24:1)
- Chloroform:isoamyl alcohol (24:1)
- 40% (w/v) acrylamide stock (38:2 acrylamide:bis-acrylamide)
- Gel elution buffer: 0.5 M ammonium acetate, 10 mM magnesium acetate, 1 mM EDTA, 0.1% SDS
- TE (see Protocol 3)
- 3 M sodium acetate (pH 5.2)

Method

1 Perform the restriction digest with 20 μg DNA and check an aliquot on an agarose gel to ensure that digestion has gone to completion.

2 Add an equal volume of phenol/chloroform, vortex mix briefly and centrifuge for 5 min at full speed in a microcentrifuge.

3 Collect the supernatant, add an equal volume of phenol:chloroform:isoamyl alcohol, and vortex and centrifuge as in step 2.

4 Collect the supernatant and add 0.1 vol 3 M sodium acetate (pH 5.2) followed by 2.5 vols 100% ethanol. Mix and centrifuge at full speed in a microcentrifuge for 20 min.

5 Decant the supernatant and wash the pellet with 70% ethanol, dry, and dissolve in 60 μl TE (pH 7.5).

6 In a microcentrifuge tube make the following reaction mix: digested DNA (approximately 100 ng), 6.5 μl 10× T4 buffer, 13 μl 10 mM DTT, and water to a final volume of 50 μl. Warm to 30 °C.

7 To another tube add the fill-in mix: 25 μl [α-^{32}P]dGTP,[a] 4 μl of combined dCTP, dATP, and TTP stock (dNTP stock 1), and 1 μl Klenow fragment (5 U/μl).

8 Add 1.25 μl T4 DNA polymerase (1 U/μl) to the resection mixture. Mix gently. After 20 s[b] add the contents to the fill in mixture. Mix gently by pipetting.

9 Incubate for 45 min at 42 °C.

10 Add: 8 μl combined dATP, dCTP, dGTP, and TTP stock, 0.5 μl Klenow fragment (5 U/μl) and 1 μl T4 DNA polymerase (1 U/μl).

11 Incubate for 1 h at 37 °C.

12 Add 1 μl EDTA (0.5 M) and incubate at 70 °C for 5 min.

13 Add 9 μl 3 M sodium acetate and 225 μl 100% ethanol. Incubate at −80 °C in a suitable container until frozen.

14 Centrifuge for 20 min at full speed in a microcentrifuge to pellet the DNA. Wash with 70% ethanol. The supernatants contain a large amount of radioactivity and should be handled with caution.

15 Dry the DNA and dissolve it in the appropriate buffer for the second restriction digest.

16 Add the second restriction enzyme and incubate for 1 h.

17 During the digest cast the following mini gel:[c]

11.55 ml H_2O;

1.50 ml 10× TBE;

1.88 ml 40% (w/v) acrylamide stock (38:2 acrylamide:bis-acrylamide);

100 µl 10% (w/v) ammonium persulphate (fresh);

20 µl TEMED.

18 Add an appropriate amount of loading dye to the sample. Load onto the gel and run so that the labelled promoter fragment is well separated from any vector fragments.

19 Once the electrophoresis is finished rinse and clean all apparatus, and properly dispose of the buffer, which will be radioactive. Remove one of the gel plates so that the gel remains attached to the other. Cover the gel and glass plate carefully with Saran wrap.

20 In a darkroom, put the gel onto a piece of X-ray film gel side down. Draw around the gel plate with a marker pen, and mark one corner of both the film and glass plate for orientation purposes. Leave the gel on the film for approximately 4 min.

21 Develop and dry the X-ray film.

22 Place the developed film on the gel. Position correctly and mark the position of the labelled promoter fragment by punching four holes through the film and into the gel using a syringe needle.

23 Cut out the probe and transfer to a microcentrifuge tube. Steps 20 and 21 can be repeated to verify that the probe has been correctly excised.

24 Macerate the gel fragment, add 450 µl gel elution buffer and vortex strongly.

25 Incubate in a suitable container at 37 °C overnight.

26 Centrifuge for 10 min at full speed in a microcentrifuge. Add 150 µl of gel elution buffer. Vortex, centrifuge, and combine the two supernatants.

27. Run the supernatant over a Sephadex G25 column (ready-to-use columns are available from Pharmacia).

28 The probe can now be used immediately or stored at −20 °C until use. Best results are obtained with freshly labelled promoter probes. As an alternative to steps 24–27, a commercially available kit can be used and the fragment eluted with water. This may be advantageous if Maxam and Gilbert sequencing is to be performed.

[a] Higher specific activities can be obtained by using more than one labelled nucleotide.

[b] The optimal resection times must be determined for each probe, but 20–30 s is a good indicative time.

[c] This can be scaled up or down depending on the apparatus used.

4.2 DNase I footprinting reactions

The footprinting reactions are simple and quick. However, the amount of protein required to produce a footprint must be determined for each protein sample, as must the amount of DNase I (*Protocol 6*).

Protocol 6

DNase I footprint reactions

Reagents

- HEN-5 buffer: 20 mM HEPES pH 7.5, 1 mM EDTA, 50 mM NaCl, 5% glycerol
- HEN-25 buffer: 20 mM HEPES pH 7.5, 1 mM EDTA, 50 mM NaCl, 25% glycerol
- 100 mM $MgCl_2$, 20 mM $CaCl_2$
- DNase I (25 U/µl)
- Stop solution: 15 mM EDTA, 0.15% SDS, 100 mM NaCl
- tRNA (2 µg/µl)
- Phenol:chloroform:isoamyl alcohol (24:24:1)

- Poly(dI-dC) (1 µg/µl)
- Formamide loading dye: 95% formamide, 4% EDTA, 0.1% bromophenol blue, 0.1% xylene cyanol
- 5× TBE: 54.0 g Tris base, 27.5 g boric acid, 20 ml 0.5 M EDTA pH8.0, water to 1 l final volume
- 40% (w/v) acrylamide stock (38:2 acrylamide: bis-acrylamide)

Method

1 Each footprinting reaction contains the following in a total volume of 15 µl: between 2 and 8 µl nuclear protein (depending on concentration), 2 µl 10 mM DTT, 1 µl Poly(dI-dC) (10 ng/µl), 4.5 µl HEN-5, and 2.5 µl promoter probe. Control tracks lacking protein contain HEN-25 instead of protein and 1 µl Poly(dI-dC) (1 µg/µl).

2 Mix gently by pipetting and incubate at room temperature for 20–30 min.

3 Prepare stocks of 1 U/µl and 0.1 U/µl DNase I in HEN-5. Store on ice.

4 Add 1.0 µl of the 100 mM $MgCl_2$, 20 mM $CaCl_2$ solution to the first footprinting reaction. Mix very gently.

5 Add the required amount of DNase I[a] and mix very gently. Incubate for 1 min.

6 Add 80 µl stop solution and mix. Add 1 µl tRNA. Repeat for the other footprinting reactions.

7 Add an equal volume of phenol/chloroform, vortex for 1 min, and centrifuge at full speed in a microcentrifuge. Transfer the upper phase to a fresh microcentrifuge tube. Do this twice for the samples that contain protein.

8 Add 250 µl −20 °C ethanol. Mix and store at −80 °C until frozen.

9 Centrifuge for 20 min at full speed in a microcentrifuge.

10 Wash with 70% ethanol and recentrifuge for 10 min.

11 Air dry and dissolve in 1.5 µl loading dye.

12 Denature at 90 °C for 3 min.

13 Load onto a pre-electrophoresed sequencing gel. If required, run Maxam and Gilbert A + G sequencing reactions (10) alongside to establish the identity of the bases in the footprinted regions.

14 Electrophorese the gel, then remove and dry it. Autoradiograph the dried gel at −80 °C overnight using an intensifying screen.

[a] This will have to be optimized for each new probe or protein preparation. As a guide, we found that DNase I in the range 0.5–2.0 µl of 0.1 U/µl is optimal for control reactions without protein. For binding reactions with 50–100 µg protein 1.0–2.0 µl 1.0 U/µl DNase I produced the best results.

5 *In vivo* footprinting

All the methods presented so far in this chapter are *in vitro* methods for the analysis of DNA–protein interactions and these techniques may not, in all cases, reflect the *in vivo* situation faithfully. By contrast, *in vivo* footprinting can provide information on the interaction of *trans*-acting factors with their cognate *cis*-acting elements *in vivo* to single nucleotide resolution.

In vivo footprinting is a time consuming process. First, dimethylsulphate (DMS) is used to produce a partial methylation of intact chromatin in living cells. These cells can be subjected to various treatments that alter promoter occupancy before methylation. The DNA is then purified, digested with a restriction endonuclease of choice, enriched for a promoter fragment of interest, and chemically cleaved with piperidine at positions of methylated guanine residues. The resulting fragments are separated on a denaturing polyacrylamide gel, transferred to a nylon membrane, and visualized by hybridization with a short single-stranded DNA probe that shares one end with a chosen restriction site ('indirect end-labelling'; *Figure 2*). The intensity of each band is compared with that obtained when using naked DNA *in vitro*. Apparent protection or enhancement of cleavage should reflect binding of DNA-binding proteins at or close to these residues.

Our *in vivo* footprinting protocol is based on that by Schulze-Lefert *et al.* (11). This protocol, although highly reliable, is relatively old, and thus a number of modifications or improvements are suggested. More recent PCR-based approaches can also be employed (12), and probe preparation by primer extension and isotachophoresis could be avoided by using end-labelled oligonucleotides. The entire process is, however, long and complicated, and we therefore suggest using the original, highly reliable, protocol that is presented here as a start point.

5.1 *In vivo* methylation of plant cells

Protocol 7 has been developed for use with plant cell cultures (11) and works best with such a homogeneous cell type. The chosen time points and/or treatments of the cells should be chosen carefully so that any changes in DNA–protein interactions are maximized. Before embarking on these experiments it should be stressed that DMS is extremely carcinogenic.

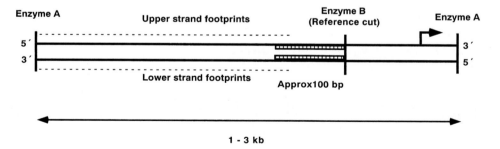

Figure 2 A schematic representation of the *in vivo* footprinting method. A 1–3 kb *in vivo* methylated DNA fragment containing the promoter region of interest is produced by the digestion of genomic DNA with restriction enzyme A. After enrichment, the DNA is cleaved at the position of methylated guanine residues. The fragments are separated on a sequencing gel and transferred to a nylon membrane. The fragments are then visualized by hybridization to a short single-stranded probe (striped boxes) that shares one end with a chosen restriction site (the reference cut—enzyme B). The intensity of each band is compared with that obtained when using naked DNA *in vitro*. Apparent protection or enhancement of cleavage should reflect binding of DNA-binding proteins at, or close to, these residues.

Gloves should always be worn and all solutions and glassware should be treated with 10 M NaOH after use.

Protocol 7

In vivo methylation of plant cells

Reagents

- DMS
- Ice-cold water

Method

1   Add DMS to the plant cells to a final concentration of 0.5% (v/v) (in our case 200 ml of parsley cell suspension culture and 1 ml DMS).
2   Swirl to mix and incubate for 1 min at 26 °C.
3   Add 10 vols of ice-cold water.
4   Mix and filter through a Buchner funnel onto filter paper.
5   Wash twice with an equal volume of ice-cold water.
6   Store the cells at −80 °C after freezing in liquid nitrogen.

5.2 Preparation of genomic DNA

Genomic DNA can be isolated by a standard SDS/phenol/chloroform method, but can also, for example, be prepared by a kit to ensure rapid isolation of good quality genomic DNA. We routinely use the Nucleon Phytopure Plant DNA extraction kit from Amersham.

5.3 Production and enrichment of promoter fragments

The next steps in the protocol are the production and enrichment of promoter fragments (*Protocol 8*). For the promoter of interest, a digest is performed with a restriction endonuclease that results in a convenient (1–3 kb) fragment of genomic DNA that includes the region of DNA to be investigated. The digested DNA is enriched for fragments in this size range by a sucrose gradient. The enriched DNA is then digested with a different restriction endonuclease to generate the 'reference cut' that defines the 3′ end of the investigated region and the start of the probe (*Figure 2*).

Protocol 8

Separation and enrichment of promoter fragments

Reagents

- Phenol:chloroform:isoamyl alcohol (25:24:1)
- Chloroform
- Isopropanol
- 70% Ethanol
- TE: 10 mM Tris–HCl pH 7.5, 1 mM EDTA
- Sucrose/NaCl solution: 20 mM Tris–HCl pH 8.0, 5 mM EDTA, 40% (w/v) sucrose, 1 M NaCl
- Sucrose solution: 20 mM Tris–HCl pH 8.0, 5 mM EDTA, 10% (w/v) sucrose
- 3 M sodium acetate pH 5.2

Protocol 8 continued

Method

1 Digest 1 mg genomic DNA for 2 h at 37 °C with 1000 U of an appropriate restriction enzyme to generate a range of fragments of about 1–3 kb.

2 Add an equal volume of phenol/chloroform, vortex for 1 min, and centrifuge at full speed in a microcentrifuge. Transfer the upper phase to a fresh microcentrifuge tube.

3 Add an equal volume of chloroform, vortex for 1 min, and centrifuge at full speed in a microcentrifuge. Transfer the upper phase to a fresh microcentrifuge tube.

4 Add 0.1 vol of 3 M sodium acetate and 0.6 vol isopropanol.

5 Mix and centrifuge for 20 min at full speed in a microcentrifuge.

6 Wash with 70% ethanol and centrifuge for 20 min at full speed in a microcentrifuge.

7 Air dry the pellet and dissolve in TE at a concentration of 1 mg/ml.

8 Using a gradient mixer prepare a linear 10–40% sucrose gradient using the sucrose and sucrose/NaCl.[a]

9 Carefully layer the DNA solution onto the top of the gradient and centrifuge at 141 000 g for 20 h at 4 °C.[b]

10 Fractionate the gradient into 0.5 ml aliquots.[c]

11 Apply 30 μl of every third or fourth fraction directly onto a 0.8% agarose gel.

12 Electrophorese and compare the size of the DNA in each fraction with size markers.[d]

13 Pool the fractions containing the DNA with the desired fragment lengths.

14 Dialyse overnight against TE.

15 Concentrate the dialysed DNA solution by several extractions with 2-butanol (1:1 v/v).

16 Extract once with phenol/chloroform, once with chloroform, and precipitate as described in steps 2–7.

17 Redissolve the DNA in TE.

[a] We use a gradient former with two 25 ml reservoirs and prepare a linear gradient in a 38 ml centrifuge tube for use in a SW28 Beckman rotor (see also *Chapter 8, Protocol 7*).

[b] We use a Beckman SW28 rotor in a suitable centrifuge and centrifuge at 28,000 rpm.

[c] The method that we use for this employs a glass capillary that is carefully inserted into the centrifuge tube and pushed down to the bottom of the gradient. The solution is drawn into a tube that is attached to the capillary and the first aliquot removed to a microcentrifuge tube. The withdrawal is stopped by a clamp, the microcentrifuge tube replaced, and the process repeated until the whole gradient has been fractionated.

[d] Confirmation that the desired fragment is present can be obtained by performing a Southern blot.

5.4 Preparation of fractionated DNA for gel electrophoresis

Both the fractionated DNA from the sucrose gradient and cloned promoter fragment DNA are prepared for electrophoresis (Protocol 9). The cloned DNA serves as a reference by showing which nucleotides show protection from or enhancement of methylation due to the binding of proteins *in vivo*.

About 500 μg of DNA is separated on a sucrose gradient. For each gel, half of the pooled DNA fractions from one sucrose gradient is digested with an appropriate restriction enzyme in order to create the reference cut. This is about 50 μg DNA.

Protocol 9

Preparation of genomic DNA sample for electrophoresis

Reagents

- Phenol:chloroform (1:1)
- Chloroform
- 10% Piperidine[a] diluted freshly with water
- Formamide loading buffer: 94% deionized formamide, 0.05% (w/v) xylene cyanol, 0.05% bromophenol blue, 10 mM EDTA pH 8.0

Method

1 Digest half of the pooled DNA (about 50 µg) with 20–30 U of an appropriate restriction enzyme for 1.5–2 h at 37 °C in order to create the 'reference cut'. This is about 50 µg DNA.

2 Add an equal volume of phenol/chloroform, vortex for 1 min, and centrifuge at full speed in a microcentrifuge. Transfer the upper phase to a fresh microcentrifuge tube.

3 Add an equal volume of chloroform, vortex for 1 min, and centrifuge at full speed in a microcentrifuge. Transfer the upper phase to a fresh microcentrifuge tube.

4 Add 0.1 vol of 3 M sodium acetate and 0.6 volume isopropanol.

5 Mix and centrifuge for 20 min at full speed in a microcentrifuge.

6 Wash with 70% ethanol and centrifuge for 20 min at full speed in a microcentrifuge.

7 Air dry the pellet and dissolve in 100 µl of 10% piperidine.

8 Incubate at 90 °C for 30 min.

9 Cool the tubes on ice.

10 Punch holes into the lid of the tubes with a syringe needle.

11 Freeze the tubes in liquid nitrogen.

12 Dry in a lyophilizer.

13 Add 50 µl water, freeze in liquid nitrogen, and dry again in a lyophilizer.

14 Redissolve in 5 µl formamide loading buffer.

15 Store at −20 °C.

[a] Piperidine is stored in dark bottles at 4 °C. The concentration is 10 M.

5.5 Preparation of cloned promoter fragment DNA for gel electrophoresis

As mentioned above, cloned promoter fragment DNA serves as a reference by showing which nucleotides show protection from or enhancement of methylation due to the binding of proteins *in vivo*. In addition, this DNA enables the reading of the genomic sequence. The cloned promoter fragment DNA is methylated *in vitro* and has the same 'reference cut' as the *in vivo* methylated genomic DNA (*Figure 2*). The DNA is then cleaved at guanine and adenine residues using the Maxam and Gilbert sequencing reactions (*Protocol 10*).

Protocol 10

Preparation of cloned promoter fragment DNA sample for electrophoresis

Reagents

- DMS buffer: 50 mM sodium cacodylate pH 8.0, 1 mM EDTA
- DMS stop: 1.5 M sodium acetate pH 7.0, 1.0 M 2-mercaptoethanol
- 10% Piperidine diluted freshly with water

- 1 mg/ml herring sperm DNA in water
- A buffer: 1.5 M NaOH
- A Stop: 1.0 M acetic acid
- TE: 10 mM Tris–HCl pH 8.0, 1 mM EDTA

Method

1 Digest 1 μg of a plasmid containing the promoter fragment of interest with the same restriction enzyme used to generate the 'reference cut' with the *in vivo* methylated DNA for 1–2 h.

2 Stop the reaction by adding 0.25 M EDTA to a final concentration of 10 mM.

3 Both a G-reaction and an A>C-reaction are performed. For the G-reaction, combine 200 μl DMS buffer, 3 μl carrier DNA, 1 μl DNA from step 2, and 1 μl DMS in a microcentrifuge tube on ice.

4 Incubate at 37 °C for 30 s.

5 Add 50 μl DMS stop and 750 μl 100% ethanol.

6 Mix and centrifuge for 20 min at full speed in a microcentrifuge.

7 For the A>C-reaction combine 80 μl A buffer, 3 μl carrier DNA, and 1 μl DNA from step 2 in a microcentrifuge tube.

8 Incubate at 90 °C for 5 min.

9 Add 150 μl A stop and 750 μl 100% ethanol.

10 Mix and centrifuge for 20 min at full speed in a microcentrifuge.

11 For both the G and A>C reaction, dry the samples, and add 100 μl 1 M piperidine.

12 Incubate at 90 °C for 30 min.

13 Cool on ice.

14 Dry in a lyophilizer.

15 Add 50 μl water, freeze in liquid nitrogen, and dry again in a lyophilizer.

16 Dissolve in 40 μl formamide loading buffer.

17 Store at −20 °C.

5.6 Preparation of a genomic filter

The DNA samples are separated on a more or less standard sequencing gel (*Protocol 11*). The percentage of the gel is dependent on the size of the fragments that are to be visualized. For fragments of 160–180 bp, we use a 6% polyacrylamide gel (19:1 acrylamide:bisacrylamide) made in 0.5× TBE and containing 7 M urea.

Protocol 11

Preparation of genomic filter

Reagents

- Nylon membrane or similar[a]
- 40% (w/v) acrylamide stock (38:2 acrylamide:bis-acrylamide)
- 6% polyacrylamide gel premix: 420 g urea, 145 ml 40% acrylamide stock, water to 1 l
- 10× TBE: 121.1 g Tris base, 51.3 g boric acid, 3.7 g EDTA, water to 1 l, pH 8.3

Method

1 Cast a polyacrylamide sequencing gel.[b]

2 Pre-run the gel in 0.5% TBE for 3 h at 3000 V.

3 Heat the DNA samples (both genomic and cloned) for 3 min at 95 °C, put on ice and then load each sample onto the gel.

4 Run the gel at 1600 V. The run time of the gel is 8–9 h.

5 Transfer the gel carefully onto one or more (depending on size) sheets of Whatmann 3MM paper.

6 Wet Gene Screen membrane (or similar) in 0.5% TBE and put on top of the gel.

7 Set up an electroblot.[c]

8 Wrap the membrane containing the DNA in Saran Wrap and irradiate with UV-light to covalently cross-link the transferred DNA to the membrane.[d]

9 Store the membrane at room temperature.

[a] We used Gene Screen from NEN, but many membranes could be suitable. This must be tested by the experimenter.

[b] We use 6% gels that are 80 cm long and 0.5 mm thick.

[c] Electroblots are run in precooled 0.5× TBE. The conditions we use are: 45 min at 35 V and 1.1 Amps at a temperature of 4 °C.

[d] The conditions are: 15 min at a 15 cm distance, λ max 306 nm (no 254 nm UV-light).

5.7 Probe synthesis

Originally, our probes were synthesized by primer extension using a M13 template (11). Now, however, we suggest the use of a directly labelled long oligonucleotide as being far more convenient (*Protocols 12* and *13*). For completeness, we also present the original protocols for probe synthesis (primer extension and isotachophoresis) as they have been used successfully. These probes should be 90–150 bases long and should be located next to the sequence to be analysed (*Figure 2*). Digestion of a primer extension-labelled probe with a restriction enzyme will release the probe, defined by the primer at the 5 prime end and the reference cut at the 3 prime end.

Protocol 12

Probe synthesis

Reagents

- Primer: 10 ng/μl in TE
- 10× Klenow buffer: 100 mM NaCl, 50 mM Mg Cl$_2$
- dNTP stock: dCTP, dGTP, TTP 10 mM of each
- 10 mM dATP
- 250 μCi dATP (10 μCi/μl)
- DNA polymerase I (2 U/μl)
- 3 M sodium acetate pH 5.2
- Restriction enzyme (the same as used for the reference cut)
- 100% Ethanol
- 70% Ethanol
- Formamide loading buffer: 94% deionized formamide, 0.05% (w/v) xylene cyanol, 0.05% bromophenol blue, 10 mM EDTA pH 8.0

Method

1 Combine the following in a microcentrifuge tube: 6 μl DNA template (1 μg/μl), 7 μl primer (10 ng/μl), 6 μl 10× Klenow buffer, 14 μl water.

2 Heat to 75 °C for 2 min.

3 Incubate at 42 °C for 30 min.

4 Add 3 μl dNTP mix, 25 μl dATP (250 μCi), 4 μl DNA polymerase I (8 U).

5 Incubate at room temperature for 7 min.

6 Add 1 μl 10 mM dATP.

7 Incubate further at room temperature for 10 min.

8 Inactivate the enzyme by heating at 70 °C for 2 min.

9 Add 25 μl water, 10 μl 10× restriction buffer, 20 U restriction enzyme.

10 Incubate at 37 °C for 1 h.

11 Stop the reaction by adding 2 μl 0.5 M EDTA.

12 Add 0.1 vol 3 M sodium acetate and 2.5 vols 100% ethanol. Mix.

13 Centrifuge for 10 min at full speed in a microcentrifuge.

14 Wash with 70% ethanol, centrifuge, and air dry.

15 Dissolve in 20 μl formamide loading buffer.

16 Pour a 6% polyacrylamide gel (in 1× TBE) with 1-cm wide slots.

17 Prerun the gel for 30 min at 500 V.

18 Denature the probe by boiling for 3 min.

19 Chill on ice, load onto the gel, and run at 500 V for 20 min.

20 When the gel has run cut out the probe and put the gel slice into a fresh microcentrifuge tube.

5.8 Probe elution

The probe is electroeluted from the gel slice by isotachophoresis. Other elution strategies are possible and may be more rapid, but the isotachophoresis protocol presented has been routinely and successfully used by us (*Protocol 13*).

Protocol 13

Isotachophoresis

Reagents

- Sephadex G50 fine
- Dialysis membrane with a 3500 cut off size
- 100 mM 6-aminocaproic acid
- TE: 10 mM Tris–HCl pH 8.0, 1 mM EDTA

Method

1 Insert a piece of ashless hardened paper into a 5-ml plastic syringe so that it acts as a support.
2 Pack degassed Sephadex G50 into the column to about halfway up the column.
3 Equilibrate the column with TE.
4 Close the column tip with a dialysis membrane by pushing a small plastic band over the column tip and dialysis membrane. Make sure that no air bubbles are trapped.
5 Remove any residual TE from the top of the column.
6 Overlay the column carefully with 100 mM 6-aminocaproic acid.
7 Fill a beaker with TE. This will act as the lower reservoir. Dip the column tip with dialysis membrane into this buffer.
8 Insert a platinum wire (the cathode) into the upper reservoir (caproic acid) and another (the anode) into the lower reservoir (TE).
9 Check that the column is functioning correctly by pre-running at 300 V for 1 min
10 Load the gel pieces onto the column and in addition a drop of 0.1% xylene cyanol as indicator.
11 Run at 300 V until the indicator dye is focused and has reached the bottom of the column.
12 Stop the isotachophoresis and elute the DNA by puncturing the dialysis membrane with a syringe needle. The peak of DNA is in front of the dye marker. Collect 500 μl fractions.
13 Centrifuge for 10 min at full speed in a microcentrifuge.
14 Count 1 μl in a scintillation counter. The probe should contain more than 10^5 Czerenkov counts.

5.9 Hybridization

Protocol 14

Hybridization

Reagents

- Hybridization mix: 89 g $Na_2HPO_4.2H_2O$, 4 ml 85% H_3PO_4, 1% BSA, 7% SDS, water to 1 l, pH 7.2
- Wash buffer: 7.12 g $Na_2HPO_4.2H_2O$, 0.32 ml 85% H_3PO_4, 1% SDS, 2 ml 0.5 M EDTA, 100 mM NaCl per 5% reduction in GC content, water to 1 l, pH 7.2

Method

1 Wet a cross-linked membrane in 0.5× TBE and roll it around a 10-ml glass pipette before inserting it into glass hybridization tube.
2 Pour the 0.5% TBE off and add 20 ml of hybridization mix.

Protocol 14 continued

3 Close the tube with the screw cap and prehybridize in an incubator[a] that continually rotates the tube for 1 h at 65 °C.

4 Pour off the hybridization mix and add 7–10 ml of hybridization mix containing the DNA probe.

5 Hybridize for 16–20 h.

6 Pour off the probe solution and rinse once with 100 ml wash buffer.

7 Wash the membrane four times with 1 l of wash buffer at 60 °C for 10 min.

8 Cover the moist filter with Saran wrap and expose to an X-ray film at −80 °C for at least 4 days using an intensifying screen.[b]

[a] We use a GFL 7601 oven.

[b] Membranes can be rehybridized with different probes after stripping with TE at 65 °C for 5 min.

6 Concluding remarks

Many protocols, even if they do work easily and reliably in one laboratory, cause trouble when first tried somewhere else. There are many reasons for this, but the protocols given in this chapter should be a good starting point for most studies. However, when running into trouble it might be helpful to check the literature for other people's experiences on DNA–protein interaction studies. As a starting point, we suggest references 2 and 13–15. Good luck!

Acknowledgements

We thank Thomas Eulgem, Jonathan Phillips, and Imre Somssich, for critical reading of the manuscript, and Richard Hooley, for his support and advice during the development of a number of these protocols.

References

1. Rushton, P. J., Hooley, H. and Lazarus, C.M. (1992). *Plant Mol. Biol.*, **19**, 891.
2. Armstrong, G. A., Weißhaar, B. and Hahlbrock, K. (1992). *Plant Cell*, **4**, 525.
3. Garner, M. M. and Revzin, A. (1981). *Nucl. Acids Res.*, **9**, 3047.
4. Fried, M. and Crothers, D. M. (1981). *Nucl. Acids Res.*, **9**, 6505.
5. Carey, J. (ed.) (1991). In *Methods in enzymology*, Vol. 208, p. 103. Academic Press, London.
6. Buratowski, S., Hahn, S., Guarente, L. and Sharp, P.A. (1989). *Cell*, **56**, 549.
7. Davidson, I., Xiao, J. H., Rosales, R., Staub, A. and Chambon, P. (1988). *Cell*, **54**, 931.
8. Korfhage, U., Trezzini, G. F., Meier, I., Hahlbrock, K. and Somssich, I. E. (1994). *Plant Cell*, **6**, 695.
9. Holdsworth, M. J. and Laties, G. G. (1989). *Planta*, **179**, 17.
10. Maniatis, T., Fritsch, E. F. and Sambrook, J. (eds) (1982). *Molecular cloning: a laboratory manual*. Cold Spring Harbor Laboratory Press, New York.
11. Schulze-Lefert, P., Dangl, J. L., Becker-Andre, M., Hahlbrock, K. and Schulz, W. (1989). *EMBO J.*, **8**, 651.
12 Hammond-Kosack, M. C. U. and Bevan, M. W. (1993). *Mol. Biol. Reporter*, **11**, 249.
13. Lane, D., Prentki, P. and Chandler, M. (1992). *Microbiol. Rev.*, **56**, 509.
14. Ceccarelli, E. and Giuliano, G. (1991). In *BioMethods (A laboratory guide for cellular and molecular plant biology)* (ed. H. P. Saluz and M. M. Becker), Birkhauser Verlag, Basel, p. 256.
15. Foster, R., Gasch, A., Kay, S. and Chua, N-H. (1992). In *Methods in Arabidopsis research* (ed. C. Koncz, N-H. Chua and J. Schell). World Scientific, Singapore.

Chapter 5
Inducible gene expression in plants

Maki Ohgishi and Takashi Aoyama

Institute for Chemical Research, Kyoto University, Uji, Kyoto 611–0011, Japan

1 Introduction

Making transgenic plants has become a popular technique in the plant science field and an increasing number of genes are now introduced into transgenic plants for the analysis of their functions. As a means of controlling transgene expression, various natural and artificial promoters have been used. Among them, constitutively active promoters such as the 35S promoter of cauliflower mosaic virus have been chosen most frequently. Ectopic over-expression of transgenes using such promoters is often sufficient to provide evidence for gene function. It is also true, however, that defined and controllable transgene expression, i.e. inducible gene expression systems, are desired for many reasons. Most simply, for genes for which ectopic expression is toxic to plants, constitutively active promoters cannot be used. In addition, it is often argued that temporal or conditional expression of transgenes would be better to establish gene function. It can be said that inducible expression of transgenes is a technique comparable to conditional mutants in classical genetics, while the latter is much more difficult for plants.

Among the advantages of inducible gene expression, the following three are the most important for basic studies. The first advantage is, as mentioned above, that inducible systems allow us to analyse the functions of genes for which ectopic expression is toxic. For genes playing important roles in fundamental biological processes, such as cell proliferation, neither mutant nor transgenic analyses with constitutive expression systems are thought to be sufficient because any abnormality in their expression would result in lethality. Secondly, in an induction experiment, the function of the transgene can be analysed by comparing a single plant, or plants with a single genetic background under non-induced and induced conditions. On the other hand, in transgenic analyses with constitutive expression systems transgenic plants are usually compared with isogenic wild-type plants. There are always two kinds of genetic differences between them: one is the T-DNA insert(s) and the other is the breakage of the genomic sequence at the site of the T-DNA insertion(s). The latter sometimes complicates the analysis of transgene function. Thirdly, and most importantly, in inducible systems it is possible to fix the starting time for transgene activation. To fully understand the events caused by the function of a gene, it is necessary to trace events along a time course. Induction systems provide an opportunity to do this.

In response to the need for inducible gene expression, various systems have been developed in plants (for reviews, see 1–3). These are largely categorized into two classes based on the origin of their components. First, there are those that consist of plant promoters responsive to specific environmental stimuli, whereas systems of the second class consist of *trans*- and *cis*-acting elements originating from heterologous organisms. In this chapter, inducible gene expression systems available in plants are listed and their characteristics are briefly discussed. Subsequently, practical tips for induction experiments with a gluco-corticoid-inducible system are described.

2 Inducible systems using plant promoters

2.1 General characteristics

Plants possess many promoters responsive to environmental stimuli. Among them, those which have been studied the best are now being used for the inducible expression of transgenes. The plant promoter-derived systems that have been used in transgenic plants or in transformed cells are listed in *Table 1*. These systems are simple and reliable because they utilize endogenous mechanisms of plants.

On the other hand, they have some disadvantages. Most importantly, the transcriptional induction of a transgene is accompanied by the endogenous responses of the host plant to the inducer. In other words, the induction is pleiotropic. Another point to note is that naturally inducible promoters often have various undesirable properties, e.g. they may be constitutively active or uninducible in particular tissues or developmental stages. It is therefore essential to understand the total characteristics of the inducible promoter to be used.

2.2 Heat-inducible systems using plant promoters

Promoters of heat shock (HS) protein genes have been studied in many plant species (for review, see ref. 19). In an experiment with transgenic *Arabidopsis*, over 40-fold induction of the ß-glucuronidase (*GUS*) gene was observed within 2 h using the promoter of the *Arabidopsis* *HSP18.2* gene (20). In addition to the quick and strong induction, an evident advantage of this system is that the inducer, heat, can be delivered to plant tissues more homogeneously compared with chemical inducers. On the other hand, HS systems are not suitable for long-term induction because the induced activities of HS promoters are usually down-regulated after several hours (20) and because long-term treatment with high temperature is deleterious for

Table 1 Inducible systems using plant promoters[a]

Inducer	Promoter	Application		Reference
		Gene	Host	
Heat	hsp70, maize	ipt	Tobacco, *Arabidopsis*	4
	Gmhsp17.5-E, soybean	ipt	Tobacco	5
		FLP	Maize	6
	Gmhsp17.6-L, soybean	FLP	*Arabidopsis*	7
		Hsp70[b]	*Arabidopsis*	8
	HS6871, soybean	tmr	Tobacco	9
	HSP18.2, *Arabidopsis*	GUS	Tobacco BY-2 cells	10
	HSP18.1, *Arabidopsis*	ARA-4	*Arabidopsis*	11
Chemicals inducing SAR	PR-1a, tobacco	δ-endotoxin	Tobacco	12
Herbicide safener	In2–2, maize	GUS	*Arabidopsis*	13
Wounding	pin2, potato	GUS	Tobacco	14,15
		ipt	Tobacco	16
Pathogen	prp1–1, potato	Barnase	Tobacco	17
Carbohydrate starvation	Amy8, rice	GUS	Tobacco, potato	18

[a]Plant promoters that have been used in transgenic plants or transformed cells are listed. Promoter analyses with reporter genes in homologous hosts are not included.
[b]An antisense construct of the *HSP70* gene was induced.

plants. One example of a successful application is the inducible somatic recombination that has been achieved in maize (6) and *Arabidopsis* (7). In those experiments, the yeast site-specific recombinase gene, *FLP*, was put under the control of a soybean HS promoter and the FLP/FRT-mediated recombination was induced by heat treatment.

2.3 Chemically-inducible systems using plant promoters

Chemicals inducing systemic acquired resistance (SAR) and herbicide safeners have been explored as inducers for inducible gene expression. SAR responses are induced by treatment with chemicals, such as salicylic acid (SA), as well as pathogen infection of a resistant plant. A class of genes including those encoding pathogenesis-related (PR) proteins are transcriptionally activated during the SAR response (21, 22). By combining the tobacco *PR-1a* promoter with SA or 2,6-dichloroisonicotinic acid (INA), an inducible system has been established in transgenic tobacco plants (1,12). Using this system, the insecticidal ∂-endotoxin of *Bacillus thurungiensis* was activated by SA and resistance against insects was successfully achieved (12). Recently, it was reported that thiadiazole-7-cabothioic acid S-methyl ester (BTH) has an even stronger effect on the *PR-1a* promoter and is much less phytotoxic than SA or INA (23–25). Moreover, BTH can cause a systemic induction because it moves systemically through the plant. These findings may encourage the application of inducible systems based on chemicals inducing SAR.

Herbicide safeners are chemicals that confer plant tolerance to certain herbicides. They induce the expression of genes encoding enzymes involved in detoxification, such as glutathione S-transferases and cytochrome P-450 mixed-function oxygenases. Maize *In2–1* and *In2–2*, which are responsive to the herbicide safener 2-chlorobenzen sulphonamide (2-CBSU) are regulated very tightly by the chemical, i.e. there is no detectable expression in non-inducing conditions (26). Using the promoter of the *In2–2* gene, an inducible system was constructed recently and examined in *Arabidopsis* (13).

2.4 Other induction systems using plant promoters

Wounding is a signal that induces the expression of genes including those encoding proteinase inhibitors and polyphenyl oxidases (27). Among the wound-inducible genes, the potato proteinase inhibitor II gene *pin2* is the best characterized at the level of transcriptional regulation (14, 15, 28). Its promoter has been used successfully for the wound-inducible expression of the bacterial isopentenyl transferase (*ipt*) gene, which confers cytokinin-mediated resistance to insects in transgenic tobacco plants (16). Other examples of plant promoter-derived systems include the pathogenesis-specific promoter of the potato *prp1-1* gene (17) and the carbohydrate starvation-inducible promoter of the rice α-amylase gene *αAmy8* (18), which have been examined in transgenic plants and suspension cultured cells, respectively.

3 Inducible systems composed of heterologous elements

3.1 General characteristics

Although endogenous plant promoters provide one means to control transgene expression, their induction is often accompanied by undesirable pleiotropic effects. Such pleiotropic effects can confuse the analysis of transgene function. To alleviate this problem, heterologous promoters have been used as inducible systems in plants. Each system of this type is essentially composed of an inducer chemical, an inducer-responsive transcription factor, and a promoter containing the *cis*-acting element recognized by the transcription factor. Systems that have been established in transgenic plants or transformed cells are listed in *Table 2*. In these systems, pleiotropic effects caused by the induction can be minimized because hetero-

Table 2 Inducible systems composed of heterologous elements[a]

Inducer	*trans*-factor	Application		
		Gene	Host	Reference
Tetracycline	TetR	GUS	Tobacco	29
		Mutant PG13	Tobacco	30
		rolB	Tobacco	31
		SAMDC	Potato	32
		rolC	Tobacco	33
		Arginine decarboxylase	Tobacco	34
		ipt	Tobacco	35
		ABP1	Tobacco	36
	TetR-VP16[b]	GUS	Tobacco	37
		SAUR-AC1	Tobacco BY-2 cells	38
		GFP	Arabidopsis	39
Glucocorticoid	GAL4-VP16-GR	LUC	Tobacco, Arabidopsis	40
		avrRpt2	Arabidopsis	41
		TIR1	Arabidopsis	42
	TetR-GR-VP16[c]	GUS	Tobacco	43
Tebufenozide	GR-VP16-HEcR	GUS	Tobacco	44
Copper ion	ACE1	GUS	Tobacco	45
		ipt	Tobacco	46
		GUS, Aspartate amino-Transferase-P2[d]	Lotus corniculatus	47
Ethanol	ALCR	Invertase	Tobacco	48
		CAT	Tobacco	49

[a]Inducible systems established in transgenic plants or transformed cells are listed.
[b]Transgene expression is down-regulated by tetracycline.
[c]Transgene expression is induced by glucocorticoid and down-regulated by tetracycline.
[d]An antisense construct of the *aspartate aminotransferase-P2* gene was induced.

logous mechanisms are not expected to have any interaction with endogenous regulatory mechanisms in plants. Another advantage of these systems is the induction of transgene expression in specific tissues, i.e. spatio-temporal gene expression, which can be achieved when the inducer-responsive transcription factor is placed under control of a tissue-specific promoter.

3.2 Inducible systems using prokaryotic repressors

The *tet* repressor (TetR), encoded by the bacterial transposon Tn10, regulates the *tet* operon, which confers resistance to tetracycline (tc) in bacteria. TetR binds to specific DNA sequences as a dimer, and the DNA-binding activity is inhibited by the binding of tc to TetR (for review, see 50). Taking advantage of this chemically-controllable DNA-binding activity of TetR, induction systems have been developed in plants. Initially, TetR itself was used as a repressor (illustrated in *Figure 1a*) in combination with a 35S promoter modified to contain TetR-binding sequences in the vicinity of the TATA box (51, 29). In the absence of tc, the TetR protein masks the TATA box region and represses promoter activity. By adding tc, the repressor is released from the promoter, resulting in induction of transcription. The system has been successfully used in transgenic tobacco and potato (see *Table 2*). In tobacco, a 500-fold induction of GUS activity was achieved (29). In tomato and *Arabidopsis*, however, the system was not functional probably because high TetR levels, which are required for tight repression, cannot be achieved (52, 53).

Subsequently, another system was developed originally for use in animal tissue culture cells (54) in which a fusion protein between TetR and the transcriptional activation domain of the herpes simplex viral protein VP16 acts as a tc-controlled transcriptional activator

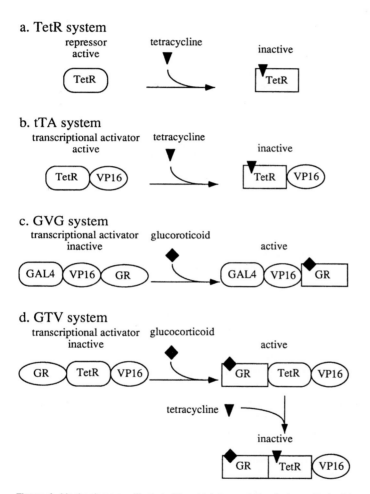

Figure 1 Mechanisms are illustrated by which transcription factors of inducible systems are activated or inactivated. TetR: the repressor protein of the Tn10 *tet* operon, VP16: the transcription activation domain of the herpes simplex viral protein VP16, GAL4: the DNA binding domain of the yeast transcription factor GAL4, GR: the hormone binding domain of the rat glucocorticoid receptor. (a) The binding of tetracycline inhibits the DNA-binding activity of TetR and makes the repressor inactive. (b) The binding of tetracycline to the TetR moiety inhibits the DNA-binding activity of the fusion protein TetR-VP16 and makes the transcriptional activator inactive. (c) The binding of glucocorticoid to the GR moiety cancels the inhibitory effect of the GR HBD and makes the transcriptional activator GAL4-VP16-GR active. (d) The binding of glucocorticoid to the GR moiety cancels the inhibitory effect of the GR HBD and makes the transcriptional activator GR-TetR-VP16 active. The binding of tetracycline to the TetR moiety inhibits the DNA-binding activity of GR-TetR-VP16 and makes the transcriptional activator inactive.

(tTA; illustrated in *Figure 1b*). In this system, tTA-activated transcription from the promoter containing TetR-recognition sequences is shut down by the addition of tc. This principle has also been shown to work in transgenic tobacco and *Arabidopsis* (37, 39). In *Arabidopsis*, tTA-activated expression of a Green Fluorescence Protein (GFP) was effectively repressed by treatment with 100 ng/ml tc (39). However, the induction of gene expression by this system is complicated because continuous treatment with tc is needed to maintain the promoter silent before induction. On the other hand, shutting down transgene expression can be easily done with this tTA-based system. This system has been successfully applied for studying the decay rates of mRNA in tobacco BY-2 culture cells (38).

TetR has a high affinity ($K_B = 10^9 M^{-1}$) to tc (55) and tc easily penetrates into cells. These assets allow the reduction of tc to concentrations as low as 1 mg/l in induction experiments. Notwithstanding, this dose of tc in hydroponic culture causes browning of the roots, root growth inhibition, and reduction of Photosystem II efficiency in tobacco (43). These toxic effects can be minimized using a less toxic tc derivative such as chlor-tc or anhydro-tc (43).

3.3 Inducible systems using steroid-hormone receptors

Steroid hormone receptors are transcription factors prevalent in the animal kingdom (for review, see 56). Because not only sensing a ligand, but also transcriptional regulation is performed by a single protein, it can be said that a steroid hormone receptor constitutes the simplest signal transduction pathway in eukaryotic cells. Moreover, steroid hormones are hydrophobic compounds that easily penetrate cells. For these reasons, steroid hormone receptors have been thought to provide ideal artificial induction systems. A system consisting of a glucocorticoid receptor (GR) and the promoter containing its *cis*-acting elements has been found to work successfully in yeast (57) and cultured tobacco cells (58). However, this system does not work in transgenic plants (59).

For use in transgenic plants, another approach was taken, in which the hormone binding domain (HBD) of GR was used in combination with the DNA-binding domain of yeast GAL4 and transcriptional activation domain of the herpes simplex viral protein VP16 (40). HBDs of steroid hormone receptors are known to function as regulatory domains *in cis* with fusion proteins, as well as with their own receptors (60). In the absence of ligands, HBDs repress the function of sterically neighbouring domains by forming a complex with multiple proteins including the heat-shock protein HSP90. Ligand binding releases the complex resulting in derepression (61, 62). In such a system, the chimeric transcription factor (GVG; illustrated in *Figure 1c*) activates transcription from the promoter containing GAL4-binding sequences only in the presence of glucocorticoid. In an experiment using a luciferase (*LUC*) gene as a reporter, over 100- and 1000-fold induction of LUC activity has been achieved in transgenic tobacco and *Arabidopsis*, respectively, when the plants were treated with 10 μM dexamethasone (DEX), a glucocorticoid derivative (40, 63). Successful applications of the system in basic studies have also been reported (41, 42).

Unlike tc, glucocorticoid is not only easily absorbed into plant cells, but it is not toxic. DEX, at the concentrations used in induction experiments (at least up to 30 μM), does not have any observable physiological effects in wild-type tobacco or *Arabidopsis*. Glucocorticoid is a well-studied biological compound and many derivatives with different properties are available from commercial sources. Another advantage of the GVG system is that it is possible to control the amount of transgene expression. The induction level can be controlled using different concentrations of DEX. In an experiment using a *LUC* reporter gene in *Arabidopsis*, induction was detectable at a concentration of 3 nM DEX or higher, and a good correlation between DEX concentrations and induction levels was obtained in the range from 3 nM to 1 μM (63).

This feature allows the analysis of dose-dependent effects of transgene products in a single genetic background.

However, it should be noted that excess amounts of the active GVG protein seem to be toxic to *Arabidopsis* plants. In some transgenic *Arabidopsis* lines that highly over-express the *GVG* gene, growth inhibition is observed when plants are treated with glucocorticoid (unpublished data). One practical way to avoid this toxicity is to separate the system into two T-DNA constructs. First, *Arabidopsis* is transformed with one of the constructs containing the *GVG* gene and a transgenic line without the inducible toxicity is established. Subsequently, a second transformation is done on the established line with the other construct containing the inducible promoter followed by the transgene of interest. By doing this, transgenic plants with the inducible gene expression and without the toxicity problems can be obtained. In addition, the first transgenic plants, which carry the *GVG* gene only, will be a negative control for the experiments.

Most recently, two other steroid hormone-inducible systems have been reported. One is a glucocorticoid-inducible and tetracycline down-regulatable system (43), and the other is an ecdysteroid agonist-inducible system (44). Although they make use of the same HBD regulatory mechanism as the GVG system, each of them has a unique and useful property. In the former system, the tTA-based shut-off system is combined to the regulatory mechanism of the GR HBD by making a chimeric transcription factor (TGV; illustrated in *Figure 1d*). In this system, glucocorticoid induces transgene expression, and tetracycline down-regulates it. The latter system has been developed as an inducible system by using the HBD of an insect ecdysone receptor. Because the induction chemical, ecdysone agonist RH5992 (tebufenozide), is currently used as a lepidoptran control agent and is not phytotoxic (44), the system is thought to be applicable safely in the field.

3.4 Other heterologous inducible systems

Other inducible systems have also been constructed using heterologous transcription factors responsive to specific chemicals. One is the system consisting of the yeast copper-responsive transcription factor ACE1 and the promoter containing its recognition sequences (45). In the presence of copper ions, the ACE1 protein becomes competent to bind the recognition sequences and to activate transcription from the promoter. Transgenic tobacco plants carrying the system fused to a *GUS* reporter gene showed 50-fold increases in GUS activity after treatment with low concentrations of copper ion (50 μM $CuSO_4$ for the nutrient solution or 0.5 μM $CuSO_4$ for a foliar spray). The system has been applied in tobacco (46), *Lotus* (47), and *Arabidopsis* (64).

Ethanol has also been used as an inducer. The *Aspergillus nidulans* transcription factor ALCR activates transcription from the promoter of the alcohol dehydrogenase gene *alcA* in response to low concentrations of ethanol (65). By making use of ALCR and the regulatory region of the *alcA* promoter, an ethanol-inducible gene expression system has been developed (48). In transgenic tobacco, 0.01% of ethanol in liquid growth media initiates the expression of the chloramphenicol acetyl transferase (*CAT*) reporter gene within 4 h (49). This system is unique in that the inducer is a chemical that is very mild to the environment and which is very safe for application in the field.

Because commonplace chemicals are used as inducers, users of these systems should take precautions to avoid unwanted induction, e.g. ethanol might be a solvent for some of the substances in the growth medium and soil might contain a low concentration of copper ions. It should also be noted that these chemicals are not completely innocuous for plants. High concentrations of ethanol or copper ion have physiological effects on plants (49, 64).

4 Practical tips for induction experiments

In this section, protocols for induction experiments using the GVG system are described. An analysis of dynamic changes caused by transgene function is one of the major purposes of induction experiments. Quick and uniform induction is desirable for this purpose. To achieve this, we are using the following procedure with transgenic *Arabidopsis* plants. Using *Protocol 1*, induction of a *LUC* reporter gene can be detected within 1 h at both the mRNA and LUC activity levels (data not shown).

Protocol 1

Short-term induction experiments

Equipment and reagents

- 0.8% agar plates containing MS salts and B5 vitamins
- PLANTCON® containers (Flow Laboratories Inc.)
- 6 × 6 cm plastic meshes with 3 mm height spacers
- DEX
- Ethanol

Method

1 Germinate *Arabidopsis* seeds on MS agar plates and grow for 3 weeks under standard conditions.

2 Transfer the plants together with agar onto a spacer-attached plastic mesh in PLANTCON® containers and put a minimal amount of water that can spread over the bottom of the container (illustrated in *Figure 2*).[a]

3 Gradually remove the top of the plastic container over several days, then keep the plants in open-air conditions for 1 day.[b]

4 Make 30 mM DEX solution in ethanol and then dilute 1000-fold directly with water to make 30 μM DEX solution.[c]

5 Exchange the water under the mesh with the 30 μM DEX solution.[d]

6 Harvest the plants after appropriate periods and analyse.

[a] Maintain the some level of water during the experiment.

[b] Avoid plant damage caused by the rapid reduction of humidity. Vigorous roots should grow under the water.

[c] Multi-step dilution by water may make the compound insoluble.

[d] Use water containing only ethanol as a negative control.

In cases where morphological changes are observed, a long-term induction should be done. Long-term experiments with *Arabidopsis* seedlings or young plants are done as indicated in *Protocol 2*. In this experiment, the level of transgene expression can be controlled with different concentrations of DEX.

In general, an important part of induction experiments is to keep the plants in good conditions. High induction levels can be achieved only in healthy plants, as in induction experiments in bacteria or yeast. Although uniform induction through a total plant is attempted in the above two experiments, it is actually difficult for two reasons. First, responses of different tissues to induction are not the same. Such differences are inevitable because of different properties of plant tissues. The other reason is due to differences in the delivery of the inducer chemical. In the former experiment, plants take up DEX from the roots and transpirational water flow in plants also delivers DEX. Although delivery is very rapid, DEX accumulates in peripheral tissues such as the leaves, as a result of transpiration. On the other hand, in the latter experiment where transpirational water flow is very slow due to the air-tight conditions in the plastic containers, DEX is thought to be delivered more equally by molecular diffusion (40).

Protocol 2

Long-term induction experiments

Equipment and reagents

- 0.8% agar medium containing MS salts and B5 vitamins
- Plastic plates
- DEX
- Ethanol

Method

1 Make a series of DEX solutions in ethanol with various concentrations (between 10- and 3000-fold dilutions of a 30 mM stock solution).[a]

2 After autoclaving the agar medium, dilute the DEX solutions directly with the medium and pour into plastic plates.[b]

3 Germinate *Arabidopsis* seeds or put young *Arabidopsis* plants on the agar plates and grow in standard conditions.

4 Transfer the plants to fresh DEX-containing agar plates every week if the experiment is continued.

[a] Minimize the final ethanol concentration in the medium because concentrations greater than 0.1% can inhibit *Arabidopsis* seed germination.

[b] Multi-step dilution by water may make the compound insoluble.

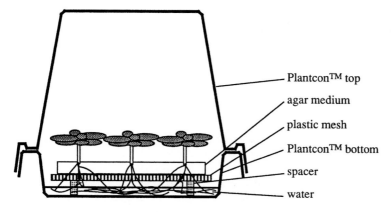

Figure 2 System for the rapid induction by glucocorticoid. Young *Arabidopsis* plants grown on an agar medium are placed together with the agar onto a spacer-attached plastic mesh in PLANTCON® containers. A minimal amount of water is added to the bottom of the container. The top is gradually removed over several days to adapt the plants to the open-air conditions. Vigorous roots should grow under the water during this period.

Glucocorticoid treatment can also be done with plants growing in pots. Induction can be done simply by pouring DEX solution (30 μM) into the pot, although it is uncertain how much DEX is trapped in the soil (63). When exposed epidermal tissues are the target of induction, spraying methods are effective and easy. In this method, plants can be sprayed with a solution containing 30 μM DEX and 0.01% Tween-20 (40). Soaking plants in DEX solution is also effective. Using a soaking solution containing 0.025% Silwet L-77 (OSI Specialties) and 5 μM DEX, the activity of the fusion protein between an *Arabidopsis* MADS box protein AP3 and the GR HBD was successfully induced in the floral meristem of *Arabidopsis* (66). These simple methods allow induction experiments with healthy plants grown under natural conditions.

Acknowledgements

Research by this laboratory has been supported in part by BRAIN, Japan. I would like to thank Dr Chris Bowler for his suggestions on the manuscript.

References

1. Ward, E. R., Ryals, J. A. and Miflin, B. J. (1993). *Plant Mol. Biol.*, **22**, 361.
2. Gatz, C. (1996). *Curr. Opin. Biotechnol.*, **7**, 168.
3. Gatz, C. (1997). *Ann. Rev. Plant Physiol. Plant Mil. Biol.*, **48**, 89.
4. Medford, J., Horgan, R., El-Sawi, Z. and Klee, H. J. (1989). *Plant Cell*, **1**, 403.
5. Ainley, W. M., McNeil, K. J., Hill, J. W., Lingle, W. L., Simpson, R. B., Brenner, M. L., Nagao, R. T. and Key, J. L. (1993). *Plant Mol. Biol.*, **22**, 13.
6. Lyznik, L. A., Hirayama, L., Rao, K. V., Abad, A. and Hodges, T. K. (1995). *Plant J.*, **8**, 177.
7. Kilby, N. J., Davies, G. J., Snaith, M. R. and Murray, J. H. A. (1995). *Plant J.*, **8**, 637.
8. Lee, J. H. and Schoffl, F. (1996). *Mol. Gen. Genet.*, **252**, 11.
9. Smart, C., Scofield, M., Bevan, M. W. and Dyer, T. A. (1991). *Plant Cell*, **3**, 647.
10. Yoshida, K., Kasai, T., Garcia, M. R. C., Sawada, S., Shoji, T., Shimizu, S., Yamazaki, K., Komeda, Y. and Shinmyo, A. (1995) *Appl. Microbiol. Biotechnol.*, **44**, 466.
11. Ueda, T., Anai, T., Tsukaya, H., Hirata, A. and Uchimiya, H. (1996). *Mol. Gen. Genet.*, **250**, 533.
12. Williams, S., Friedrich, L., Dincher, S., Carozzi, N., Kessmann, H., Ward, E. and Ryals, J. (1992). *Bio/Technol.*, **10**, 540.
13. De Veylder, L., Van Montagu, M., Inze, D. (1997). *Plant Cell Physiol.*, **38**, 568.
14. Thornburg, R. G., An, G., Cleveland, T. E., Johnson, R. and Ryan, C. A. (1987). *Proc. Natl Acad. Sci. USA*, **84**, 744.
15. Keil, M., Sanchez-Serrano, J. J. and Willmitzer, L. (1989). *EMBO J.*, **8**, 1323.
16. Smigocki, A., Neal, J. W. Jr, McCanna, I. and Douglass, L. (1993). *Plant Mol. Biol.*, **23**, 325.
17. Strittmmatter, G., Janssens, J., Opsomer, C. and Botterman, J. (1995). *Bio/Technol.*, **13**, 1085.
18. Chan, M-T., Chao, Y-C. and Yu, S-M. (1994). *J. Biol. Chem.*, **269**, 17635.
19. Nagao, R. T. and Gurley, W. B. (1999). In *Inducible gene expression in plants* (ed. P. H. S. Reynolds), p. 97. CABI Publishing, Wallingford.
20. Takahashi, T., Naito, S. and Komeda, Y. (1992). *Plant J.*, **2**, 751.
21. Ward, E. R., Ukens, S. J., Williams, S. C., Dincher, S. S., Wiederhold, D. L., Alexander, D. C., Ahl-Goy, P., Metraux J-P. and Ryals, J. A. (1991). *Plant Cell*, **3**, 1085.
22. Ukens, S., Mauchi-mani, B., Moyer, M., Potter, S., Williams, S., Dincher, S., Chandler, D., Slusarenko, A., Ward, E. and Ryals, J. (1992). *Plant Cell*, **4**, 645.
23. Gorlach, J., Volrath, S., Knauf-Beiter, G., Hengy, G., Beckhove, U., Kogel, K-H., Oostendrop, M., Staub, T., Ward, E., Kessmann, H. and Ryals, J. (1996). *Plant Cell*, **8**, 629.
24. Friedrich, L., Lawton, K. A., Ruess, W., Masner, W., Specker, N.,Rella, M, G, Meier, B., Dincher, S., Staub, T., Uknes, S., Metraux, J-P., Kessmann, H. and Ryals, J. (1996). *Plant J.*, **10**, 61.
25. Lawton, K. A., Friedrich, L., Hunt, M., Weymann, K., Delaney, T., Kessmann, H., Staub, T. and Ryals, J. (1996). *Plant J.*, **10**, 71.
26. Hershey, H. P. and Stoner, T. D. (1991). *Plant Mol. Biol.*, **17**, 679.
27. Schaller, A., Bergey, D. R. and Ryan, C. A. (1995). *Plant Cell*, **7**, 1893.
28. Lorberth, R., Dammann, C., Ebneth, M., Amati, S. and Sanchez-Serrano, J. J. (1992). *Plant J.*, **2**, 477.

29. Gatz, C., Frohberg, C. and Wendenburg, R. (1992). *Plant J.*, **2**, 397.
30. Rieping, M., Fritz, M. Prat, S. and Gatz, C. (1994). *Plant Cell*, **6**, 1087.
31. Roder, F. T., Schmulling, T. and Gatz, C. (1994). *Mol. Gen. Genet.*, **243**, 32.
32. Kumar, A., Taylor, M. A., Arif, S. A. M. and Davis, H. V. (1995). *Plant J.*, **9**, 147.
33. Faiss, M., Strnad, M., Redig, P., Dolezal, K., Hanus, J., Van Onckelen, H. and Schmulling, T. (1996). *Plant J.*, **10**, 33.
34. Masgrau, C., Altabella, T., Farras, R., Flores, D., Thompson, A. J., Besford, R. T. and Tiburcio, A. F. (1997). *Plant J.*, **11**, 465.
35. Faiss, M., Zalubilova, J., Strnad, M. and Schmulling, T. (1997). *Plant J.*, **12**, 401.
36. Jones, A. M., Im, K-H., Savka, M. A., Wu, M-J., DeWitt, G., Shillito, R. and Binns, A. N. (1998). *Science*, **282**, 1114.
37. Weinmann, P., Gossen, M., Hillen, W., Bujard, H. and Gatz, C. (1994). *Plant J.*, **5**, 559.
38. Gil, P. and Green, P. J. (1995). *EMBO J.*, **15**, 1678.
39. Love J., Scott A. C. and Thompson W. F. (2000). *Plant J.*, **21**, 579.
40. Aoyama, T. and Chua, N-H. (1997). *Plant J.*, **11**, 605.
41. McNellis, T. W., Mudgett, M. B., Li, K., Aoyama, T., Horvath, D., Chua, N-H. and Staskawicz, B. J. (1998). *Plant J.*, **14**, 247.
42. Gray, W. M., del Pozo, J. C., Walker, L., Hobbie, L., Risseeuw, E., Banks, T., Crosby, W. L., Yang, M., Ma, H. and Estelle, M. (1999) *Genes Dev.*, **13**, 1678.
43. Bohner, S., Lenk, I., Rieping, M., Herold, M. and Gatz, C. (1999). *Plant J.*, **19**, 87.
44. Martinez, A., Sparks, C., Hart, C. A., Thompson, J. and Jepson, I. (1999). *Plant J.*, **19**, 97.
45. Mett, V. L., Lockhead, L. P. and Reynolds, P. H. S. (1993). *Proc. Natl Acad. Sci. USA*, **90**, 4567.
46. McKenzie, M. J., Mett, V. L., Reynolds, P. H. S. and Farnden, K. J. F. (1998). *Plant Physiol.*, **116**, 969.
47. Mett, V. L., Podivinsky, E., Tennant, A. M., Lockhead, L. P., Jones, W. T. and Reynolds, P. H. S. (1996). *Transgenic Res.*, **5**, 105.
48. Caddick, M. X., Greenland, A. J., Jepson, I., Krause, K-P., Qu, N., Riddell, K. V., Salter, M. G., Schuch, W., Sonnewald, U. and Tomsett, A. B. (1998). *Nature Biotechnol.*, **16**, 177.
49. Salter M. G., Paine, J. A., Riddell, K. V., Jepson, I., Greenland, A. J., Caddick, M. X. and Tomsett, A. B. (1998). *Plant J.*, **16**, 127.
50. Hillen, W. and Berens, C. (1994). *Ann. Rev. Microbiol.*, **48**, 345.
51. Gatz, C. and Quail, P. H. (1988). *Proc. Natl Acad. Sci. USA*, **85**, 1394.
52. Corlett, J. E., Myatt, S. C. and Thompson, A. J. (1996). *Plant Cell Env.*, **19**, 447.
53. Gatz, C. (1999). In *Inducible gene expression in plants* (ed. P. H. S. Reynolds), p. 11. CABI Publishing, Wallingford.
54. Gossen, M. and Bujard, H. (1992). *Proc. Natl Acad. Sci. USA*, **89**, 5547.
55. Takahashi, M., Altschmied, L. and Hillen, W. (1986). *J. Mol. Biol.*, **87**, 341.
56. Laudet, V., Hanni, C., Coll, J., Catzeflis, F. and Stehelin, D. (1992). *EMBO J.*, **11**, 1003.
57. Schena, M. and Yamamoto, K. R. (1988). *Science*, **241**, 965.
58. Schena, M., Lloyd, A. M. and Davis, R. W. (1991). *Proc. Natl Acad. Sci. USA*, **88**, 10421.
59. Lloyd, A. M., Schena, M., Walbot, V. and Davis, R. W. (1994). *Science*, **266**, 436.
60. Picard, D., Salser, S. J. and Yamamoto, K. R. (1988). *Cell*, **54**, 1073
61. Beato, M. (1989). *Cell*, **56**, 335.
62. Picard, D. (1994). *Curr. Opin. Biotechnol.* **5**, 511.
63. Aoyama, T. (1999). In *Inducible gene expression in plants* (ed. P. H. S. Reynolds), p. 43. CABI Publishing, Wallingford.
64. Mett, V. L. and Reynolds, P. H. S. (1999). In *Inducible gene expression in plants* (ed. P. H. S. Reynolds), p. 61. CABI Publishing, Wallingford.
65. Felenbok, B., Sequeval, D., Mathieu, M., Sibley, S., Gwynne, D. I. and Davies, R. W. (1988). *Gene*, **73**, 385.
66. Sablowski, R. W. M. and Meyerowitz, E. M. (1998). *Cell*, **92**, 93.

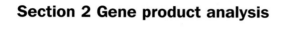

Section 2 Gene product analysis

Chapter 6
Expression of recombinant proteins

Avital Yahalom and Daniel A. Chamovitz
Department of Plant Sciences, Tel Aviv University, Tel Aviv 69978, Israel

1 Introduction

Complicated developmental processes can be readily dissected through elegant genetic analyses. However, genetic approaches that iimportant genes are only first steps in elucidating gene function. With the completion of the sequencing of the Arabidopsis and rice genomes, plant biologists will find themselves needing to analyse numerous putative proteins. Understanding the cellular and biochemical basis of a particular developmental phenomenon, or more particularly, a cellular and biochemical description of the gene product, necessitates the generation of new research tools, the most important of which is the generation of recombinant protein. Recombinant proteins are needed for the generation of both mono- and polyclonal antibodies, for affinity purification of antibodies, for various *in situ* and *in vitro* studies, and structural studies.

Numerous protocols and books have been published concerning the expression of recombinant proteins (1). This chapter intends to give a general overview of protein expression with a few basic protocols.

2 Preliminary considerations

Recombinant protein production can be a simple and painless process for some, while for others it can be the demise of a project. Unfortunately, much of this is dependent on serendipity—the particular qualities of your favorite protein. Much heartache can be avoided if you attempt only the minimum needed for your research. This minimum is dictated by answering the question: 'Why do you need the recombinant protein?' The answer to this question leads to several secondary questions and determines which expression and purification system to use, and in general, how much time you will put into purifying the protein.

2.1 Does the protein need to be soluble?

Perhaps the biggest problem of recombinant protein expression is getting soluble protein. Weeks of work can be wasted trying to get protein from the inclusion bodies to the soluble fractions, and there are many technical variations that can aid in producing soluble protein (*Protocol 1*). However, an insoluble protein is not necessarily deleterious. For example, if you are expressing a protein in order to generate antibodies (*Protocol 8*), the protein does not need to be soluble (*Protocol 7*). This is especially true, if the antibodies are needed primarily for use in immunoblot ('western') analysis. As the protein detected by immunoblot on a membrane is denatured, antibodies generated against denatured proteins may be more sensitive. As the purification of proteins from inclusion bodies is quick and easy (*Protocol 6*), attempting to isolate large amounts of soluble protein is unnecessary and can delay a project.

However, if the antibodies are needed to detect native proteins (e.g. in immunoprecipitation), then more effort should be invested in getting soluble protein. However, this too can be side-tracked by selecting for antibodies against native epitopes through affinity purification.

Recombinant proteins for use in biochemical studies (*Protocol 9*) need to be soluble. However, as biochemical studies usually need less protein than for immunizations, it is usually possible to isolate small amounts of soluble protein, although much of the protein may be insoluble. Thus, even if only a very small fraction is soluble, through scaling up the purification, it is possible to isolate the needed quantities of protein.

In general, maintaining protein solubility can be enhanced by reducing recombinant protein expression levels by:

- lowering IPTG concentrations;
- shortening induction times;
- changing host strains;
- growing the cells at lower temperatures.

Alternatively, expressing the protein as a fusion-protein with a very soluble protein or expressing only putative soluble regions of the protein can enhance solubility.

2.2 Does the protein have to be in its native state?

A soluble protein does not imply native state. The addition of various post-translational modifications can alter the structure of a protein, which may affect its activity or antigenicity. The choice of the proper expression system can allow for protein modifications.

2.3 How pure does the protein need to be?

While, ideally, all experiments would be carried out with 100% homogeneous protein, this is often difficult. However, the problem of purification can be side stepped. For example, rather than putting effort into purifying an antigen to 100% homogeneity, it is possible to immunize with a relatively pure antigen (say 80% pure), purified through expression system A, and then affinity purifying the antibodies over the antigen purified from expression system B. In this way, the only common protein in the two systems is your recombinant protein, while the antibodies against the contaminating proteins from A will not be recovered. For biochemical analyses, the design of experiments with proper controls can negate the effects of any contaminating proteins.

3 Vector choices

A large number of expression systems are available from numerous companies (*Table 1*). It is possible to express a recombinant protein from any vector that contains a cloning site downstream from an inducible promoter. Even simple pUC18 is a potential expression vector, expressing the protein as a fusion with the first 20 amino acids of lacZ. The more advanced plasmids have several improvements that allow for tight control of expression induction, easy purification of the protein as a fusion protein, epitope tags for the identification of the protein, cleavage of the fusion protein, and sites for modification of the protein by kinases. As it is impossible to know *a priori* which expression system will work best for a particular protein, the guiding principle in vector choice should be to attempt to express the protein in at least two different systems at the same time. This is simplified as most vectors are available in three cloning reading frames and most vectors have similar multiple cloning sites. If more than one system works, this can have the added advantage of having available protein for affinity purification. A new cloning technology, 'recombinational cloning', provides an elegant method for moving a gene of interest simultaneously into several expression vectors (3).

Table 1 Cloning systems for expression of recombinant proteins

Fusion protein/peptide	Vector system	Supplier	Miscellaneous (see ref. 2)
Glutathione S-transferase (GST)	pGEX, pET41, 42	Pharmacia, Novagen	GST is rather large (24 kD), which is advantageous for identifying the recombinant protein, but has several disadvantages. Large proteins are often expressed at lower levels, and if used as an antigen, the GST is also antigenic. This can be side-stepped by either cleaving the fusion protein, or affinity purifying the antibodies over a second non-GST fusion protein.
Histidine tag	pET14–16, 19–42 Various pBAD pRSET	Novagen Invitrogen	The HIS tag has little effect on the size of the recombinant protein. Can also be used as an epitope tag.
Thioredoxin	pET-32 pTrxFus	Novagen Invitrogen	Increased solubility of fusion protein
Cellulose binding domain (CBD)	pET34–38	Novagen	Very simple and inexpensive purification of the recombinant protein.
T7-tag	Various pET	Novagen	Epitope tag
S-tag	Various pET	Novagen	Epitope tag
Myc tag	PBAD/*Myc*-His	Invitrogen	Epitope tag
PKA site	pET33	Novagen	*In vitro* phosphorylation
CAMP-dependent kinase site	PGEX-2TK	Pharmacia	*In vitro* phosphorylation
Biotinylated peptide	PinPont Xa	Promega	

An older maltose binding protein (MBP) expression system is also widely used. However, it should be noted that MBP is highly antigenic, and a number of labs have purified good anti-MBP antibodies, but no specific antibodies against their protein. While not endorsing any specific company, we have had great success using a mixture of GST- and HIS-tagged proteins (4–6).

3.1 Considerations for construction of plasmids

Unfortunately, many recombinant proteins are deleterious for the host cell. This toxicity can be overcome by placing the target gene under the control of an inducible promoter. The most common example is the IPTG inducible *lac* promoter found in numerous vectors. However, the *lac* promoter is leaky and highly toxic recombinant proteins cannot be expressed in this system. The pET and other vectors bypass this shortfall by putting the recombinant gene under the control of the T7 promoter, which is not recognized by the *E. coli* RNA polymerase. Therefore, virtually no expression occurs until a source of T7 polymerase is provided. The efficient use of this system necessitates the use of correct *E. coli* strains (see below).

Most recombinant proteins can be expressed to high levels in *E. coli*. However, some proteins are very toxic to the cells and are expressed poorly. Other proteins may be unstable and subject to degradation. Thus, the conditions for optimal expression of individual proteins must be determined empirically. Deletion of potential toxic regions such as large hydrophobic regions from the protein can solve this problem for some proteins. Others, particularly smaller proteins are better expressed as a fusion with a larger soluble protein such as GST.

4 Host cell choices

4.1 E. coli

The expression and purification of recombinant proteins is most readily accomplished in *E. coli*. The extensive understanding of the regulation of gene expression in *E. coli*, and of its genetics and physiology, has led to the development of numerous expression systems that are used on all levels, from small labs, to large biofermentors, for the production of recombinant proteins. However, as *E. coli* does not add certain post-translational modifications to proteins, correct folding and activity may be affected, representing the major drawback of protein expression in *E. coli*.

Using the wrong *E. coli* strain can ruin an expression project. A suitable host strain containing the genetic traits necessary for a particular expression system must be used. There are numerous *E. coli* strains available, and this information is available in the appendices of most commercial catalogues. For example, as mentioned above, several expression systems utilize the T7 promoter for regulating transgene expression. To fully utilize this system, the plasmid is constructed in an *E. coli* strain not containing a source of T7 RNA polymerase, such as DH5α, and then transferred to a strain containing the polymerase on an inducible promoter, such as BL21(DE3)pLys, for the expression of the recombinant protein.

Most strains also carry mutations in proteases such as *ompT* and *lon*. Strains carrying the *trxB* mutation are defective in thioredoxin reductase, which can increase protein solubility and proper folding. Efforts have been made to produce *E. coli* strains that over-express cocktails of chaperones, such as *groE*, *dnaK*, and *clpB*, which should also increase protein solubility (7, 8). Such strains may soon be commercially available.

4.2 Eukaryotic expression systems

When proper post-translational modifications for the protein are important, a eukaryotic expression system must be adopted. While giving the advantage of a more native protein, these systems are often labour intensive and require some degree of expertise. There are at least three different eukaryotic expression systems—baculovirus, yeast, and plants.

4.2.1 Baculovirus

The baculovirus system is well documented and high-level protein expression, up to 100 mg of recombinant protein per 10^9 cells, is readily accomplished (9). Baculoviruses comprise a family of DNA viruses isolated from insects. *Autographa californica* nuclear polyhedrosis virus (AcMNPV) is the prototype baculovirus strain most commonly used in expression vector systems. Glycosylation, myristilation, phosphorylation, addition of fatty acids, and other modifications occur in baculovirus-infected cells. However, baculovirus expression necessitates the establishment of an insect cell culture system, which needs constant attention (transferring cells, etc.).

4.2.2 Yeast

Several yeast strains can be very useful for expression and analysis of plant proteins. The yeast *Saccharomyces cerevisiae*, *Schizosaccharomyces pombe*, and *Pichia pastoris* are well characterized genetically, which has led to their development as bioreactors for protein expression. Various inducible expression systems have been commercialized for all three yeast species. Single-celled yeast provide the convenience of quick growth and ease of handling and storage associated with working with *E. coli*, while still allowing for many eukaryotic post-translational modifications found in other eukaryotic cell culture-based systems, without the trouble of a cell culture system. Of the three, *P. pastoris* has been more widely used for producing high levels of active recombinant protein (10).

As this book deals specifically with molecular plant biology, it should be obvious that, ultimately, we should study proteins expressed in plants and not in bacteria. New viral-based transient expression strategies may enable the widespread use of plants as a host for the expression of recombinant proteins. Tobamovirus derivatives have been used to produce large quantities of foreign proteins in plants (11). Such systems have yet to be commercialized.

5 Expression of recombinant proteins

In general, it is advisable to do all manipulation of proteins at low temperature (0–4 °C). If the protein of interest is very sensitive to proteolysis, it is recommended that protease inhibitors are added and that proteins are kept at a low temperature (0–4 °C). However, from our experience in many cases this is not necessary.

Protocol 1

Induction and initial analysis of fusion proteins

Equipment and reagents
- Sorvall centrifuge GSA and SS34 rotors
- Microcentrifuge
- Mini vertical electrophoresis unit
- LB medium: Tryptone 10g/l; yeast extract 5g/l; NaCl 10g/l
- 20% sterile glucose
- IPTG (Isopropyl-β-D-thiogalactoside) 100 mM
- 10% Triton X-100
- PBS: 140 mM NaCl; 2.7 mM KCl; 10 mM Na_2HPO_4; 1.8 mM $K_2H_2PO_4$ (pH 7.3)
- 2× SDS sample buffer: 0.1 M Tris-HCl ph 6.8; 20% glycerol; 4% SDS, 200 mM DTT, bromophenol blue
- French press
- Coomassie Brilliant Blue R-250 or similar

Method

1 Grow pGEX- or pET-containing appropriate *E. coli* cells overnight at 37 °C in 5 ml LB + 2% glucose[a] and the appropriate antibiotic.[b]

2 Dilute the overnight culture 1:100 into 200 ml LB + 2% glucose[a] and the appropriate antibiotic, and grow to $O.D_{600}$ of 0.6–1.0 at 20–37 °C.[c]

3 Add IPTG to a final concentration of 0.1 mM and grow for 3 h more.[d]

4 Pellet the cells by centrifugation for 10 minutes at 3000 g.

5 Discard the supernatant and resuspend the bacterial pellet in 20 ml (1/10 the culture volume) PBS, put in ice.

6 Disrupt the cells by French press.[e]

7 Add Triton X-100 to a final concentration of 1% to the 20 ml lysed cell suspension and mix for 30 min at room temperature.

8 Centrifuge the suspension at 12 000 g for 10 min.

9 Transfer the supernatant to a new tube and keep the pellet.

10 Resuspend the pellet with a volume of PBS equal to the volume of the supernatant.

Protocol 1 continued

11 Take 20 μl from each fraction (soluble and non-soluble) and boil for 5 min with 20 μl 2× SDS sample buffer. Run 20 μl of each on SDS-PAGE. As a control run the appropriate control protein.[b]

12 Stain the gel or blot it to a membrane and probe the membrane with anti-GST or anti-His antibodies.

13 If the protein is soluble,[c, d] proceed to *Protocol 4 or 5*; if it is insoluble, proceed to either *Protocol 2 or 3*.

[a] For His-fusion proteins from pET vectors usually there is no need to add the glucose.

[b] As a control, prepare a second culture containing the appropriate empty vector.

[c] It is possible to increase the soluble fraction by lowering the growing temperature from 37 °C to 28–20 °C.

[d] The amount of IPTG can be lowered to increase expression and/or solubility of the fusion protein.

[e] We found that disruption with a French press is much better and more consistent than sonication.

Protocol 2

Solubilization of GST-fusion proteins

It was suggested that common denaturants such as 4–8 M guanidine-HCl, 4–8 M urea, detergents, or N-lauryl sarkosyl can solubilize proteins from inclusion bodies. From our experience, detergents such as Triton X-100, NP40, or Tween 20 gave lousy results. N-lauryl sarkosyl solubilized proteins very well, but even at low concentrations (such as 0.25%), N-lauryl sarkosyl inhibits the binding of GST to glutathione sepharose beads. It is possible to solubilize the proteins to some extent with 6 M guanidine-HCl or 8 M urea, and to bind the protein to glutathione-beads in the presence of these denaturants.

Equipment and reagents

- Sorvall centrifuge GSA and SS34 rotors
- Microcentrifuge
- Mini vertical electrophoresis unit

- Redissolving buffer: 8 M urea, 10 mM β-mercaptoethanol, PBS (*Protocol 1*)

Method

1 Add the pellet-PBS suspension from *Protocol 1* to a pre-weighed tube.

2 Centrifuge the suspension at 12 000 g for 10 min and discard the supernatant.

3 Weigh the tube and subtract the weight of the empty tube to determine the weight of the pellet.

4 Slowly add by stirring, redissolving buffer to a final concentration of 1–2 ml/mg pellet.

5 Stir the mixture for approximately 2 h.

6 Centrifuge the mixture for 10 min at 12 000 g and collect the supernatant.

7 For binding to glutathione beads, continue with *Protocol 4*.

8 In step 2 of *Protocol 4*, wash the glutathione-sepharose beads once with redissolving buffer and then with PBS.

9 Continue as in *Protocol 4*.

Protocol 3

Solubilization of HIS-fusion proteins

Equipment and reagents

- Sorvall centrifuge GSA and SS34 rotors
- Microcentrifuge
- Mini vertical electrophoresis unit
- 6 M guanidine-HCl in 0.1M Na-phosphate, 0.1M Tris pH 8.0

Method

1 Add the pellet-PBS suspension from *Protocol 1* to a pre-weighed tube.

2 Centrifuge the suspension at 12 000 g for 10 min and discard the supernatant.

3 Weigh the tube and subtract the weight of the empty tube to determine the weight of the pellet.

4 Resuspend the pellet (1–2 ml/mg pellet) in 6 M guanidine-HCl.

5 Incubate the mixture on an orbital shaker for approximately 2 h.

6 Centrifuge the mixture for 10 min at 12 000 g and collect the supernatant.

7 For binding to nickel beads, continue with *Protocol 5*, except prewash the nickel beads with 6 M guanidine-HCl.

8 Continue as in *Protocol 5*.

Protocol 4

Binding of soluble GST-fusion proteins to glutathione sepharose

Equipment and reagents

- Sorvall centrifuge GSA and SS34 rotors
- Microcentrifuge
- Mini vertical electrophoresis unit
- PBS (see *Protocol 1*)
- Glutathione-sepharose beads, 50% in PBS: prepare the slurry by pelleting glutathione-sepharose beads at 3000 g, washing the beads with PBS, and resuspending the beads in 0.75 vols PBS
- 2× SDS sample buffer
- Glutathione elution buffer: 10 mM reduced glutathione in 50 mM Tris-HCl (pH 8.0)

Method

1 Incubate in two microcentrifuge tubes 200–500 μl supernatant from *Protocol 1* with 20–30 μl glutathione-sepharose slurry for 1 h at room temperature.

2 Wash the beads five times by centrifugation at 5000 g and resuspending the pellet with 1 ml PBS. Do not add PBS after last wash.

3 Add 20 μl of 2× SDS sample buffer to the beads in one microfuge tube and boil it for 5 min.

4 Incubate the second microtube for 1 h at room temperature with 20μl glutathione elution buffer.

5 Pellet the beads at 12 000 g for 1 min, and mix the supernatant with 20 μl 2× SDS sample buffer in a new tube and boil for 5 min.

Protocol 4 continued

6 To check the efficiency of the GST binding and elution by glutathione buffer, analyse both samples by SDS-PAGE.[a,b,c]

[a] Empty vector expressing GST should be used as a control.

[b] If the fusion protein binds to the glutathione-sepharose beads, the procedure can be scaled up to several litres.

[c] The fusion protein can also be released from the beads by digestion with the appropriate protease, if a protease site is present between GST and the cloned gene.

Protocol 5

Binding of soluble His-fusion proteins to nickel beads

Equipment and reagents

- Sorvall centrifuge GSA and SS34 rotors
- Microcentrifuge
- Mini vertical electrophoresis unit
- PBS (see *Protocol 1*)
- Nickel beads, 50% in PBS.
- 6 M guanidine-HCl (6 M GuHCl; 0.1 M Na-phosphate, 0.01 M Tris-HCl pH 8.0)
- 2× SDS sample buffer

- Buffer A: with 8 M urea, 0.1 mM phosphate buffer, 0.01 M Tris-HCl, pH 8.0
- Buffer B: with 8 M urea, 0.1 mM phosphate buffer, 0.01 M Tris-HCl, pH 6.3
- Buffer C: with 8 M urea, 0.1 mM phosphate buffer , 0.1 M EDTA, 0.01 M Tris-HCl, pH 5.7
- Imidizole buffer: 500 mM imidazole in 50 mM Na phosphate, 300 mM NaCl, 10% glycerol, pH 6

Method

1 Add 500 μl from the supernatant in *Protocol 1* to 50 μl of washed nickel beads.

2 Incubate the suspension on an orbital shaker for 1–2 h at room temperature.

3 Wash the beads five times by centrifugation at 5000 g and resuspending the pellet with 1 ml 6 M guanidine-HCl buffer.

4 Wash the beads three times with buffer A and three times with buffer B.

5 Elute the fusion protein by incubating the beads with 20 μl of either buffer C or imidazole buffer.[a,b]

6 Pellet the beads at 12 000 g for 1 min, mix the supernatant with 20 μl 2× SDS sample buffer in a new tube, and boil for 5 min.

7 Add 20 μl 2× SDS sample buffer to the eluted beads and boil for 5 min.

8 Analyse both samples by SDS-PAGE.[c,d]

[a] Do not boil sample with imidazole as it will partially hydrolyse acid labile bonds. Instead incubate the mixture 10 min at 37 °C just before loading the gel.

[b] To get a cleaner protein sample it is possible to load the sample on a small nickel column (in 1 ml syringe), and to elute the protein with a step or linear gradient of imidazole (0–500 mM).

[c] If the fusion protein binds to the nickel beads, the procedure can be scaled up by increasing the amount of nickel beads and volume of protein supernatant.

[d] The fusion protein can also be released from the beads by digestion with the appropriate protease, if a protease site is present between the HIS tag and the cloned gene.

Protocol 6

Partial purification of inclusion bodies by differential centrifugation

Differential centrifugation is a simple method to partially clean the inclusion bodies. In the first step of centrifugation, most of the cell DNA is pelleted. As different proteins will pellet at different g-forces, it is possible to select the cleanest fraction that still contains the recombinant protein.

Equipment and reagents

- Sorvall centrifuge GSA and SS34 rotors
- Microcentrifuge
- Mini vertical electrophoresis unit
- PBS (see *Protocol 1*)
- 2× SDS-sample buffer

Method

1 Follow *Protocol 1* to step 6.

2 Centrifuge the supernatant at 2000 g for 10 min.

3 Remove the supernatant to a new tube and keep the pellet on ice.

4 Centrifuge the supernatant at 4000 g for 10 min.

5 Repeat steps 2 and 3 four times, with increasing centrifugal forces: 6000, 8000, 10 000, and 12 000 g.

6 Resuspend the pellets in a volume of PBS equal to the volume of the supernatant.

7 Take 20 μl from each fraction and boil it with 20 μl 2× SDS-sample buffer.

8 Analyse proteins by SDS-PAGE and Coomassie blue staining.

Protocol 7

Preparing insoluble protein for affinity purification of anti-serum

The fact that a protein is not soluble may be an advantage, because the inclusion bodies might contain mainly the recombitant protein. If the protein in the inclusion bodies from *Protocol 6* is fairly clean, it is possible to solubilize the protein in coupling buffer containing 1% N-lauryl sarkosyl.

Equipment and reagents

- Sorvall centrifuge GSA and SS34 rotors
- Microcentrifuge
- Mini vertical electrophoresis unit
- Coupling buffer: 0.2 M $NaHCO_3$, 0.5 M NaCl containing 1% N-lauryl sarkosyl pH 8.0
- 0.45 μm membrane

Method

1 Centrifuge pellet-PBS suspension from *Protocol 6* at 12 000 g for 10 min and discard the supernatant.

2 Slowly add by stirring, coupling buffer containing 1% N-lauryl sarcosyl.

3 Stir the mixture for approximately 2 h at room temperature.

4 Centrifuge the mixture for 10 min at 12 000 g and collect the supernatant.

Protocol 7 continued

5 Filter the supernatant through a 0.45 μm membrane.

6 For binding to NHS-activated sepharose or other media, follow the manufacturer's instructions.

If a binding step (either to nickel or GST beads) is necessary following solubilization with denaturating agents as in *Protocol 2 and 3* (8 M urea or 6 M guanidine-HCl) it is still possible to very gradually dialyse out the denaturing agent against the appropriate coupling buffer containing 1% N-lauryl sarkosyl.

Protocol 8

Preparing insoluble protein for immunization of rabbits

Equipment and reagents

- Sorvall centrifuge GSA and SS34 rotors
- Microcentrifuge
- Mini vertical electrophoresis unit
- Coomasie Brilliant Blue R-250
- Destaining solution: 50% methanol, 10% acetic acid
- Double distilled H_2O

- PBS (see *Protocol 1*)
- Surgical blade
- 2–2.5 ml Louer lock syringes
- Needles (nos 18–21)
- Ponceau Red
- 3MM Whatman paper
- dimethylsulfoxide (DMSO)

Methods

1 Separate the insoluble protein (*Protocol 6*) on a preparative SDS-PAGE gel.

Method A: in gel preparation

2 Stain the gel for 10 min.

3 Destain the gel for 30 min.

4 Wash the gel with several changes of H_2O or PBS until the background is transparent.

5 Cut the protein band from the gel.

6 Trim it to small pieces.

7 Dry the gel pieces overnight in a hood.

8 Grind the gel pieces in a pestle and mortar.

9 Place in a 2.5 ml syringe and add 500 μl of PBS.

10 Pass the gel pieces through a 2.5 ml syringe without needle and then pass the gels through needle no. 18, followed by needles no. 19, 20, and 21.

11 This gel suspension can be used for injection.

Method B: membrane preparation

2 Blot the gel to a nitrocellulose membrane (do NOT use PVDF[a]).

3 Stain the nitrocellulose membrane with Ponceau Red.

4 Cut the protein strip and wash it with PBS several times to remove the Ponceau.

Protocol 8 continued

5   Dry the protein membrane strip on 3MM Whatman paper and put the membrane pieces in a clean microcentrifuge tube.[b]

6   Add 200μl DMSO to dissolve the membrane. This suspension is ready for injection.

[a] PVDF is not dissolved by DMSO.

[b] It is very important that the membrane is completely dry, as even a drop of water will cause an aggregation of the membrane.

Protocol 9

Pull down assay

In order to assay for protein–protein interactions by pull down assay, both proteins must be soluble, either as recombinant proteins or after *in vitro* translation. One of the proteins must be 'tagged' to a motif that can be easily used for selective binding to a resin, such as GST or a poly-His tag. In addition, it must be possible to detect the proteins, either through antibodies against the recombinant protein, antibodies against the protein tag, or radioactive labelling of the proteins.

Equipment and reagents

- Sorvall centrifuge GSA and SS34 rotors
- Microcentrifuge
- Mini vertical electrophoresis unit
- TTBS buffer : 0.1% Tween 20 in TBS

Method

1   Couple the recombinant protein to glutathione sepharose beads as in *Protocol 4*.

2   Prepare microcentrifuge tubes containing TTBS and increasing amounts of the second protein (0, 25, 50, 75, and 100 ng).

3   Incubate 20–40 μl of coupled protein for 1 h at room temperature in buffer containing TTBS and increasing amounts of the second protein.

4   Pellet the beads at 12 000 g for 1 min and discard the supernatant.

5   Wash the beads five times with 1 ml TTBS as in *Protocol 4*.

6   Extract the proteins by boiling with SDS-sample buffer.

7   Run the samples on SDS-PAGE.

8   Blot the gel and probe with antibodies against the second protein.

References

1. Tuan, R. S. (ed.) (1997). *Recombinant Gene Expression Protocols*. Humana Press, Totowa.
2. Parts reprinted from the Novagen 2000 catalog with the permission of Novagen, Inc.
3. Walhout, A. J. M., Sordella, R., Lu, X., Lu, X., Hartley, J. L., Temple, J. F., Brasch, M. A., Thierry-Mieg, N. and Vidal, M. (2000). *Science*, **287**, 116.
4. Karniol, B., Yahalom, T., Kwok, S. F., Tsuge, T., Matsui, M., Deng, X-W. and Chamovitz, D. A. (1998). *FEBS Lett.*, **439**, 173.
5. Karniol, B., Malec, P. and Chamovitz, D. A. (1999). *Plant Cell*, **11**, 839.
6. Freilich, S., Oron, E., Kapp, Y., Nevo-Caspi, Y., Orgad, S., Segal, D. and Chamovitz, D. A. (1999). *Curr. Biol.*, **9**, 1187.
7. Goloubinoff, P., Gatenby, A.A. and Lorimer, G.H. (1989). *Nature*, **337**, 44.

8. Mogk, A., Tomoyasu, T., Goloubinoff, P., Rudiger, S., Roder, D., Langen, H. and Bukau, B. (1999). *EMBO J.*, **18**, 6934.
9. King, L. A. and Possee, R. D. (1992). In *The Baculovirus Expression System: a laboratory manual.* Chapman and Hall, London.
10. Cereghino, J. L. and Cregg, J. M. (2000). *FEMS Microbiol. Rev.*, **24**, 45.
11. Kumagai, M. H., Donson, J., della-Cioppa, G. and Grill, L. K. (2000). *Gene*, **245**, 169.

Chapter 7

Import of proteins into isolated chloroplasts and thylakoid membranes

Colin Robinson and Alexandra Mant
Department of Biological Sciences, University of Warwick, Gibbet Hill Road,
Coventry CV4 7AL, UK

1 Introduction

The biogenesis of the chloroplast is a complex process, involving a great deal of protein traffic. The majority of chloroplast proteins (around 80%) are imported from the cytosol, as they are encoded by nuclear genes (reviewed in refs 1 and 2). These proteins must be both specifically targeted into the chloroplast and accurately sorted to their final sites of function within the organelle. With the exception of a subset of outer envelope proteins, which have no cleavable targeting peptides (3), all imported proteins analysed to date are initially synthesized as larger precursors containing amino-terminal presequences. The presequences have been shown to contain information specifying targeting into the chloroplast, and in some cases, localization within the organelle.

This sorting of imported proteins is particularly interesting, due to the structural complexity of the chloroplast, which consists of three membrane compartments (outer and inner envelopes, and the thylakoid membrane) and three soluble compartments (the intermembrane space, the stroma, and the thylakoid lumen). Each of these compartments contains cytosolically-synthesized proteins, and intensive efforts have been made to understand the mechanisms that ensure correct targeting and intra-organellar routing of imported proteins. These studies have depended totally on the development of efficient *in vitro* assays for protein translocation across the envelope and thylakoid membranes. The aim of this article is to describe in detail basic protocols for the import of *in vitro*-synthesized proteins into both intact chloroplasts and isolated thylakoids. *Protocols 1* and *2* describe how to isolate intact chloroplasts from a homogenate of pea or barley seedlings. The methods essentially rely on differential centrifugation to separate the intact from broken plastids, and from other small organelles such as mitochondria. Assays for protein uptake into chloroplasts or isolated thylakoids generally take the form of mixing a suspension of organelles in buffer with a sample of *in vitro*-translated, radiolabelled precursor protein, followed by incubation in the light and fractionation to locate the imported protein. *Protocols 3, 4*, and *5* explain how to prepare *in vitro*-translated proteins starting from a sample of purified plasmid DNA. *Protocols 6* and *7* are basic descriptions of SDS-PAGE analysis of proteins, which are intended only as starting points, as many readers may prefer to use their own methods for SDS-PAGE, which are often tailored to suit an individual laboratory's electrophoresis equipment. The latter part of the article (*Protocols 8, 9*, and *10*) describes basic assays for the import of precursor proteins into intact chloroplasts and isolated thylakoid membranes.

1.1 Choice of plants

The correct choice of plant is of great importance—only a few higher plant species yield intact chloroplasts capable of efficient protein import. The majority of species contain phenolic compounds in the vacuole, which become oxidized upon disruption of the tissue and are believed to inactivate isolated organelles. Researchers who wish to study the import of proteins into chloroplasts, tend to favour a relevant plant species that is already being grown in their laboratory for other biochemical or genetic experiments. Hence, a group which studies the function of photosystem I protein subunits in barley (*Hordeum vulgare*) and grows these plants routinely, would probably choose to import their photosystem I precursor proteins into barley chloroplasts. Mainly for this reason, a variety of plant species are used for chloroplast protein import experiments, including barley, wheat (*Triticum aestivum*), maize (*Zea mays*), spinach (*Spinacea oleracea*), Arabidopsis (*Arabidopsis thaliana*), and pea (*Pisum sativum*).

Seedlings from dwarf pea varieties (*Pisum sativum*) most commonly provide the starting material for the isolation of intact chloroplasts. If the reader has no preference for a particular plant species, then we suggest that he or she start with pea seedlings. This is because pea chloroplasts are easily and rapidly isolated, and will import precursor proteins from a variety of monocot and dicot species, although the uptake efficiency can vary between proteins. Furthermore, pea seedlings grow rapidly and in simple conditions, which enables the researcher to carry out an experiment each day of the week, if so desired. The pea variety 'Feltham First' is excellent, when available. Alternatively, we have found 'Kelvedon Wonder' to be a high quality substitute for 'Feltham First' ('Kelvedon Wonder' may be purchased from Nickerson-Zwaan).

Spinach, the favourite species for photosynthetic measurements, is another abundant source of suitable chloroplasts. The intact chloroplasts import proteins and route them to the thylakoids very efficiently. However, the isolated thylakoid membranes translocate proteins rather poorly, which is a disadvantage. A benefit of using spinach is that intact chloroplasts can be prepared from leaves that have been stored for a short time after harvesting, e.g. spinach purchased in a market. However, great care should be taken to buy genuine *Spinacia oleracea*, as various leafy vegetables are often mistakenly sold under the guise of spinach, in both shops and markets, and these substitutes yield very poor quality chloroplasts. To rely on a local market for a suitable supply of leaves is not always possible, so laboratories often grow their own spinach plants. Growth conditions for spinach are more complicated than for pea. In order to obtain good quality leaves of adequate size, the plants should be grown hydroponically. It can take between 6 weeks and 2 months for the plants to reach the stage where leaves may be harvested. As this article is intended as in introduction to the subject, we feel the reader would be best advised to begin with pea, which is so easy to grow.

The availability of countless cDNA clones from the Arabidopsis Biological Resource Center (http://aims.cps.msu.edu/aims/) encoding chloroplast proteins, combined with the relative ease with which *A. thaliana* may be genetically manipulated, means that it would be highly desirable to isolate import-competent plastids from this species. Recently, protocols have been developed for the import of proteins into *Arabidopsis thaliana* chloroplasts—descriptions of growth conditions and isolation methods may be found in refs 4 and 5. The small size of the plants means that they are grown for around a month before there is enough tissue to yield sufficient chloroplasts for experiments. It is possible that chloroplasts from leaves of this age may not import proteins as efficiently as the chloroplasts from very young, greening tissue, like that provided by pea or barley seedlings. There are additional difficulties caused by the breakdown of glucosinolates into isothiocyanates and other chemicals harmful to chloroplasts, which occurs when *A. thaliana* tissue is disrupted. At the time of writing, it appears that *A. thaliana* chloroplasts import proteins less efficiently than pea and we are not yet aware of any reports of protein uptake using isolated thylakoid membranes, but undoubtedly the number of studies published on this subject will increase before too long.

In the following sections we present a protocol for isolating intact chloroplasts from pea seedlings. A second protocol for isolation of chloroplasts from barley seedlings is also included. It is slightly more complicated than the first protocol, but we feel it is worth sharing as a representative method for isolating import-competent chloroplasts from a monocotyledonous plant.

1.2 Growth conditions

Peas may be planted in compost (e.g. Levington's multipurpose compost) or vermiculite. Do not cover the seeds with a very thick layer of compost or vermiculite, because it will only delay germination; about 1 cm of material on top of the seeds is plenty. It is not the exact age of the seedlings *per se* that is important, but rather the exact stage of their development. The seedlings should be harvested 2–3 days after the emergence of the first leaves, when they are still appressed to one another. If the leaves have completely opened out and tendrils are visible, then the seedlings have grown past the optimum stage for isolating import-competent chloroplasts. Only young leaves should be used, because isolated chloroplasts from mature tissue show little or no protein uptake *in vitro*. The growth regime should be a 12 h photoperiod at $20 \pm 2\,°C$. The light intensity must be relatively low (40–50 μE m^2/s) to minimize the formation of starch grains, which cause chloroplast rupture during the isolation procedure. Under these conditions, it should take 7–9 days for the peas to germinate and grow, but this will depend on the exact conditions of growth facilities in individual laboratories. As an extra precaution against excessive starch deposits, pea leaves should ideally be harvested about an hour after the lights come on (starch grains increase in size during the day). It is also thought that chloroplast import efficiency follows a circadian rhythm, and that import efficiency is greatest in the morning. The trays of plants in the foreground of *Figure 1* show pea seedlings at the perfect stage for chloroplast isolation, around 2–3 days after the emergence of the first leaves.

We plant barley in vermiculite and grow at $20 \pm 2\,°C$ under a 12 h photoperiod. The light intensity is 75 μE m^2/s. Again, it is the stage of development of the plants that matters. Using these growth conditions, it should take 7–8 days for the plants to become 10 cm tall, ready for harvesting.

2 Isolation of intact chloroplasts

There are several important considerations to bear in mind when isolating chloroplasts. Pea leaves begin to deteriorate as soon as they are harvested, so do not start cutting leaves until everything needed for the chloroplast isolation procedure has been prepared. As soon as the tissue has been homogenized, the protocol should be worked through rapidly because the chloroplasts' ability to import proteins decreases over time. All tubes, beakers, media, adaptors for centrifuge rotors, and, of course, the chloroplasts themselves should be ice-cold. In particular, the grinding medium should be placed in a $-20\,°C$ freezer until it forms an icy slurry (shaking a bottle of super-cooled buffer will convert the liquid into an instant slush). In our laboratory we use a Polytron homogenizer (from Kinematica AG) to grind the leaves. Other types of blender will serve just as well, providing the grinding blades are very sharp and frothing, which is difficult to avoid and detrimental to the intactness of the chloroplasts, is kept to an absolute minimum. It is vitally important to reserve a set of centrifuge tubes to be used only for chloroplast isolation, and avoid cleaning these tubes with any sort of detergent or abrasive. Finally, great care must be exercised in handling chloroplasts; their

Figure 1 Growth of pea seedlings for chloroplast isolation. The seed trays in the foreground contain pea seedlings (var. Kelvedon Wonder) at the best growth stage for isolation of import-competent chloroplasts: 2–3 days after leaf emergence (8–9 days old under our greenhouse conditions). The larger seedlings behind are 11 days old and are suitable for other purposes, such as purification of thylakoid lumen proteins.

large size means that they are prone to lysis from shearing forces, such as those caused by pipetting. When it is necessary to pipette a suspension of chloroplasts, widen the bore of disposable tips to at least 1 mm by cutting off the ends with scissors.

Over the years we have gradually eliminated the more complicated buffers from our isolation procedure for pea chloroplasts without any loss in the yield or quality of the intact organelles. *Protocol 1* is, therefore, a streamlined method, which will save time and consumables money. Barley chloroplasts are more difficult to isolate than their pea counterparts. We work with barley infrequently and have not had the opportunity to simplify that protocol, so we recommend that the experimenter follows the method described in *Protocol 2*. Protocols for the isolation of chloroplasts for other biochemical analyses can be found in *Chapter 8*.

Protocol 1

Isolation of intact chloroplasts from pea seedlings

Equipment and reagents

- 100 ml or 50 ml centrifuge tubes, with rounded, not pointed bottoms[a]
- 15 ml Corex centrifuge tubes
- Polytron homogenizer (Kinematica AG) or blender of similar efficiency, together with a vessel for homogenization[b]
- Muslin or Miracloth (Calbiochem)
- Refrigerated centrifuge with swing-out rotor, e.g. Sorvall HB-4

- 5× HEPES-sorbitol (5× HS): 250 mM HEPES-KOH pH 8.0, 1.65 M sorbitol[c]
- 1× HEPES-Sorbitol (HS): dilute 5× HS with distilled water
- 35% Percoll (Pharmacia) in HS (mix 9 ml distilled water, 7 ml Percoll, and 4 ml 5× HS to give 20 ml of 35% Percoll)
- 80% v/v acetone in distilled water

Protocol 1 continued

Method

1 Harvest leaves from pea seedlings, avoiding the lower leaves and stem, and mix with an icy slush of HS. Use 100 ml HS per 20 g leaves.

2 Homogenize the leaves for the minimum time necessary to produce a suspension of small, evenly-sized leaf fragments. Two 3 s bursts from a Polytron set at 75% full speed should be sufficient.

3 Pour the homogenate onto either eight layers of muslin or two layers of Miracloth, resting over a pre-cooled beaker. To prevent the muslin or Miracloth falling into the beaker, it may be secured with an elastic band. If using Miracloth, thoroughly moisten it with a small quantity of HS before filtering the homogenate. When the majority of the filtrate has passed by gravity through the muslin or Miracloth, gather up the edges of the material, and apply very gentle pressure to encourage the remaining filtrate into the beaker. If there are bubbles on the surface of the material or if the plant debris is squeezed dry, then the pressure applied was too great.

4 Pour the suspension into the pre-cooled centrifuge tubes. Centrifuge at 4000 g for 1 min in a swing-out rotor. Discard the supernatant in one smooth motion and, while holding the tube upside down, quickly wipe around the inside with a tissue, to remove any froth. At this stage, the pellet is fairly firm.

5 Resuspend each pellet very gently in a small volume HS (4 ml), using a cotton swab or a paint-brush that has been briefly dipped in HS before contacting the pellet.

6 Layer each resuspended sample on a Percoll pad, comprising an equal volume of 35% Percoll in HS in a 15 ml Corex tube. This is made easier by tilting the tube at 30° to the vertical and gently running the chloroplast suspension down the lower side of the tube.

7 Centrifuge at 1400 g for 8 min once the operating speed is attained in a swing-out rotor, with the brake off. Intact chloroplasts are pelleted, whereas lysed and aggregated organelles fail to penetrate through the Percoll pad (illustrated in *Figure 2*).

8 Remove the Percoll and green debris using a glass Pasteur pipette. Place the end just above the layer of lysed chloroplasts and remove them in a long, billowing string. The intact chloroplast pellet is easily disturbed at this stage, so take care removing the last drops of supernatant.

9 Wash the pellet of chloroplasts by filling up the Corex tubes with HS to remove residual Percoll. Centrifuge at 4000 g for 1 min in a swing-out rotor. Remove the supernatant by pipetting, taking care not to disturb the pellet. Resuspend the chloroplast pellet in 1 ml HS.[d]

10 Measure the chlorophyll concentration of the suspension. A rapid estimate (6) may be obtained by adding 5 μl chloroplast suspension to 1 ml 80% acetone (do three replicates). Incubate on the bench for 1 min.

11 Centrifuge the tubes for 2 min in a microcentrifuge (top speed), to pellet protein precipitates.

12 Transfer the supernatant to a glass or quartz cuvette, and read the absorbance at 652 nm, against an 80% acetone blank. The chlorophyll concentration (mg/ml) is given by the formula: chlorophyll concentration = $A_{652} \times 5.6$. Adjust the concentration of the chloroplast suspension using HS to 1 mg/ml with respect to chlorophyll and use the chloroplasts as soon as possible.

[a] 100 ml tubes are preferable to 50 ml tubes, simply because there will be fewer samples to handle and therefore less time spent resuspending pellets. Some researchers also suggest that shortening the distance the organelles travel during centrifugation will reduce damage to them. Using this line of reasoning, it would be better to half-fill large tubes with homogenate than completely fill small tubes. However, rotors such as HB-4 (Sorvall), which carry 50 ml tubes are more widespread than rotors, which carry 100 ml tubes, so we suggest that the experimenter use these if a swing-out rotor for 100 ml tubes is not available.

Protocol 1 continued

b Do not use glass beakers in combination with a Polytron homogenizer. Use either a homogenization vessel designed for the Polytron and available from Kinematica AG or general laboratory equipment suppliers, or a plastic beaker. Do not allow the rotating blades to touch the bottom of the container, because this contact will damage the container and blunt the blades.

c The 5× HS stock may be stored in a −20 °C freezer, as HEPES is unstable to autoclaving.

d Check the intactness of the organelles by phase-contrast microscopy; intact organelles appear bright green, often with a surrounding halo, whereas broken chloroplasts appear darker and more opaque. The majority of the organelles should be intact, although 50% intactness should give reasonable results.

Protocol 2

Isolation of intact chloroplasts from barley seedlings[a]

Equipment and reagents

- Polytron or other homogenizer with very sharp blades, plus a vessel (see *Protocol 1*)[a]
- Miracloth or 30 μM pore nylon mesh
- Refrigerated centrifuge; with rotors equivalent to Sorvall GSA, SS34 and HB-4
- 2 × Grinding buffer (2 × GB): 100 mM HEPES-KOH pH 7.5, 660 mM sorbitol, 10 mM sodium ascorbate, 0.5% (w/v) BSA (fraction V), 4 mM EDTA, 2 mM $MgCl_2$, 2 mM $MnCl_2$

- 1 × Grinding buffer (GB): dilute 2 × GB with distilled water
- Percoll gradient: dissolve 2 mg glutathione in 35 ml 2 × GB and mix with 35 ml Percoll
- HS (see *Protocol 1*)
- 80% v/v acetone in distilled water

Method

1 Prepare the Percoll gradients by dividing the Percoll mixture described in the equipment and reagents section between two clear centrifuge tubes (e.g. Sorvall SS34 type). Centrifuge the tubes in a SS34 rotor or equivalent for 40 min at 47 800 g (at r_{max}). Store the gradients on ice.

2 Harvest the barley seedlings by pulling each one out of its leaf sheath, and cutting the **lower** half of the seedling into 5 mm long sections with scissors, avoiding the very pale part at the base. The harvested seedlings should be stored on ice until ready for grinding.

3 Put the seedlings into ice-cold GB and homogenize them in batches of around 10 g tissue/100 ml GB. Approximately, 100 g leaf tissue can be obtained from a tray of 7 day-old seedlings, 10 cm in height. As with peas, do not over-homogenize.

4 Strain the homogenate through two layers of Miracloth or controlled-pore nylon mesh, premoistened with GB. Pour the filtrate into 300 ml centrifuge bottles and centrifuge at 3200 g for 8 min in a Sorvall GSA rotor or equivalent.

5 Discard the supernatant and gently resuspend each pellet in 3–4 ml GB, by gentle swirling of the bottle.

6 Layer the suspension onto a Percoll gradient (prepared in step 1), and centrifuge at 7700 g in a Sorvall HB-4 rotor or equivalent, for 15 min without the brake. There will be an upper layer of lysed chloroplasts, a lower layer of intact organelles and a pellet of starch grains on the bottom of the tube.

Protocol 2 continued

7   Remove the lower green layer of intact chloroplasts, transfer it to a fresh SS34 tube or similar, estimate its volume, and dilute it with 3 vols of GB. Centrifuge at 3200 g for 6 min in a SS34 rotor or equivalent, to remove the Percoll.

8   Discard the supernatant, and resuspend each pellet, with a paintbrush if necessary, in 2 ml HS. Pool the pellets and dilute them with HS until the tube is full (the exact volume is not important). Centrifuge at 1200 g for 6 min in a SS34 rotor or similar, and resuspend the pellet in 500 μl HS.

9   Determine the chlorophyll concentration, as in *Protocol 1*, steps 10–12, and adjust the concentration of the suspension to 1 mg/ml with respect to chlorophyll.[b]

[a] The method is modified from that described by Dahlin and Cline (7).

[b] If barley and pea chloroplasts (prepared as described) are compared under the microscope, the barley organelles will be smaller and appear less green than those from pea.

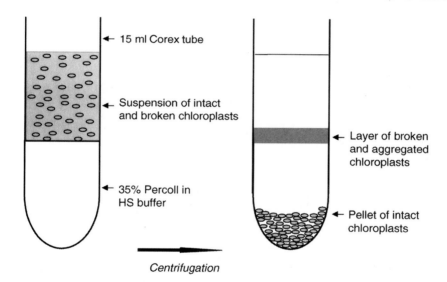

Centrifugation

Figure 2 Percoll purification of intact chloroplasts. A suspension of broken and intact pea chloroplasts (*Protocol 1*, step 5) is layered upon 35% Percoll in HS buffer. After centrifugation in a swing-out rotor, the intact chloroplasts form a loose pellet, and the remaining debris can be withdrawn using a Pasteur pipette and discarded.

3 *In vitro* synthesis of nuclear-encoded chloroplast proteins

A number of methods are available for the generation *in vitro* of synthetic mRNA following the isolation of a full-length cDNA clone (8). The transcription protocol described in *Protocol 3*, which can be used with SP6, T7, or T3 RNA polymerase, is useful because the transcription products can be used directly to programme a cell-free translation system, without first phenol- or salt-extracting the RNA. Try the protocol first without linearizing the vector, but in some cases this may be necessary. If possible, avoid using restriction enzymes, which produce

4-base 3′ protruding ends (*Apa* I, *Kpn* I, *Pst* I, *Sac* I, *Sph* I), as these linearized templates produce inhibitory non-coding RNA (9). Where these enzymes must be used, fill in the resulting overhangs with Klenow DNA polymerase.

Protocol 3

In vitro transcription of cDNA

Equipment and reagents

- Nuclease-free plastic ware[a]
- Transcription mix containing the following nuclease-free reagents: 40 mM Tris-HCl pH 7.5, 6 mM $MgCl_2$, 2 mM spermidine, 10 mM DTT, 0.5 mM ATP, CTP and UTP 50 μM GTP (the NTPs should be pH 7–8), and 100 μg/ml BSA
- Water bath set to 37 °C
- RNase inhibitor (RNasin, Promega)
- Monomethyl cap: $m^7G(5')ppp(5')G$ (Promega or Pharmacia)
- RNA polymerase (SP6: Gibco, 15 U/μl; T7: Gibco, 50 U/μl or T3: Stratagene, 50 U/μl)

Method

1 Prepare high quality plasmid DNA in an appropriate transcription vector, either CsCl-purified or from a midiprep kit (e.g. Qiagen or Promega), at a concentration of 1 μg/μl in water or 10 mM Tris-HCl pH 8.0.

2 Mix at room temperature (cold temperatures can cause DNA to precipitate in the presence of spermidine): 2 μl DNA, 15.5 μl transcription mix, 20 U RNasin, 0.1 U monomethyl cap, and 1 μl of the appropriate RNA polymerase.

3 Incubate for 30 min at 37 °C, then add 1 μl of 10 mM GTP and continue the incubation for a further 30 min. The products of the transcription reaction may be frozen at -80 °C until required.

[a] Wear gloves when working with RNA to prevent nucleases on the skin from entering the transcription reaction (see also *Chapter 1*).

Generally, the best method for the synthesis of nuclear-encoded chloroplast protein precursors is the wheatgerm lysate system, described by Anderson *et al.* (10). The reticulocyte lysate system may also be used and should be tried when wheatgerm lysates fail to translate large mRNA templates. However, several groups have found that reticulocyte lysate can lyse chloroplasts, if used in high concentrations. Both wheatgerm and reticulocyte lysates are available commercially (e.g. Promega or Amersham).

For optimal results, follow the manufacturer's instructions when using wheatgerm lysate or rabbit reticulocyte lysate. *Protocols 4* and *5* are based on the instructions supplied with Promega translation systems. Highly detailed guides to translation using their products may also be found on the Internet at www.promega.com. In our laboratory, we find that about 1 μl transcription products (from *Protocol 3*) per 25 μl translation reaction gives good results, using 30 μCi [^{35}S] methionine, or 25 μCi [^{3}H] leucine. We selected these isotope amounts as a compromise between achieving high levels of incorporation into translated proteins, on the one hand, and reducing the amount of money spent on radioactive isotopes on the other. [^{35}S] cysteine may also be used, but suffers twin drawbacks: namely, high expense and chemical instability, meaning that aliquots should ideally be stored under liquid nitrogen. It is advisable to optimize the translation reaction with respect to the concentration of transcription

Concentration of K⁺, mM / (volume of mRNA mix per 25 µl translation)

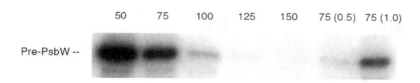

Figure 3 Variation of *in vitro* translation efficiency with K⁺ ion concentration. Pre-PsbW mRNA was used to programme a wheatgerm extract translation system in the presence of ^{35}S-methionine. After A 1 h incubation period, the products were analysed by SDS-PAGE and fluorography. In the first five tracks, the volume of transcription mixture was held at 1.5 µl per 25 µl translation reaction, and the concentration of K⁺ ions was varied from 50 mM to 150 mM. In the last two tracks, the K⁺ ion concentration was held at 75 mM, and the volume of transcription mixture varied to 0.5 or 1.0 µl per 25 µl translation reaction. PsbW is a subunit of photosystem II, located in the thylakoid membrane.

products (serially dilute the mRNA template first, and add the same volume to each sample); optimal concentrations of potassium and magnesium can also vary with different mRNA species. *Figure 3* illustrates how drastically translation efficiency may vary with ion concentration in the wheatgerm mixture.

Protocol 4 explains how to prepare three 25 µl translation reactions labelled with [^{35}S] methionine: the sample of interest, one negative control and one positive control using a template supplied by the wheatgerm lysate manufacturer. Volumes may be scaled-up for larger reactions. 0.5 ml microcentrifuge tubes may be substituted for 1.5 ml tubes, if preferred. Wear gloves to protect against accidental contamination with radioisotope and, importantly, to prevent nucleases from entering the translation reaction.

Protocol 4

In vitro translation of [^{35}S] methionine-labelled precursor proteins using wheatgerm lysate[a]

Equipment and reagents

- Nuclease-free plastic-ware
- Water bath set to 25 °C
- Nuclease-free water
- RNase inhibitor (RNasin, Promega)
- L-[^{35}S] methionine: Redivue™ 1000 Ci mmol/10 mCi ml (Amersham)[b]
- Wheatgerm lysate ('Wheat Germ Extract' from Promega)

- Amino acid mixture, minus methionine (the mixture contains 1 mM of each amino acid except methionine, and is supplied with 'Wheat Germ Extract' from Promega)
- Products from an *in vitro* transcription reaction (*Protocol 3*)
- Brome Mosaic Virus (BMV) mRNA as a positive control for translation (supplied with 'Wheat Germ Extract' from Promega)

Method

1. Thaw all frozen reagents slowly on ice.[c]
2. Remove 45 µl wheatgerm lysate and transfer it into an ice-cold 1.5 ml microcentrifuge tube. Centrifuge at 2000 g for 5 min at 4 °C in a microcentrifuge to sediment aggregated protein.

Protocol 4 continued

3   Remove the supernatant with a pipette and transfer it to a fresh microcentrifuge tube on ice. For three 25 μl translation reactions, a total of 37.5 μl of supernatant is required.

4   Assemble the following components on ice, in a 1.5 ml microcentrifuge tube, making sure that the RNasin is the first item to be added: 0.5 μl (20 U) RNasin, 6 μl nuclease-free water, 2 μl amino acid mixture minus methionine, 3 μl (30 μCi) [35S] methionine, 1 μl products from a transcription reaction (*Protocol 3*), and 12.5 μl wheatgerm lysate supernatant. For the negative control,[d] substitute 1 μl nuclease-free water for transcription products; for the positive control[e] add 1 μl BMV mRNA instead of the experimenter's own transcription products. After each addition, gently mix the components with a pipette tip, but do not blow bubbles in the mixture.

5   Incubate the translation reactions at 25 °C for 60 min. Promega suggests extending the incubation time to 120 min if the experimenter is not satisfied with the yield from a 60 min incubation.

6   Analyse the translation reaction using SDS-PAGE (*Protocol 6*).[f]

[a]   The same protocol may be used with other radiolabelled amino acids, such as [3H] leucine, but take care to use the corresponding amino acid mixture when setting up the translation reactions.

[b]   This grade of [35S] methionine is suggested because it is very high quality, shows increased stability during storage and is visible in the microcentrifuge tube, due to the addition of a dye.

[c]   The first time a tube of wheatgerm lysate is thawed, it is wise to divide it into smaller, single-use aliquots before re-freezing the remainder, because wheatgerm lysate loses activity after several rounds of freezing and thawing.

[d]   The negative control allows the experimenter to assess whether there is background incorporation of radiolabelled amino acids.

[e]   There should be four main viral proteins visible in the positive control, which have the following sizes: 109, 94, 35, and 20 kDa.

[f]   If the experimenter's mRNA template is not translated at all, or poorly compared to the positive control, then try varying the concentration of transcription reaction added to the translation reaction. If mRNA templates contain large stretches of GC nucleotides, then regions of secondary structure can form, which inhibit translation. This secondary structure can sometimes be reduced by heating an aliquot of the transcription reaction from *Protocol 3* to 67 °C for 10 min, followed by immediate cooling on ice. Centrifuge the tube for a few seconds to return condensed water to the bottom of the tube, before adding the transcription products to a translation reaction. If neither of these suggestions improve translation efficiency, vary the potassium or magnesium concentrations in the translation reaction. Promega supplies 1 M potassium acetate with their 'Wheat Germ Extract'. In the translation reactions described above, the final potassium acetate concentration is about 50 mM. The optimal potassium concentration for a given mRNA template should lie in the range 50–200 mM. The final magnesium acetate concentration is about 2 mM in the translation reactions described above. The optimal magnesium concentration for most mRNA templates should lie in the range 2–5 mM.

Protocol 5 explains how to prepare *in vitro*-translated proteins using rabbit reticulocyte lysate. This system should be tried if the protein of interest fails to translate in a wheatgerm system. Of the many proteins in our laboratory that we have tried to translate and import into chloroplasts, less than 5% translated so poorly in wheatgerm lysate that we used rabbit reticulocyte lysate instead.

Protocol 5

In vitro translation of [^{35}S] methionine-labelled precursor proteins using rabbit reticulocyte lysate[a]

Equipment and reagents

- Nuclease-free plastic-ware
- Water bath set to 30 °C
- Nuclease-free water
- RNase inhibitor (RNasin, Promega)
- L-[^{35}S] methionine: Redivue™ 1000 Ci mmol/10 mCi/ml (Amersham)
- Rabbit reticulocyte lysate (Promega)

- Amino acid mixture, minus methionine (the mixture contains 1 mM of each amino acid except methionine, and is supplied with rabbit reticulocyte lysate from Promega)
- Products from an *in vitro* transcription reaction (*Protocol 3*)

Method

1 Thaw all frozen reagents slowly on ice.[b]

2 Assemble the following components on ice, in a 1.5 ml microcentrifuge tube, ensuring that the RNasin is the first item to be added: 0.5 µl (20 unit) RNasin, 3.5 µl nuclease-free water, 0.5 µl amino acid mixture minus methionine, 2 µl (20 µCi) [^{35}S] methionine, 1 µl products from a transcription reaction (*Protocol 3*), and 17.5 µl rabbit reticulocyte lysate. For a negative control, set up a similar sample, substituting 1 µl nuclease-free water for the transcription products. After each addition, gently mix the components with a pipette tip, but do not blow bubbles in the mixture.

3 Incubate the translation reactions at 30 °C for 90 min.

4 Analyse the translation reaction using SDS-PAGE (*Protocol 6*).[c]

[a] The same protocol may be used with other radiolabelled amino acids, such as [^{3}H] leucine, but take care to use the corresponding amino acid mixture when setting up the translation reactions.

[b] Like wheatgerm lysate, rabbit reticulocyte lysate should not be subjected to multiple freeze-thaw cycles. The first time a tube of rabbit reticulocyte lysate is defrosted, we suggest that it is divided into single-use aliquots before being returned to the −80 °C freezer. Rabbit reticulocyte lysate is inactivated by storage on dry ice, so it is preferable to freeze unused lysate in liquid nitrogen.

[c] If the mRNA template is not translated, then try varying the concentration of transcription products added to the translation reaction. Promega also suggest heating an aliquot of mRNA template to 65 °C for 3 min, followed by immediate cooling in an ice/water bath, to denature any secondary structure in the mRNA. The concentration of rabbit reticulocyte lysate in each translation reaction may be varied between 50 and 70% v/v of the total sample volume. Promega advise that use of concentrations lower than 70% may require the experimenter to optimize the translation reaction with respect to potassium and magnesium concentrations (see *Protocol 4*).[f]

3.1 Analysis of proteins by SDS-PAGE

The results of the *in vitro* translation reaction should be analysed to check that the protein is of the expected size. SDS-polyacrylamide gel electrophoresis (SDS-PAGE; 11) is described in *Protocol 6*, both as a technique to analyse the results of translation reactions, and the results of chloroplast or isolated thylakoid import assays. *Protocol 7* describes a variant of SDS-PAGE particularly well suited to the analysis of small (e.g. 2–14 kDa) proteins (12).

Electrophoresis apparatus usually consists of two reservoirs of buffer, the upper containing a cathode and the lower containing an anode. A sandwich, consisting of the polyacrylamide gel between two glass plates is clamped in position between the two buffer reservoirs, such that the top of the sandwich is immersed in the cathode buffer, while the bottom end is immersed in anode buffer. After the protein samples have been loaded onto the top of the gel, the two electrodes are connected to a power supply, and either a constant current or constant potential difference is applied to the electrodes. The proteins are denatured and negatively charged, by incubation in sample buffer containing the strongly anionic detergent SDS, and migrate through the gel towards the anode at a rate which is inversely related to their size, but which may also be affected by other factors, such as degree of hydrophobicity. After electrophoresis, the gel is removed from its glass plate sandwich and submersed in a fixing solution, which immobilizes the proteins in the gel. At this stage, the proteins may also be stained using Coomassie Brilliant Blue R-250 dye. Further incubation of the gel in a destaining solution removes background colour and reveals blue bands of separated proteins. Using Coomassie staining, it is not possible to see the relatively small quantities of protein synthesized by a small *in vitro* translation reaction, like those described in *Protocols* 4 and 5. Instead, these radiolabelled proteins are visualized by autoradiography. For isotopes like [^{35}S] and [^{3}H], which emit only weak radiation, the signal is best measured either by fluorography, or by a phosphorimager. In the case of fluorography, the gel is incubated in an organic scintillant, which amplifies the radioisotope signal. The gel is dried on a sheet of filter paper, under a vacuum, and exposed to X-ray film in a cassette, which is usually developed after a period of a few hours to a few weeks, depending on the strength of the radioactive signal. For analysis by a phosphorimager, the gel is dried (again, under a vacuum) after fixing and staining, and placed in a cassette so that the dried gel is in contact with a phosphor screen. After a period of exposure (usually 0.5–2 days), the screen is 'read' by a laser scanner. One benefit of using a phosphorimager to analyse gels is that the protein bands may be quantified using image analysis software supplied with the scanner. However, the phosphorimager cassettes can be extremely costly.

There are countless suitable commercially available and home-made types of electrophoresis apparatus of different shapes and sizes. In our laboratory, we use gel plates that are 16.5 cm (width) $\times$ 14.5 cm (height), and combine these with 1 mm thickness spacers and gel combs. We find these suitable for the analysis of the results of chloroplast import assays and *in vitro* translations, because the gels are large enough to provide good resolution, but not so large that they require cooling apparatus during electrophoresis, nor that they are especially fragile during the drying process. It is likely that the reader will have his or her own preferred electrophoresis apparatus, but by way of illustration in the protocols, we will refer to a good electrophoresis set sold by CBS Scientific Company Inc. (www.cbssci.com), known as 'CBS Lite'. The gel mixtures described are to make one gel in a plate assembly 16.5 cm (width) $\times$ 14.5 cm (height) with 1 mm thickness spacers, plus a little extra in case of spillage.

Protocol 7 describes the preparation of one Tricine gel, cast between plates 16.5 cm (width) $\times$ 14.5 cm (height), using 1 mm thickness spacers and comb.

Protocol 6

Basic method for SDS-polyacrylamide gel electrophoresis (SDS-PAGE) analysis of proteins

Equipment and reagents

- Electrophoresis power supply
- Pair of glass gel plates, front and back (e.g. 16.5 × 14.5 cm from CBS Scientific)
- Spacers 1 mm in thickness (e.g. from CBS Scientific)
- Comb 1 mm in thickness (e.g. 12 wells from CBS Scientific)
- Gasket to seal the gel plates (e.g. Gel Wrap from CBS Scientific)
- Clips to clamp the gel plates and spacers together
- Hamilton 100 μl syringe or similar
- Vacuum gel dryer
- X-ray cassette
- X-ray film (e.g. Kodak Biomax™ ML)
- Buffer for resolving gel: 3 M Tris-HCl pH 8.8
- Buffer for stacking gel: 0.5 M Tris-HCl pH 6.8
- Acrylamide[a] solution 1: 30% w/v acrylamide, 0.8% w/v bis-acrylamide (N′N′-methylene-bisacrylamide)
- 10% w/v SDS
- 10% w/v ammonium persulphate (APS), freshly made
- TEMED[a] (N, N, N′, N′-tetramethylethylenediamine)

- Water-saturated butan-1-ol (mix equal volumes of butan-1-ol and distilled water in a bottle with a leak-proof cap, shake the bottle gently for about 1 min, then allow the butan-1-ol to separate from the water. Water-saturated butan-1-ol is the upper layer)
- SB (sample buffer): 125 mM Tris-HCl pH 6.8, 4% w/v SDS, 20% w/v glycerol, 10% v/v β-mercaptoethanol;[a] colour with a few grains of bromophenol blue until the buffer is blue, but not opaque
- Protein markers (e.g. Pharmacia Low Molecular Weight Markers 94–14.4 kDa), made up according to the manufacturer's protocol
- Laemmli buffer: 25 mM Tris, 192 mM glycine (pH 8.3), 0.1% w/v SDS
- Fixing solution: 10% v/v acetic acid, 45% v/v methanol, 0.25% w/v Coomassie Brilliant Blue R-250 (filter the solution through Whatman paper before use, to remove undissolved dye particles)
- Destaining solution: 7% v/v acetic acid, 40% v/v methanol
- Fluorography reagent (e.g. Amplify™ from Amersham)

Method

1 Clean the glass plates, spacers, and gel comb. Wash with detergent and a sponge, then rinse with distilled water and polish with ethanol.

2 Place the spacers at the left and right edges of the back, un-notched plate, and gently rest the notched plate on top, to form a 'sandwich'. Seal the sandwich with a rubber gasket or Gel Wrap.

3 Place two strong clips on each of the left and right edges of the sandwich so that the spacers and sealing gasket are held securely in place, and two clips on the bottom edge.

4 Prepare the stacking gel mixture (6% w/v acrylamide final concentration). Mix in a beaker: 1 ml acrylamide solution 1, 1.25 ml 0.5 M Tris-HCl pH 6.8, 2.7 ml distilled water, 50 μl 10% w/v SDS and 25 μl 10% w/v APS. Put to one side.

5 Prepare the resolving gel mixture (15% w/v acrylamide final concentration).[b] Mix in a beaker: 10 ml acrylamide solution 1, 2.5 ml 3 M Tris-HCl pH 8.8, 7.3 ml distilled water, 200 μl 10% w/v SDS, and 83 μl 10% w/v APS.

6 Add 8.3 μl TEMED to the resolving gel mixture to initiate polymerization. Swirl gently to mix, then either pour or pipette the mixture between the glass plates of the sandwich. Fill the sandwich to within 2.5 cm of the upper edge of the notched front plate.

Protocol 6 continued

7 Stand the sandwich upright and pipette 1 ml of water-saturated butan-1-ol on top of the gel. Gently insert the comb between the gel plates (without touching the gel mixture or butan-1-ol). The addition of butan-1-ol helps to smooth the upper surface of the gel, excludes air to promote polymerization, and will also serve as a useful indicator of polymerization. When the gel has polymerized, a small amount of water is excluded from the gel matrix, and it forms a distinctly visible layer between the surface of the polymerized gel and the layer of butan-1-ol. Wait an extra 30 min after the appearance of this water of polymerization, before proceeding to the next step.

8 Remove the comb and pour off the water and butan-1-ol from between the gel plates. Carefully rinse the butan-1-ol away with distilled water and then dry between the plates with a small piece of Whatman filter paper, without touching the surface of the gel.

9 Add 6 μl TEMED to the stacking gel mixture, mix, then pour or pipette the mixture between the gel plates, on top of the polymerized resolving gel. Fill the sandwich to the top edge of the notched plate.

10 Carefully insert the gel comb, making sure there is at least 1 cm between the bottom of the teeth and the top of the resolving gel. At this stage, there is a danger of acrylamide solution being splashed into the face, so wear safety glasses.

11 Allow the stacking gel to polymerize. This usually takes about 30–60 min, depending on the temperature of the laboratory.

12 Prepare the protein samples for analysis. Mix the sample with an equal volume of SB. In the case of *in vitro* translated proteins, remove a 2 μl sample and mix with 2 μl SB in any size of microcentrifuge tube.

13 Boil the sample for 5 min.[c]

14 Centrifuge the sample for 30 s at 15 000 g at room temperature in a microcentrifuge, to return condensed water from the lid to the bottom of the tube.

15 Peel away the sealing gasket from around the edges of the gel sandwich, but keep the clips in place, to prevent the plates from separating.

16 Remove the comb from the polymerized stacking gel.

17 Immediately assemble the gel sandwich in the electrophoresis apparatus, fill the upper and lower reservoirs with 1 × Laemmli buffer and flush out the wells of the stacking gel with Laemmli buffer, using a Hamilton syringe to ensure there are no air bubbles or unpolymerized acrylamide in the wells. Check that there are no air bubbles trapped underneath the bottom edge of the gel.

18 Load the protein sample and an aliquot of protein markers (see manufacturer's instructions for the quantity) into adjacent wells, using a Hamilton syringe. Rinse the syringe between loading samples. Put equal volumes of SB in any unused wells, to prevent spreading of the proteins into adjacent lanes during electrophoresis.

19 Put the lid on the electrophoresis apparatus and connect the power supply. Run the gel at a constant current of 30 mA.[d] In our laboratory it takes between 2 and 3 h for the dye-front to reach the bottom of the gel.[e] A 15% w/v polyacrylamide gel can be used to analyse proteins in the 90–13 kDa size range, but it is more useful in the 40–13 kDa size range. Proteins smaller than about 12 kDa usually run too close to the dye-front to be resolved, so then a Tricine gel is recommended (*Protocol 7*).

20 Disconnect the power supply, dispose of the buffers from the top and bottom reservoirs and remove the gel sandwich from the electrophoresis apparatus.

Protocol 6 continued

21 Remove the spacers and separate the gel plates. Carefully submerse the gel in 200 ml fixing solution containing Coomassie Brilliant Blue R-250. Use a larger volume of fixing solution if the gel is not able to move freely in the solution.

22 Incubate at room temperature on a rotating platform (e.g. 30 rpm). The gel will be adequately fixed in 30 min, but staining may take up to 90 min. The gel is fully stained when both the stacking and resolving gels are deep blue in colour, and it is just possible to make out even deeper blue protein bands against the background.

23 Pour off the staining solution and add 50 ml destaining solution. Incubate at room temperature on a rotating platform for 5 min.

24 Pour off the now blue destaining solution and add 200 ml of clean destaining solution, or sufficient that the gel is able to float in the solution. Incubate at room temperature on a rotating platform. Changing the destaining solution every 30 min should result in a light background and deep blue stained proteins after about 3 h. Alternatively, put 400 ml destain on the gel and leave it on the rotating platform overnight.

25 Pour off the final lot of destaining solution and incubate the gel in fluorographic reagent, according to the manufacturer's instructions.[f]

26 Dry the gel on filter paper under vacuum, according to the equipment manufacturer's instructions.

27 Expose the dried gel to X-ray film in a cassette, following the film manufacturer's instructions.

[a] Acrylamide is a potent neurotoxin. *Always* wear gloves and protective clothing when handling solutions of acrylamide. Even polymerized acrylamide contains small quantities of unpolymerized material, so should be handled with caution. TEMED and β-mercaptoethanol are also harmful. Use in a fume hood.

[b] Acrylamide concentrations can easily be varied between 8 and 20% w/v, by altering the proportions of water and acrylamide solution 1 in the resolving gel mixture. For example, a 12.5% w/v gel, which is useful for separating proteins between about 15 and 65 kDa in size, can be made by using 9 ml distilled water and 8.3 ml acrylamide solution 1, instead of the volumes described in step 5. The lower percentage gels are more fragile, so take care handling them. Above about 17% w/v acrylamide (final concentration), there is a greater risk of the gel cracking during drying. We prefer to use Tricine gels, rather than using standard 17% w/v or 20% w/v gels, because they are excellent at resolving size differences between small proteins.

[c] Be aware that some proteins are unstable to boiling in SDS. When testing a translation reaction for the first time, it is wise to boil one aliquot of the reaction, and heat another for 20 min at 40 °C, to test whether boiling causes the protein of interest to aggregate.

[d] For the CBS Lite gel apparatus described, 220 V is a sensible maximum potential difference when running a gel at constant current. If potential differences become too great, there is a risk that the gel will over-heat and the plates crack.

[e] Unincorporated radiolabelled amino acids run shortly behind the dye-front. So if the experimenter wishes to minimize radioisotope contamination of the electrophoresis apparatus, then he or she should not allow the dye-front to run off the gel.

[f] Adequate destaining is important to remove background labelling of the gel by unincorporated radiolabelled amino acids. Fluorographic reagents are often expensive, but this is offset by the fact that they can be re-used a number of times. However, their life span is greatly reduced if gels have not been properly destained. This is because the unincorporated radiolabelled molecules leach out into the reagent and cause an increase in the background signal for subsequent users of the reagent.

Protocol 7

Basic method for the Tricine gel variation of SDS-PAGE

Equipment and reagents

- Electrophoresis power supply
- Pair of glass gel plates, front and back (e.g. 16.5 × 14.5 cm from CBS Scientific)
- Spacers 1 mm in thickness (e.g. from CBS Scientific)
- Comb 1 mm in thickness (e.g. 12 wells from CBS Scientific)
- Gasket to seal the gel plates (e.g. Gel Wrap from CBS Scientific)
- Clips to clamp the gel plates and spacers together
- 3 M Tris-HCl pH 8.45
- Acrylamide solution 2: 48% w/v acrylamide, 1.5% w/v bis-acrylamide

- 10% w/v SDS
- 10% w/v ammonium persulphate (APS), freshly made
- TEMED (N, N, N', N'-tetramethylethylenediamine)
- Water-saturated butan-1-ol (see *Protocol 6*)
- TSB (Tricine gel sample buffer): 50 mM Tris-HCl pH 6.8, 4% w/v SDS, 12% v/v glycerol, 24% w/v urea, 2% v/v β-mercaptoethanol, bromophenol blue (see SB *Protocol 6*)
- Anode buffer: 200 mM Tris-HCl pH 8.9
- Cathode buffer: 100 mM Tris, 100 mM Tricine, 0.1% w/v SDS

Method

1 Assemble a glass plate sandwich, as described in *Protocol 6*, steps 1–3.

2 Prepare the stacking gel mixture. Mix in a beaker: 0.5 ml acrylamide solution 2, 1.55 ml 3 M Tris-HCl pH 8.45, 4.15 ml distilled water, 47 μl 10% w/v SDS, and 50 μl 10% w/v APS. Put to one side.

3 Prepare the resolving gel mixture. Mix in a beaker: 5 ml acrylamide solution 2, 5 ml 3 M Tris-HCl pH 8.45, 4.95 ml distilled water, 150 μl 10% w/v SDS, and 50 μl 10% w/v APS.

4 Add 20 μl TEMED to the resolving gel mixture to initiate polymerization. Swirl gently to mix, then either pour, or pipette the mixture between the glass plates of the sandwich. Fill the sandwich to within 2.5 cm of the upper edge of the notched front plate.

5 Follow *Protocol 6*, steps 7 and 8.

6 Add 15 μl TEMED to the stacking gel mixture to initiate polymerization. Proceed as for *Protocol 6*, steps 9–18, except that TSB should be used instead of SB for preparing protein samples.

7 Put the lid on the electrophoresis apparatus and connect the power supply. Use a constant potential difference of 30 V, while the samples run from the sample wells into the stacking gel. Immediately the samples have completely entered the stacking gel, increase the potential difference to 90 V. When the dye-front reaches the interface between stacking and resolving gel, adjust the supply to a constant 50 V, which should correspond to about 30 mA. The starting current should not exceed 35 mA, so reduce the potential difference if this is the case. In our laboratory, the gels take approximately 16 h to run, using the conditions and equipment described in this protocol. During this time, using a constant potential difference, the current should drop from 30 mA to around 8 mA. Towards the end of a run, it may be necessary to increase the potential difference, to prevent the current dropping below 8 mA.

8 Follow *Protocol 6*, steps 20–27.

4 Import of proteins into isolated chloroplasts

Reconstitution of chloroplast protein import is usually straightforward, providing that intact chloroplasts are isolated as described in *Protocols 1* or *2*, and sufficiently labelled precursor proteins can be prepared. Every precursor protein that we have tested so far has been successfully imported into isolated chloroplasts, although with varying efficiencies. Import takes place post-translationally and therefore the precursors are incubated with chloroplasts after translation is complete. Ideally, freshly prepared chloroplasts should be incubated with fresh translation mixtures, but we have found most translation mixtures can still be imported after freezing at $-80\,^{\circ}C$. However, several rounds of freeze-thawing usually lead to a rapid loss of import-competence and this is especially relevant for membrane proteins. Another point to consider is the distribution of labelled amino acids in the precursor protein—make sure there is at least one in the mature protein, so the protein may still be detected after import and processing.

The basic import assay conditions, modified from original methods (13) are given below in *Protocol 8*. ATP is required for both binding and import into chloroplasts (14, 15); GTP may also be required by the import machinery (16). It is present in the translation mixture and will also be generated in the stroma by photophosphorylation when the chloroplasts are illuminated. However, we generally add ATP to the assay and so pre-incubate the chloroplasts at $25\,^{\circ}C$ for 10 min to allow the stromal ATP concentration to reach the optimal level. A suggested arrangement for setting up an illuminated water bath is shown in *Figure 4*.

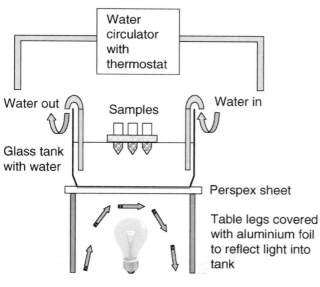

Figure 4 Suggested arrangement for an illuminated water bath. A light source is placed under a glass tank, supported on a transparent Perspex sheet. The degree of illumination is determined by the distance of the light from the base of the tank and the depth of the water in the tank. These should be fixed so that the light intensity can be measured. To prevent the hot light bulb from melting the Perspex sheet, an electric fan may be used to ventilate the air around the light. The temperature of the water bath may be controlled by using a thermostatic water circulator. If this is unavailable, then it will be necessary to monitor the water temperature closely during the course of the experiment and add some cold water if necessary.

At all times, when handling intact chloroplasts, use wide-bore pipette tips (minimum internal diameter 1 mm), made by cutting off the ends with scissors.

Protocol 8

Import of proteins into isolated chloroplasts: basic assay

Equipment and reagents

- Illuminated water bath, set at 25 °C
- HS (as in *Protocol 1*)
- 60 mM methionine in 2 × HS[a]
- 60 mM MgATP pH 7.0–8.0 in HS
- Translation mixture (see *Protocols 4* and *5*)
- SB: SDS-PAGE sample buffer (see *Protocol 6*)

Method

1 Prepare intact chloroplasts (*Protocol 1* or *2*) and *in vitro*-synthesized proteins (*Protocol 4* or *5*).

2 Mix the following components in a 1.5 ml microcentrifuge tube and pre-incubate for 10 min at 25 °C: 50 μl chloroplasts (equivalent to 50 μg chlorophyll),[b] 20 μl MgATP, and 55 μl HS.

3 Mix and then add the following to the chloroplast suspension after the pre-incubation: 12.5 μl translation mixture and 12.5 μl methionine.

4 Incubate at 25 °C for 20–60 min[c] in an illuminated water bath, at an intensity of 300 μE m^2/s. Gently shake the tubes or briefly pipette the samples with wide-bore tips if the chloroplasts appear to be settling out of suspension.

5 After incubation, dilute the assay with 1 ml ice-cold HS, and pellet the chloroplasts by centrifugation at 2000 g for 3 min at 4 °C in a microcentrifuge. *Gently* resuspend the chloroplasts in a small volume of ice-cold HS (suggested 120 μl), ready for fractionation, which is described in *Protocol 9*.

6 Remove 30 μl (equivalent to one quarter of the assay), add to SB, and boil for 5 min.[d] This sample contains imported proteins, plus any precursor molecules that are bound to the chloroplast surface. Freeze the sample in liquid nitrogen or dry ice, and store it at −80 °C until it can be analysed by SDS-PAGE.

7 Proceed to *Protocol 9* if the experimenter wishes to fractionate the remaining 90 μl chloroplasts into stroma and thylakoid samples.

[a] Unlabelled methionine is included in the import incubation mixture to prevent the high specific activity, labelled methionine in the translation mix from being incorporated into protein by the chloroplast protein synthesis machinery. When using other radiolabelled amino acids, substitute the corresponding 'cold' amino acid for methionine.

[b] The values for the amount of chloroplasts and translation mixture in the assay are intended as a guide; however, too high a chlorophyll concentration may lead to overloading problems during subsequent SDS-PAGE analysis, and the concentration of translation mixture should not exceed 10% of the total assay volume.

[c] Most of the precursor proteins that we work with are imported during a 20 min incubation. However, a few precursor proteins are imported more slowly, and may take as long as 60 min before the levels of radiolabelled protein inside the chloroplasts reach their maximum. We suggest that the experimenter carry out a time-course of protein import if he or she is interested in finding out the optimum period of time for incubation.

[d] Membrane proteins in particular may aggregate during boiling. Therefore, it is wise to test some *in vitro*-synthesized precursor in a separate experiment to make sure it will survive boiling. An alternative to boiling is to incubate the sample with SB for 10–20 min at 40 °C.

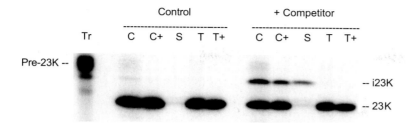

Figure 5 Import of pre-23K into intact pea chloroplasts. Pre-23K mRNA was translated in a Promega wheatgerm lysate, labelled with ^{35}S-methionine. The translation mixture (*Tr*) was incubated with intact pea chloroplasts for 20 min, in the absence (*Control*) and presence (+ *Competitor*) of *E. coli* over-expressed pre-23K protein, and the two samples were then fractionated according to *Protocol 5*. In the Control sample, pre-23K is processed to 23K, seen in the total (*C*) and thermolysin-treated (*C+*) chloroplast fractions. Lysis of the protease-treated chloroplasts yields a stromal fraction (*S*) and thylakoid membranes (*T*). The 23K protein is resistant to thermolysin digestion of the thylakoids (*T+*), demonstrating that it has been translocated into the thylakoid lumen. In the presence of an excess of unlabelled pre-23K (+ *Competitor*), translocation of the radiolabelled 23K into the thylakoid lumen is slowed down, demonstrated by the appearance of intermediate 23K (*i23K*) in the stromal fraction, and a reduction in the amount of protease-protected 23K in the thylakoid lumen.

The following method is designed to locate processing intermediates and the mature polypeptide of a thylakoid lumen protein; results from a typical import assay are shown in *Figure 5*. The strategy is to produce five fractions: total, washed chloroplasts (described above in *Protocol 4*, step 6), protease-treated chloroplasts, stromal extract, total thylakoids (with some envelope contamination), and protease-treated thylakoids. The latter fraction will demonstrate whether or not the import substrate is protected from protease degradation, within the thylakoid lumen. This method is intended as a guide, which may be adapted to the experimenter's own requirements. An entry into the literature of protein targeting to the chloroplast envelope is provided in the following reviews: Gray and Row (17) and Heins *et al.* (3).

Protocol 9

Fractionation of chloroplasts after an import assay

Reagents

- HS (as in *Protocol 1*)
- 100 mM $CaCl_2$
- Thermolysin (Sigma type X protease): 2 mg/ml in HS; also 2 mg/ml in HM
- Thermolysin mixture 1: [12.5 μl 2 mg/ml thermolysin in HS, 6.25 μl 100 mM $CaCl_2$, 141.25 μl HS] multiplied by the number of samples to be digested with protease
- Thermolysin mixture 1a (optional)[a]: [25 μl 2 mg/ml thermolysin in HS, 6.25 μl 100 mM $CaCl_2$, 128. 75 μl] multiplied by the number of samples to be digested with protease

- Thermolysin mixture 2: [10 μl 2 mg/ml thermolysin in HM, 2.5 μl 100 mM $CaCl_2$, 87.5 μl HM] multiplied by the number of samples to be digested with protease
- HSE: HS containing 50 mM EDTA
- HM: 10 mM HEPES-KOH pH 8.0, 5 mM $MgCl_2$
- HME: 10 mM HEPES-KOH pH 8.0, 5 mM $MgCl_2$, 10 mM EDTA
- 500 mM EDTA pH 8.0
- SB: SDS-PAGE sample buffer (see *Protocol 6*)

141

Protocol 9 continued

Method

1 Protease-treat the remaining 90 μl chloroplasts from *Protocol 8*, steps 6 and 7 with thermolysin in a 1.5 ml microcentrifuge tube.[a] A suggested method is to add 160 μl thermolysin mixture 1 to the chloroplasts, giving a final volume of 250 μl, and incubate for 40 min on ice.

2 Stop the protease digestion by adding 50 μl HSE to chelate the Ca^{2+} ions. Mix well, but gently, and remove 100 μl (one-third of the remaining sample) to a clean microcentrifuge tube.

3 Centrifuge this sample at 2000 g for 3 min at 4 °C in a microcentrifuge and resuspend the pellet in 15 μl HSE and 15 μl SB. Boil 5 min and then store at −80 °C.[b]

4 Centrifuge the remaining 200 μl sample at 2000 g for 3 min at 4 °C in a microcentrifuge and resuspend the pellet in 60 μl HME using uncut pipette tips, to lyse the chloroplasts. The EDTA is present to prevent proteolysis of stromal protein from any remaining thermolysin activity.

5 Incubate the chloroplast lysate for 5 min on ice. Centrifuge the sample for 5 min at 4 °C and 18 000 g in a microcentrifuge, to generate a stromal supernatant and a thylakoid pellet.[c,d] If a maximum setting of only 15 000 g is available, then this will suffice. Subsequent centrifugation steps in this protocol should be carried out at 18 000 g or 15 000 g, as available.

6 Immediately remove the stromal supernatant, and add it directly to 15 μl boiling SB. Boil for 5 min and then store at −80 °C. This stromal fraction is equivalent to two aliquots of thylakoid membranes, not one, so remember to load the gel accordingly.

7 Resuspend the thylakoid pellet in 200 μl HM and divide into two equal aliquots in fresh microcentrifuge tubes.

8 Centrifuge at top speed for 5 min at 4 °C in a microcentrifuge. Remove the supernatants.

9 Resuspend one of the pellets in 15 μl HM and 15 μl SB to give the total thylakoid fraction, boil 5 min, then store at −80 °C.

10 Resuspend the second pellet in 100 μl thermolysin mixture 2. Incubate for 40 min on ice.[e]

11 Stop the protease digestion by the addition of 2 μl 500 mM EDTA.

12 Centrifuge at top speed for 5 min at 4 °C in a microcentrifuge, and resuspend the pellet in 15 μl HME and 15 μl SB before boiling for 5 min. Store the sample at −80 °C.

13 Analyse all the fractions by SDS-PAGE, loading equivalent proportions of each sample. One basic method for preparing and running SDS-PAGE gels is described in *Protocol 7*. Tricine gels (12) are recommended when the mature protein is between 2 and 14 kDa in size. A very basic method for preparing and running a Tricine gel is given in *Protocol 8*.

14 For analysis by autoradiography, it is suggested that the gels are amplified with a fluorographic reagent (e.g. Amplify™ from Amersham) before drying them—this can be especially useful for detecting [³H]-labelled proteins.

15 Dry gels under a vacuum on filter paper, rather than between two sheets of plastic, as the plastic will reduce the signal intensity reaching the X-ray film during exposure in a cassette.

[a] Thermolysin is the preferred protease for digesting the outer surface of intact chloroplasts, because it does not alter the envelope permeability (18). However, a control digestion of the precursor protein alone should be carried out in parallel with the assay, to make sure thermolysin will degrade the unimported protein. Occasionally, a higher concentration of thermolysin is required to digest unimported precursor proteins. In this case, we suggest trying thermolysin mixture 1a. Thermolysin should be dissolved in import buffer (HS or HM) at 2 mg/ml and should be stored at −80 °C in single-use aliquots. Calcium is the cofactor required for thermolysin activity, and should only be added to the protease immediately before use.

b Supernatants from washes and protease digestions may also be kept for analysis.

c After lysis of the chloroplasts, all the membranes are pelleted by centrifugation; hence a small proportion of the total will be envelopes. However, protease-treatment of the entire sample in *Protocol 9, step 1* gets around this potential problem, by digesting unimported precursor proteins attached to the envelope. If the experimenter wishes to separate envelopes from thylakoids post-import and prior to protease digestion, a quick method is described in ref. 19.

d Make sure each fraction is removed to a clean tube to avoid potential problems with radiolabelled precursor residue, which can smear around the top of the microcentrifuge tube during the import assay and find its way into the sample buffer later, during boiling.

e The fractionation method may also be used for the analysis of thylakoid membrane proteins. However, a range of proteases and concentrations should be tested to find a combination that will completely digest non-inserted precursor protein, but at the same time give defined and reproducible degradation patterns for protein integrated in the thylakoid membrane. Chemical extraction techniques (e.g. ref. 20) are another useful means of testing whether a protein is membrane-associated or not.

5 Import of proteins into isolated thylakoid membranes

5.1 The import pathway for thylakoid lumen proteins

Most thylakoid lumen proteins of higher plants and green algae, such as *Chlamydomonas reinhardtii*, are nuclear-encoded and are synthesized in the cytosol as precursors with an N-terminal bipartite presequence. After import into the chloroplast, the first portion of the presequence, termed the envelope transit peptide, is usually, but not always cleaved off, leaving a signal peptide in tandem with the mature protein domain. All thylakoid signal peptides are superficially very similar in terms of their predicted domain structures, but in fact there are two subsets that specifically interact with distinct translocases in the thylakoid membrane. After transport across the thylakoid membrane, the signal peptide is removed by thylakoidal processing peptidase. The first four thylakoid lumen proteins to be studied were plastocyanin (PC) and the 16, 23, and 33 kDa proteins of the oxygen-evolving complex (16K, 23K, 33K). It turned out that PC and 33K are translocated across the thylakoid membrane by a Sec translocase (2), which is almost certainly inherited from a cyanobacterial progenitor of the chloroplast. On the other hand, 16K and 23K are transported by a ΔpH-dependent translocase, one component of which has been cloned in maize and has recently been shown to have a prokaryotic counterpart (21, 22). Nearly all thylakoid lumen proteins studied subsequently have been assigned to one or other of the pathways. To a certain extent, the two pathways may be distinguished by chloroplast import assays, but much more detailed analyses can be carried out by using isolated thylakoids, because conditions such as presence/absence of stromal extract, nucleoside triphosphates (NTPs), etc., are readily manipulated.

5.2 The basic import assay

What follows is a simple, light-driven assay for importing precursor proteins into isolated thylakoids, with some suggestions on how to alter conditions to test for the involvement of the Sec or ΔpH-dependent translocases. Two points should be emphasized:

1. The most efficient *in vitro* import to date has been obtained with isolated pea thylakoids.

2. For import to take place *via* the ΔpH-dependent pathway, the sole energy requirement (as might be expected), is the thylakoidal ΔpH. ATP or other NTPs are not required, nor is

stromal extract (the envelope transit peptide does not have to be removed from pre-16K and pre-23K prior to import into isolated thylakoids). Transport *via* the Sec pathway requires stromal SecA, and ATP as well as the thylakoid-associated Sec apparatus. Additional soluble factors for some Sec substrates cannot be ruled out at this stage. The thylakoidal ΔpH is not a prerequisite for transport by the Sec pathway, but may stimulate the transport of some proteins under certain *in vitro* conditions. Thus, for any new thylakoid lumen protein, it is recommended to try import in the presence and absence of stromal extract first of all.

Protocol 10

Import of proteins into isolated thylakoids

Equipment and reagents

- Illuminated water bath set to 25 °C
- HM (see *Protocol 9*)
- 100 mM CaCl$_2$
- Thermolysin: 2 mg/ml in HM
- Thermolysin mixture 2 (see *Protocol 9*)
- 500 mM EDTA pH 8.0
- Translation mixture (see *Protocols 4* or *5*)
- SB (see *Protocol 6*)

Method

1 Prepare a pellet of intact pea or barley chloroplasts as described in *Protocol 1*, steps 1–9 or *Protocol 2*, steps 1–8.

2 Lyse the chloroplasts by resuspending the pellet in ice-cold HM at a chlorophyll concentration of 1 mg/ml.[a] Leave on ice for 5 min to ensure complete lysis.

3 Centrifuge at 18 000 g for 5 min at 4 °C in a microcentrifuge to generate a stromal supernatant and a thylakoid pellet. Keep the stromal extract on ice until required.

4 Wash the thylakoids twice in 1 ml ice-cold HM, taking care not to blow bubbles in the viscous suspension. As a starting point, allow 20 μg chlorophyll per assay: for ease of handling, wash only as many membranes as are needed for the experiment, plus a little extra in case of losses. Large samples (e.g. thylakoids equivalent to 1 mg chlorophyll) are difficult to wash efficiently in a microcentrifuge tube; scaling up to larger tubes and centrifuges greatly extends the time taken to prepare the membranes, and is likely to lead to a deterioration in their quality.

5 Resuspend the final pellet in either HM or stromal extract, to a concentration of 0.5 mg/ml chlorophyll.

6 Set up the import incubation in a 1.5 ml microcentrifuge tube containing: 40 μl thylakoid suspension, 5 μl translation mixture, and 5 μl HM.

7 Incubate at 25 °C for 20 min under illumination (300 μE/m^2/s).

8 Wash the thylakoids after the incubation, by adding 1 ml ice-cold HM to the sample. Centrifuge for 5 min at 18 000 g at 4 °C in a microcentrifuge. However, the experimenter may be interested to look at a sample of the unwashed membranes. If this is the case, remove one-third of the incubation before the wash step and add it directly to an equal volume of SB, boil for 5 min, and store at −80 °C.

9 Resuspend the pellet from step 7 in a small volume of HM (e.g. 40 μl).

10 Divide the suspension equally between two clean microcentrifuge tubes. To one, add an equal volume (e.g. 20 μl) of SB, boil 5 min, and store at −80 °C.

Protocol 10 continued

11 Centrifuge the other sample for 5 min at 18 000 g at 4 °C in a microcentrifuge. Remove the supernatant.

12 Add 100 µl thermolysin mixture 2 and incubate 40 min on ice.

13 Stop the digestion by the addition of EDTA to a final concentration of 10 mM.

14 Pellet the membranes at 18 000 g for 5 min at 4 °C in a microcentrifuge, then resuspend the pellet in HM and 10 mM EDTA before quickly adding boiling SB (to ensure rapid inactivation of the protease).

15 Analyse all the samples by SDS-PAGE (*Protocols 6* or *7*).

[a] The amount of stromal extract added to the assay may be increased by lysing a more concentrated suspension of chloroplasts, e.g. at 2 mg/ml chlorophyll.

5.3 Some variations on the basic assay

Nucleoside triphosphates, the majority of which are contributed by the translation mixture, may be removed by the enzyme apyrase (Sigma, grade VI); a recent example of the method is described in ref. 23. Alternatively, stromal extract and translation mixture can be gel filtered on small, disposable Sephadex columns (obtained from Pharmacia), which then enables the researcher to add back specific NTPs, analogues, etc. An example of this technique is described in ref. 24.

The thylakoidal ΔpH may be dissipated by 2 µM nigericin (25). Note that this proton ionophore requires K^+ ions to function, but if the HM import buffer pH has previously been adjusted with KOH, there should be sufficient K^+ already present. Otherwise, add 10 mM KCl to the assay.

The translocation factor SecA can be inhibited by 10 mM sodium azide, especially in the presence of low concentrations of ATP (26). However, sensitivity to azide or a lack of it, is not enough on its own to assign a protein to a particular translocation pathway, especially in the light of results reported by Leheny *et al.* (27) who have shown that 16K translocation is inhibited by azide, despite the general acceptance that this protein is a substrate for the ΔpH-dependent pathway.

References

1. Cline, K. and Henry, R. (1996). *Ann. Rev. Cell Dev. Biol.*, **12**, 1.
2. Robinson, C. and Mant, A. (1997). *Trends Plant Sci.*, **2**, 431.
3. Heins, L., Collinson, I. and Soll, J. (1998). *Trends Plant Sci.*, **3**, 56.
4. Chen, L-J. and Li, H. (1998). *Plant J.*, **16**, 33.
5. Rensink, W. A., Pilon, M. and Weisbeek, P. (1998). *Plant Physiol.*, **118**, 691.
6. Hipkins, M. F. and Baker, N. R. (1986) In *Photosynthesis energy transduction, a practical approach* (ed. M. F. Hipkins and N. R. Baker), p. 51. IRL Press, Oxford.
7. Dahlin, C. and Cline, K. (1991). *Plant Cell.*, **3**, 1131.
8. Melton, D. A., Krieg, P., Rabagliciti, M. R., Maniatis, T., Zinn, K. and Green, M. R. (1984). *Nucl. Acids Res.*, **12**, 7035.
9. Schenborn, E. T. and Mierendorf, R. C. (1985). *Nucl. Acids Res.*, **13**, 6223.
10. Anderson, C. W., Straus, J. W. and Dudock, B. S. (1983). In *Methods in enzymology*. (ed. R. Wu, L. Grossman, and K. Moldave), Vol. 101, p. 635. Academic Press, London.
11. Laemmli, U.K. (1970). *Nature*, **227**, 680.
12. Schägger, H. and von Jagow, G. (1987). *Anal. Biochem.*, **166**, 368.

13. Grossman, A. R., Bartlett, S. G., Schmidt, G. W., Mullett, J. E. and Chua, N-H. (1982). *J. Biol. Chem.*, **257**, 1558.
14. Olsen, L. J., Theg, S. M., Selman, B. R. and Keegstra, K. (1989). *J. Biol. Chem.*, **264**, 6724.
15. Theg, S. M., Bauerle, C. B., Olsen, L. J., Selman, B. R. and Keegstra, K. (1989). *J. Biol. Chem.*, **264**, 6730.
16. Kessler, F., Blobel, G., Patel, H. A. and Schnell, D. J. (1994). *Science*, **266**, 1035.
17. Gray, J. C. and Row, P. E. (1995). *Trends Cell Biol.*, **5**, 243.
18. Cline, K., Werner-Washburne, M., Andrews, J. and Keegstra, K. (1984). *Plant Physiol.*, **75**, 675.
19. Mant, A. and Robinson, C. (1998). *FEBS Lett.*, **423**, 183.
20. Breyton, C., de Vitry, C. and Popot, J-L. (1994). *J. Biol. Chem.*, **269**, 7597.
21. Settles, A. M. and Martienssen, R. (1998). *Trends Cell Biol.*, **8**, 494.
22. Dalbey, R. E. and Robinson, C. (1999). *Trends Biochem. Sci.*, **24**, 17.
23. Thompson, S. J., Robinson, C. and Mant, A. (1999). *J. Biol. Chem.*, **274**, 4059.
24. Mant, A., Schmidt, I., Herrmann, R. G., Robinson, C. and Klösgen, R. B. (1995). *J. Biol. Chem.*, **276**, 23275.
25. Mills, J. D. (1986) In *Photosynthesis energy transduction, a practical approach* (ed. M. F. Hipkins and N. R. Baker), p. 143. IRL Press, Oxford.
26. Knott, T. G. and Robinson, C. (1994). *J. Biol. Chem.*, **269**, 7843.
27. Leheny, E. A., Teter, S. A. and Theg, S. M. (1998). *Plant Physiol.*, **116**, 805.

Chapter 8

Fractionation of plant tissue for biochemical analyses

Matthew J. Terry and Lorraine E. Williams

School of Biological Sciences, University of Southampton, Bassett Crescent East, Southampton SO16 7PX, UK

1 Introduction

The range of biochemical techniques that can be used to address problems in plant biology is vast. Essentially, all the biochemical approaches developed using microbial or animal systems are equally applicable to plant research and have, indeed, been used for many years. However, the biochemical analysis of plant tissue does require some special considerations. In particular, homogenization of plant tissue can be exceedingly problematic. This is the result of the combination of a tough exterior cell wall and a large internal vacuole, with one of the properties of the latter being to act as a repository of substances that are likely to be deleterious to the protein or compound being investigated. A second consideration is that plants produce a diverse range of compounds that are not generally found in other systems. Many of these are present in the plastids, organelles that are also unique to the plant kingdom. With these considerations in mind, the aim of this chapter is to provide a series of methods that will act as a starting point for the analysis of plant cell fractions and, as such, will be suitable for researchers with training in other disciplines who wish to undertake an analysis of, for example, a new mutant or transgenic plant.

Many of the techniques used for investigating the biochemistry of plants have not changed substantially over a number of years and certainly pre-date many of the molecular techniques described in these volumes. This being the case, there are already many authoritative and comprehensive reviews of the major issues involved in the biochemical analysis of plant tissue (1–3). In the present chapter we will focus as much as possible on recent developments in this field. We will discuss the most appropriate homogenization conditions for the isolation of biochemically-active cell components, outline protocols for isolating enriched fractions of many compartments within the cell, and provide examples of biochemical assays specific for these particular cellular compartments.

2 Homogenization of plant tissue
2.1 General considerations

Isolation of uncontaminated, intact organelles or pure homogeneous membrane fractions is very difficult to accomplish. However, the preparation of enriched material suitable for biochemical analysis can be achieved by relatively straightforward procedures and is often all that is necessary for measuring or even determining the cellular location of a particular enzyme activity or protein. Concerning the homogenization of plant tissue there are two main factors that are crucial for the isolation of biochemically-active material. The first is the choice of the homogenization protocol, which is generally a compromise between yield and the harshness of the procedure. Certain biochemical analyses require the isolation of an intact organelle and, therefore, the cell must be disrupted without breaking the organelle. One option is to prepare protoplasts that can subsequently be ruptured to isolate the organelles (for detailed methods see ref. 4). Although this approach can have advantages in terms of gentleness, it is not as well suited for analyses requiring high yields.

Alternatively, tissue can be homogenized to disrupt the cell wall allowing the organelles to be released. The harsh procedures required to break open cells often result in the release of secondary metabolites (e.g. phenolics) and hydrolytic enzymes that are normally compartmentalized in the vacuoles and plastids. These can have deleterious effects and thus it is advantageous to minimize their release. Non-mechanical methods such as razor blade chopping or grinding with a pestle and mortar are fairly gentle procedures often producing less damage. However, these methods can be less efficient in disrupting the tissue and may compromise yield. Mechanical blenders and grinders can be used (e.g. Waring blender, Polytron, etc.), but these devices can be too harsh for tissues known to contain high levels of phenolics.

Care must be taken over the speed and length of time of homogenization and also in maintaining low temperatures. The method of homogenization can directly affect the final distribution of subcellular components, influencing their partitioning during differential centrifugation. An additional point to remember is that the isolation of biochemically-active tissue will depend not only on the choice of homogenization conditions, but also on how quickly this tissue is removed from other cell debris. This will minimize the time in which the required sample is in contact with damaging substances released during homogenization.

Another important consideration is the choice of homogenization medium. Many enzymes can lose activity irreversibly if exposed to extremes of pH and thus the homogenization medium should be buffered. Usually, the pH is maintained between pH 7 and 8 to minimize hydrolytic-enzyme activity, and to offset the acidic pH of the vacuole. The pH of the homogenization buffer should be determined after the addition of protectants (as these can alter the pH) and also at 4 °C as some buffers are temperature sensitive. Osmoticum should be present to reduce bursting of organelles and sufficient protectants must be added to combat the deleterious effects of phenolics and degradative enzymes. It may be necessary to add these fresh to the homogenization medium to ensure full activity. A suitable ratio of homogenization media to tissue must be used to ensure an adequate level of protectants and to dilute the harmful components released from disrupted intracellular organelles.

Table 1 gives a list of protectants that can be added to homogenization media with guideline concentrations (see also refs 5 and 6). Do not assume that all putative protectants are beneficial; for example, sulphydryl reagents and protease inhibitors may also affect the enzyme activity to be studied. Some of the ingredients in the homogenization media are toxic and suitable precautions should be taken at all times when preparing cell fractions. It is also recommended that water that has been distilled and deionized is used. It is important that all homogenization stock solutions containing sucrose or sorbitol are stored at −20 °C. These can be defrosted overnight prior to use.

2.2 Homogenization procedures and conditions

Homogenization conditions will depend greatly on the choice of tissue. For example, mature tissue will often require more vigorous homogenization methods while younger or etiolated tissue is softer and relatively simple to homogenize. Three different homogenization protocols are described in this chapter. In this section we give the methods for preparing

Table 1 Protectants for homogenization of plant tissue

Protectant[a]	Action	Concentration[b]
Protein protectants		
DTE/ DTT	Sulphydryl reagents	1–5 mM
β-Mercaptoethanol	Sulphydryl reagent	10–15 mM
Cysteine	Sulphydryl reagent	5 mM
PMSF[c]	Serine- and cysteine-protease inhibitor	1–5 mM[d]
Leupeptin	Serine- and cysteine-protease inhibitor	1–5 μM
Chymostatin	Cysteine-protease inhibitor	20 μg/ml[e]
EDTA	Metallo-protease inhibitor	1 mM
Bestatin	Aminopeptidase inhibitor	10 μM[d]
Pepstatin A	Acid-protease inhibitor	3 μg/ml
BSA	Protease substrate	0.1–0.5% (w/v)
Phospholipid protectants		
EGTA	Phospholipase inhibitor	5 mM
Choline and ethanolamine	Phospholipase substrate	4% (w/v)
Glycerol-1-phosphate	Phosphatidic acid	
	Phosphatase inhibitor	10 mM
Defatted BSA	Prevents acyl-hydrolase activation	0.1–0.5% (w/v)
Protectants against oxidation (caused by phenolics)		
Ascorbate	Anti-oxidant	1 mM
Thiourea	Phenolic oxidase inhibitor	2 mM
Potassium metabisulfite	Phenolic oxidase inhibitor	2 mM
PVP	Adsorbs phenolics	0.5–2% (w/v)

[a]Unless otherwise stated all reagents can be obtained from Sigma, although the products of other chemical companies are equally suitable.
[b]Guideline concentration; actual concentrations should be determined empirically.
[c]A non-toxic alternative, AEBSF [4-(2-aminoethyl)—benzene sulphonylfluoride] hydrochloride, is now available.
[d]Prepare stock solution in ethanol or methanol.
[e]Prepare stock solution in DMSO.

microsomal fractions from two different types of tissues: dark-grown *Ricinus* cotyledons (a small scale fairly gentle procedure; *Protocol 1*; ref. 7) and red beetroots, a highly vacuolated material (a larger scale harsher procedure; *Protocol 2*; refs 8 and 9). In section 3.1.1, we provide a protocol designed specifically for isolating developing chloroplasts (also a small scale, gentle procedure; *Protocol 3*).

A flow diagram showing the general procedure for preparing a microsomal membrane preparation is shown in *Figure 1*. The homogenate is first filtered through muslin to remove the majority of the cell wall debris. This is followed by differential centrifugation whereby low speed centrifugation is used to pellet the majority of the starch grains and unbroken organelles. The supernatant is then centrifuged at higher speed to obtain the microsomal fraction, which is a mixed membrane fraction containing plasma membrane, tonoplast, Golgi, endoplasmic reticulum (ER), and microbodies. The amount of mitochondrial and plastid membranes will depend on the disruption of these organelles during the initial preparative steps. All procedures must be carried out at 1–4 °C, and it is important to work quickly to reduce the time between the initial homogenization and suspension of the membrane fraction in protective medium. These precautions will reduce the damaging effects of hydrolytic enzymes.

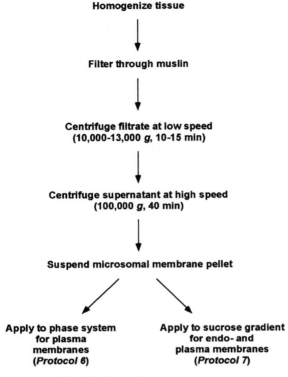

Figure 1 Flow diagram showing the general procedure for isolating microsomal membranes for further fractionation by phase partitioning or sucrose density gradient centrifugation. For further details of this procedure refer to *Protocols 1* and *2*.

Protocol 1

Small-scale, non-mechanized preparation of microsomal tissue from dark-grown *Ricinus* cotyledons

Reagents[a]

- Homogenization medium: 250 mM sorbitol, 25 mM Hepes-Bis-Tris propane (BTP) pH 7.4, 3 mM EGTA, 1 mM dithiothreitol (DTT) added just before use
- Phenylmethylsulphonyl fluoride (PMSF): 400 mM stock solution in ethanol

- Suspension buffer: 250 mM sorbitol, 2.5 mM Hepes-BTP pH 7.4, 10% (v/v) glycerol, 1 mM DTT added just before use

Method

1. All procedures should be performed at 1–4 °C.
2. Pre-chill pestle and mortar, beakers, and centrifuge tubes.
3. Harvest tissue as quickly as possible directly into ice-cold distilled water that is continuously aerated using a pump.

4 When all the tissue has been harvested take it out of water, quickly blot dry, and weigh. Return the tissue to ice-cold water while you make up the homogenization medium.

5 Measure out the ice-cold homogenization medium (10 ml per g cotyledons). Add solid DTT and 2.5 μl of 400 mM PMSF per 1 ml of homogenization medium to give final concentrations of 2 and 1 mM, respectively. Stir for 5 min and return to ice.

6 Place cotyledon tissue in the chilled mortar, add enough homogenization medium to cover the tissue, and then chop the material using scissors. Add half of the remaining homogenization medium and grind gently until there are no intact cotyledons remaining.

7 Filter the homogenate through four layers of muslin and collect the filtrate in an ice-cold beaker. Return the tissue from the muslin to the pestle and mortar. Add remaining homogenization medium, grind more thoroughly, and filter as before.

8 Pool filtrate and centrifuge at 10000 g for 10 min in 30–50 ml centrifuge tubes.[b] Without disturbing the pellet, transfer the supernatant to clean centrifuge tubes and centrifuge at 100000 g for 40 min.

9 Suspend pellets using a paintbrush initially, followed by gentle agitation with a pipette. Use approximately 30–50 μl suspension medium per g of starting material. Aim for 1–2 mg membrane protein/ml suspension medium.

10 Transfer to microcentrifuge tubes and either use immediately or freeze rapidly in liquid nitrogen and store at −80 °C.

[a]Unless otherwise stated all reagents can be obtained from Sigma, although the products of other chemical companies are equally suitable.

[b]The centrifuge tubes used for all protocols will depend on the centrifuges and rotors available in the laboratory, and in most cases the volumes used can easily be adjusted as required.

Protocol 2

Large-scale preparation of microsomal tissue from beetroots using mechanized homogenization

Equipment and reagents

- Waring blender
- Homogenization medium: 250 mM sucrose, 70 mM Tris–HCl pH 8.0, 2 mM EDTA, 5 mM MgSO$_4$, 0.5% (w/v) polyvinylpyrrolidone (PVP; mol. wt. 40,000), 10% (v/v) glycerol, 1 mM DTT added just before use

- PMSF: 400 mM stock solution in ethanol
- β-mercaptoethanol
- Suspension buffer: 250 mM sucrose, 2 mM BTP-Mes pH 7.0, 10% (v/v) glycerol, 1 mM DTT added just before use.

Method

All procedures should be performed at 1–4 °C in a chemical hood.

1 Pre-chill beakers, blender, centrifuge tubes, and glass homogenizer.

2 Place 375 ml ice-cold homogenization medium in a beaker. Add solid DTT to give a final concentration of 4 mM and 0.5 μl of 400 mM PMSF per 1 ml of homogenization medium to give a final concentration of 0.2 mM. Add 375 μl β-mercaptoethanol to give a final concentration of 15 mM, stir for 5 min to dissolve and return to ice.

3 Working as quickly as possible, peel beet roots using knife (use approximately 0.5 kg). Cut beet into approximately 0.5 cm thick slices, dice into cubes, and add to ice-cold homogenization medium. Add beet to beaker until the volume is doubled, i.e. 750 ml mark is reached.

4 Place beaker containing the tissue on ice and vacuum infiltrate the tissue with homogenization medium for 10 min by placing tissue in a vacuum dessicator attached to a water pump to provide the vacuum (this step is recommended, but not essential).

5 Homogenize the beet in a Waring blender in three batches ensuring that the medium and beet are equally divided. Use three 20 s bursts on high speed.

6 Filter homogenate through four layers of muslin, then filter again.

7 Centrifuge at 13 000 g for 15 min in 30–50 ml centrifuge tubes.

8 Without disturbing the pellet, remove the supernatant and centrifuge at 100 000 g for 40 min in 30–50 ml centrifuge tubes.

9 Suspend the pellets in 3 ml suspension medium using a paintbrush initially, followed by gentle pipetting up and down in a 1 ml pipette tip that has been widened by cutting off the narrow end. If the membranes are not fully suspended a glass homogenizer can be used.

10 Transfer to microcentrifuge tubes and either use immediately or freeze rapidly in liquid nitrogen and store at −80 °C.

3 Biochemical analysis of organelles

3.1 Etioplasts and developing chloroplasts

In this section we will concentrate on the isolation of etioplasts and developing chloroplasts. For information on organelles such as the microbodies (including peroxisomes and glyoxysomes), the reader is referred to other sources (3, 6). Similarly, other plastid types (including amyloplasts and leucoplasts) have been covered previously (10, 3; see also Chapter 7 for alternative protocols).

As with all purification procedures the choice of methods for isolating plastids will depend on the particular tissue, and the required quality of the plastid preparation with respect to both purity and plastid integrity. Most procedures for isolating chloroplasts were originally developed for mature chloroplasts from fully developed leaves and a number of reviews have thoroughly covered the important aspects of these methods (11, 12). However, while these chloroplast preparations are ideally suited to the study of photosynthetic reactions, they are less suitable for the analysis of biosynthetic pathways or, indeed, the complex biochemistry involved in the biogenesis of the photosynthetic machinery itself. The reasons for this include the difficulty in obtaining reproducible tissue samples on an appropriate time scale, the toughness of mature leaves and the high levels of starch present in these plastids. However, it should be noted that for species such as *Arabidopsis*, using fully developed leaf tissue may be the only option for obtaining sufficient starting material. For many biochemical analyses, a sensible alternative to mature chloroplasts is to use developing chloroplasts from dark-grown seedlings that have been transferred to a standardized white light source for a set period (most commonly 4–20 h). The growth conditions are simple to replicate and the cotyledons or young leaves from these plants are easy to homogenize and biochemically active.

The purity and integrity of the plastids required is also an important consideration. For many biochemical analyses, the amount of tissue may well be limiting, for example, when comparing wild-type seedlings with mutants or transgenic lines. In such cases, the maximum yield of plastids is the most important factor. For experiments investigating the localization of the process or protein in question, the purity of the plastid preparation is paramount. The analysis of processes taking place on the outer envelope membrane requires the isolation of intact plastids and methods for isolating intact organelles specifically for protein import experiments are given in Chapter 7.

3.1.1 Isolation of etioplasts and developing chloroplasts

Protocol 3 describes a simple method for isolating an enriched fraction of etioplasts or developing chloroplasts from cucumber cotyledons based on differential centrifugation. The protocol was originally developed by Pardo *et al.* (13) for the investigation of tetrapyrrole synthesis and was subsequently modified by replacing sucrose with sorbitol (14). The final pellet includes both intact and broken chloroplasts, but is devoid of mitochondrial contamination as judged by the complete absence of succinic dehydrogenase activity (13). However, this preparation is contaminated with microbodies (13). The protocol is suitable for use with a wide range of tissues, although minor modifications are recommended such as the ratio of homogenization buffer to tissue used and the protectants included. For example, 0.5% (w/v) soluble PVP and 142 mM β-mercaptoethanol were included in the homogenization buffer in order to obtain plastids from etiolated tomato seedlings suitable for phytochrome chromophore biosynthesis assays (15). It is important to remember that for etioplast preparations all procedures should be performed under green safelight, primarily to prevent protochlorphyllide reduction, but also to eliminate photoreceptor-mediated effects.

Protocol 3

Protocol for isolation of crude etioplasts and developing chloroplasts from cucumber cotyledons

Reagents

- Homogenization stock solution: 1 M sorbitol, 40 mM N-tris(hydroxymethyl)methyl-2-aminoethanesulfonic acid (TES)-20 mM Hepes-NaOH pH 7.7, 2 mM $MgCl_2$, 2 mM EDTA (free acid), 2 mM EDTA (di-Na salt)
- Bovine serum albumin (BSA) and cysteine

Method

1. Grow cucumber seedlings in moist vermiculite in the dark for 7 days at 28 °C using plastic wrap to keep the humidity high. For developing chloroplasts, irradiate 6-day-old seedlings with white fluorescent light for 20 h.

2. Prepare homogenization buffer from stock solution by diluting two-fold with H_2O. Add BSA and cysteine to final concentrations of 0.2% (w/v) and 5 mM, respectively, and adjust back to pH 7.7.

3. Harvest cotyledons by hand, quickly remove the majority of seed coats and homogenize in a pre-chilled mortar and pestle in 4 ml/g fresh weight ice-cold homogenization buffer. For etioplast preparations, these and subsequent steps should be performed under green safelight. The sample should be kept chilled (4 °C) throughout the isolation procedure.

4. Filter homogenate through four layers of muslin.

5. Centrifuge for 1 min at 8000 g in 30–50 ml centrifuge tubes.

6. Discard supernatant and wipe away any starch from the side of the tube with a tissue.

7. Gently suspend pellet in homogenization medium (~1 ml/g fresh weight) using a paintbrush to tease the pellet away from the centrifuge tube.

Protocol 3 continued

8 Centrifuge for 1 min at 100 g to remove unbroken cells and other debris.

9 Centrifuge supernatant for 2 min at 1500 g in 30–50 ml centrifuge tubes.

10 Wash the plastids by resuspending the pellet(s) in 10 ml of the assay buffer to be used and centrifuge again for 2 min at 1500 g.

3.1.2 Purification of intact etioplasts and developing chloroplasts

For some experiments it will be important to separate intact from broken plastids. Intact plastids are plastids in which the envelope membranes are complete and undamaged and the simplest method to prepare these is by centrifugation through silica sols (16). A protocol for the purification of intact plastids by centrifugation through 45% Percoll (Pharmacia) is shown below (*Protocol 4*). This protocol, based on the method of Fuesler *et al.* (14), has the benefit that contaminating material of other origins will also be removed. A slightly more sophisticated alternative is to use an 80% Percoll cushion under the 45% layer. The intact plastids can then be collected directly from the interface with no need for resuspension. In addition, any remaining nuclei will pass into the 80% layer.

Protocol 4

Protocol for separating intact etioplasts and developing chloroplasts

Reagents

- Suspension buffer: 0.5 M sorbitol, 20 mM TES-20 mM Hepes-NaOH pH 7.7, 1 mM MgCl$_2$, 1 mM EDTA (free acid), 1 mM EDTA (di-Na salt), 0.2% (w/v) BSA, and 5 mM cysteine

- Percoll (Pharmacia)

Method

1 Using a paintbrush, suspend the crude plastid pellet (from *Protocol 3*) in ~0.3 ml/g fresh weight tissue of suspension buffer (the homogenization medium of *Protocol 3*).

2 Gently layer 5 ml onto 35 ml of homogenization medium containing 45% (w/v) Percoll in a 50 ml centrifuge tube.

3 Centrifuge for 5 min at 6000 g in a swing-out rotor. The intact plastids will form a pellet, while the broken and damaged plastids will form a broad band near the top of the centrifuge tube.

4 For intact plastids, discard supernatant and remove excess Percoll by suspending in a small volume of the assay buffer to be used and centrifuge for 2 min at 1500 g.

5 If required, collect the upper band containing the broken and damaged plastids using a Pasteur pipette and dilute this sample into 25 ml assay buffer. The broken plastids can then be obtained by centrifuging at 27 000 g for 5 min in a 30–50 ml centrifuge tube.

The intactness of chloroplasts prepared as described in *Protocol 4* can be estimated in a number of ways (12). One commonly used assay is the latency of the stromal enzyme gluconate-6-phosphate dehydrogenase (17). The activity of this enzyme can be assayed quite simply by following the reduction of NADP$^+$ by the change in absorbance at 340 nm. A protocol based on the modification of this assay by Lee *et al.* (18) is given below (*Protocol 5*).

Alternatively, ferricyanide can be used as an electron receptor in the Hill reaction (see ref. 12 for protocol), although this is more suitable for mature chloroplasts.

Protocol 5

Estimating the intactness of developing chloroplasts

Reagents

- Assay solution: 0.3 M sucrose, 20 mM Tes-NaOH pH 7.5, 10 mM $MgCl_2$, 0.5 mM $NADP^+$
- Triton X-100
- 200 mM gluconate-6-phosphate

Method

1 Put 0.5 ml of assay solution into a cuvette and monitor absorbance at 340 nm.

2 Add a small aliquot of plastid suspension containing approximately 100 µg of plastid protein, and allow sample to equilibrate and the baseline to settle to a new value.

3 Add gluconate-6-phosphate (2 mM final concentration), mix and record the increase in absorbance at 340 nm.

4 After 5 min add 0.1% Triton X-100 to lyse plastids, mix, and continue to follow absorbance changes.

5 Determine the rates of gluconate-6-phosphate dehydrogenase activity in the original sample (activity in the absence of Triton X-100 = I) and the total gluconate-6-phosphate dehydrogenase activity (activity in the presence of Triton X-100 = T). The concentration (c) of NADPH can be calculated using a molar absorption (extinction) coefficient for NADPH at 340 nm (ε_{340}) of 6220 $M^{-1}cm^{-1}$ and the equation $A = \varepsilon_{340}cd$, where A is absorbance and d the light path in centimetres (usually 1 cm).

6 Calculate the intactness (or latency) of the plastid preparation using the following equation:
% Intact = $[(T - I)/T] \times 100$

3.1.3 Localization within plastids

For the biochemical analyses of plastid processes it may be sufficient to demonstrate that the activity is plastid localized or it may be necessary to resolve the localization further by fractionation of the isolated plastids. For localizing an activity to plastids the most common approach is to demonstrate that the activity correlates with plastid markers and not with markers for other organelles or membranes. The use of marker assays is dealt with later in this chapter (see section 6.2). However, there are alternative approaches to localization of a biochemical activity in plastids. If the substrates for the reaction of interest are unable to pass across an intact plastid envelope then it is possible to demonstrate that this activity is plastid localized by performing a simple latency experiment. In this experiment, a plastid-localized enzyme activity will be dependent on the presence of a detergent that permits substrate accessibility (the method is essentially described in Protocol 5 for gluconate-6-phosphate dehydrogenase). Activity released from other compartments during homogenization will not be latent as it will not be membrane bound. A different approach, though similar in principle, is to change the activity measured by manipulating substrate concentrations within the plastid. This method was used for the phytochrome chromophore biosynthesis enzyme, phytochromobilin synthase (19). Since phytochromobilin synthase requires NADPH as an energy source, etioplasts were incubated with the triose phosphates, 3-phosphoglycerate and dihydroxyacetone phosphate, to cause a net consumption or production of NADPH, respectively, within the plastid. As the experiment was performed in the presence of an external NADPH scavenging system (oxidized glutathione and glutathione reductase), the observed changes in phytochromobilin synthase activity could only be the result of its localization within the plastid (19).

3.1.4 Fractionation of chloroplasts

Isolated chloroplasts can be fractionated quite easily into three components: the thylakoid membranes, the envelope membranes, and a fraction representative of stromal proteins. It is always best to start with purified chloroplasts, which are then lysed either by osmotic shock or by freeze-thaw cycles (20). For freeze-thawing, the chloroplasts should be incubated in a hyper-osmotic sucrose solution prior to lysis as this reduces the formation of mixed membrane populations. The fractions can then be separated using discontinuous sucrose gradients, a technique pioneered by Douce *et al.* (21). More recently, these methods have been modified to produce fractions of greater purity. To increase the purity of thylakoid membranes, the thylakoids isolated from the sucrose gradient are washed twice in 330 mM sorbitol in 10 mM tricine-NaOH pH 7.8 (22). Envelope membranes can be isolated using a flotation technique in which the lysed chloroplasts in 1.3 M sucrose are overlayed with buffers containing 1.2 and 0.4 M sucrose, and centrifuged at 112 000 g for 5 h (23). The envelope membranes, free of thylakoid contamination, are then recovered from the 1.2/0.4 M interface and can be purified on a sucrose gradient as before. It is also possible to separate the inner and outer envelope membranes using discontinuous sucrose gradients (24).

For many purposes this level of purity is not required and if the amount of sample is limiting, a simpler method may be suitable. One method, based on the work of Smeekens *et al.* (25), has been used successfully to study the intraplastidic localization of the haem biosynthesis enzyme, ferrochelatase (26). It relies on the fact that the density of envelope membranes is much lower than thylakoid membranes and requires only two differential centrifugation steps. First chloroplasts are lysed in a minimum volume of lysis buffer (50 mM Hepes-KOH pH 8.0) by incubating for 10 min on ice. The thylakoid membranes are then pelleted by centrifugation at 12 000 g for 5 min, while the envelope membranes are collected by centrifuging at 75 000 g for 45 min. The supernatant constitutes the stromal fraction.

Use of markers is important for determining the degree of contamination of the membrane fractions. Envelope contamination of thylakoid fractions is particularly common. Chlorophyll is an ideal marker for thylakoid membranes (see *Protocol 11*), but envelope markers are more problematic. The most commonly used is 1,2-diacylglycerol 3-β-galactosyltransferase (27). This assay follows the synthesis of galactolipids from UDP-[^{14}C]galactose, which are then quantified by scintillation counting. A suitable alternative to this assay is to use antibodies recognizing envelope-specific proteins such as the triose-phosphate translocator (28). In this case antibodies to chlorophyll *a/b*-binding proteins (e.g. 29) can be used for thylakoid marker analysis (see section 6.2.2.2 for further details).

3.2 Mitochondria

The techniques and conditions used for the purification of mitochondria closely parallel those described above for plastids. It is possible to produce an enriched mitochondrial fraction by differential centrifugation, which will be suitable for many purposes (30). For example, this method employs a low speed centrifugation (1500 g for 20 min), which removes nuclei, plastids, intact cells and cell debris, followed by a higher speed centrifugation (12 000 g for 20 min), which pellets the intact mitochondria, but not other cellular membranes. For purer mitochondrial preparations Percoll gradients can be used (31). The exact procedure depends on the starting material, and there is considerable variation between photosynthetic or non-photosynthetic tissue.

3.3 Nuclei

The isolation of a crude nuclei preparation is again relatively straight forward, requiring only filtration and differential centrifugation and a range of methods and conditions have been described by Jackson (32). A simple method is given by Datta *et al.* (33). Nuclei were isolated by filtration through a nylon mesh followed by centrifugation at 1900 g for 10 min. This crude nuclei preparation was then further purified using a discontinuous Percoll gradient with the intact nuclei collected at the interface of the 25 and 50% Percoll layers (33). Further information on the isolation of nuclei can be found in Chapter 4.

4 Biochemical analysis of plasma membranes and intracellular membranes

Sub-fractionation of plasma- and intracellular membranes can be achieved by differential and density gradient centrifugation, phase partitioning, and free-flow electrophoresis, each method having particular advantages and disadvantages (34). Since the equipment for free-flow electrophoresis is not generally accessible this method will not be discussed here. The starting material for the isolation of plasma and intracellular membranes is a total microsomal membrane fraction. Methods for preparing this fraction have been described fully in *Protocols 1* and *2*.

4.1 Isolation of plasma membrane using aqueous two-phase partitioning

Aqueous two-phase partitioning is now a routine procedure that can be used to isolate plasma membrane vesicles with up to 95% purity; this is compared to about 50–75% achieved for sucrose gradient centrifugation. Two-phase partitioning separates membranes on the basis of hydrophobicity and surface charge, rather than density and gives rise to a population of plasma membrane vesicles orientated predominantly cytoplasmic side-in (for a more detailed account see refs. 5, 35). Cytoplasmic-side out vesicles can be obtained using a freeze/thaw procedure (36) or by treatment with the detergent Brij 58 (37). Phase partitioning is temperature-dependent and should therefore be carried out at 4 °C. The procedure involves mixing a microsomal fraction with a solution of defined ionic composition and pH, containing appropriate concentrations of two aqueous polymers, Dextran, and polyethylene glycol (PEG). The polymer and ionic concentration can be adjusted to give optimum separation (34). The phase system is allowed to separate and the upper, plasma membrane-enriched fraction is collected and transferred to fresh lower phase and the procedure repeated. An outline of this protocol is shown in *Figure 2*. If an intracellular membrane fraction is to be prepared from the same microsomal fraction, fresh upper phase can be added to replace the initial upper phase, and this procedure can be repeated alongside the preparation of the plasma membrane fraction (*Figure 2*).

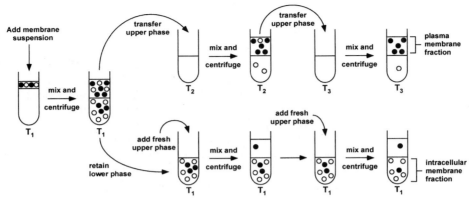

Figure 2 Flow diagram showing the procedure for purifying plasma membranes by aqueous two-phase partitioning. For further details of this procedure refer to *Protocol 6*.

Protocol 6

Isolation of plasma membrane using aqueous two-phase partitioning[a]

Reagents

- Suspension buffer: 250 mM sorbitol,[b] 10% (v/v) glycerol, 2 mM Tris-2-(N-morpholino)ethanesulfonic acid (Mes) pH 7.5, and 1 mM dithioerythritol (DTE)
- Phase-suspension medium: 250 mM sorbitol, 5 mM K phosphate buffer pH 7.8, and 1 mM DTE
- 40% (w/w) PEG (mol. wt. 3350) in H_2O
- 20% (w/w) Dextran T-500 (Pharmacia). Put magnetic stirrer in beaker and layer 22 g Dextran onto 84.25 g H_2O. Cover with

parafilm and stir overnight to dissolve. Do not attempt to dissolve too quickly otherwise large insoluble lumps will form. The weight allows for the water content of the Dextran and this should be checked using a polarimeter (specific rotation +199° ml/g/dm, measured at 589 nm) and adjusted accordingly

- 200 mM K phosphate buffer pH 7.8
- 120 mM KCl
- Brij 58 (Sigma P 5884)

Method

1 Prepare a bulk 300 g phase system to provide fresh upper and lower phases during the phasing procedure. A 6.5% phase-system contains: 250 mM sorbitol, 1 mM DTE, 5 mM K phosphate buffer pH 7.8, 3 mM KCl, 6.5% PEG, 6.5% Dextran. PEG and Dextran concentrations may be altered if required. Weigh beaker and stirrer, add 13.66 g sorbitol and 0.046 g DTE, and dissolve in approximately 45 ml H_2O. Add 7.5 ml 200 mM K phosphate buffer pH 7.8, 7.5 ml 120 mM KCl, 48.75 g 40% PEG, 97.5 g 20% Dextran, and make contents up to 300 g with H_2O and stir.

2 Transfer to separating funnel, mix thoroughly, and leave in cold room overnight to allow separation into upper and lower phases.

3 Next day remove stopper and carefully collect lower phase so as not to disturb interface. Discard interface and collect upper phase.

4 The microsomal fraction should be suspended in phase-suspension medium before loading onto the phase system. Either suspend directly following preparation or pellet the microsomal

Protocol 6 continued

fraction by diluting in suspension buffer and centrifuging at 100 000 g for 40 min in 30–50 ml centrifuge tubes, and then suspend the pellet in phase-suspension medium. Suspend approximately 5 mg protein in 0.4 ml of phase-suspension medium for each 8 g phase system. (NB. It is possible to use multiples of the 8 g system, generally we use 32 g systems).

5 To carry out phase partitioning using an 8 g system, three tubes (T_1, T_2, and T_3) of ~10 ml are required for one complete phase procedure.

6 Prepare the first tube (T_1) by weight. This can be prepared the night before, covered with parafilm and stored in a cold room overnight. Add 0.364 g sorbitol, 0.0012 g DTE, and approximately 0.9 ml H_2O. Cover with parafilm and vortex to dissolve. Add 0.2 g 120 mM KCl, 0.2 g 200 mM K phosphate buffer, 1.3 g 40% PEG, and 2.6 g 20% Dextran. Make contents up to 7 g with H_2O.

7 For tubes T_2 and T_3, add 4 g of lower phase (from the bulk phase system, step 1).

8 Allow tubes to equilibrate on ice. Take tube T_1, tare (zero) it on the balance, and add the microsomal fraction (which must contain 0.4 ml of phase-suspension medium). Make up to 1 g with H_2O (the tube will now contain a total of 8 g of solution).

9 Cover tube with parafilm and mix thoroughly by inverting 20–30 times and centrifuge at 2500 g for 5 min in a swing-out rotor.

10 Noting the weight, transfer 95% of upper phase to tube T_2 avoiding the interface. Make up to 4 g with fresh upper phase (see step 2 of *Figure 2*).

11 If the lower phase material (intracellular membrane) is to be enriched, replace the upper phase of T_1 with fresh upper phase (see step 2 of *Figure 2*).

12 Mix tubes T_1 and T_2 20–30 times as before, and centrifuge at 2500 g for 5 min.

13 Repeat steps 10, 11, and 12 (substituting tube T_3 for T_2).[c]

14 Dilute upper phase from tube T_3 and lower phase from tube T_1 approximately 10-fold with suspension buffer, mix thoroughly, and centrifuge at 100 000 g for 40 min.

15 Resuspend pellets in suspension buffer. Use immediately or freeze rapidly in liquid nitrogen and store at −80 °C.

16 If plasma membrane vesicles are to be reorientated to give a predominantly inside-out preparation, treat the freshly phased vesicles (i.e. not frozen after phasing if possible) with the detergent Brij 58 at a final concentration of 0.025% (w/v).[d] Incubate the vesicles for 5 min on ice then pellet by centrifugation at 100 000 g in a 30–50 ml centrifuge tube to remove detergent. If reorientation is not complete, the Brij concentration may be increased to 0.05%. Check for protein lost in the supernatant when deciding on the detergent concentration to use.

[a]See *Figure 2* for flow diagram of this protocol.

[b]Sorbitol can be replaced with sucrose.

[c]A fourth cycle can be included if further plasma membrane purity is required.

[d]See ref. 37 for further details.

4.2 Sub-fractionation of membranes using density gradient centrifugation

Microsomal membranes can be fractionated on the basis of size and density by centrifugation through continuous or discontinuous density gradients. Sucrose is the most commonly used material, but Percoll and glycerol can also be used. Continuous gradients should be produced using a gradient generator. These generally consist of two chambers connected at their base through a valved channel or tap [see Price (38) for more detailed information]. Initially, information on the separation of membranes on continuous gradients can be used to construct discontinuous gradients that are simpler to prepare. These can be used to prepare fractions enriched in tonoplast, ER, Golgi and plasma membrane that may be sufficient for biochemical analysis. However, the fractions are never pure and the degree of separation and level of cross-contamination varies with different plant species and tissues. The density of the enriched fractions varies for different plant species, different tissues within a plant and tissues at different developmental stages. It is therefore advisable to carry out a thorough analysis of the membrane distribution following fractionation, rather than assuming a particular location based on published values.

Isolation of enriched ER fractions has been aided by the use of a magnesium-induced density shift. This technique is based on the observation that ribosomes dissociate from the ER in the absence of Mg^{2+} resulting in a decrease in the buoyant density of the rough ER (39). The Mg^{2+} concentration is critical and it is particularly important to establish that the density shift is due to ribosomal removal and not non-specific membrane aggregation. Therefore, the separation of other membranes should also be monitored. When we have used this method we have included 3 mM free Mg^{2+} in the homogenization medium (taking into account the presence of chelating agents) and 3 mM $MgCl_2$ in the suspension medium and gradient solutions. It is advisable to determine the appropriate concentration experimentally using these concentrations as a starting guideline.

Protocol 7

Fractionation of microsomal membranes using a continuous sucrose gradient

Reagents and equipment

- Gradient generator
- Peristaltic pump with thin tubing attached to input and output valves
- Suspension buffer: 250 mM sucrose, 2 mM Tris-Mes pH 7.0, 10% (v/v) glycerol, 2 mM DTT added just before use
- Low density sucrose solution (LDSS): 22% (w/w) sucrose in 2 mM Tris-Mes pH 7.0, 2 mM

DTT added just before use; for magnesium gradients include 3 mM $MgCl_2$
- High density sucrose solution (HDSS): 50% (w/w) sucrose in 2 mM Tris-Mes pH 7.0, 2 mM DTT added just before use; for magnesium gradients include 3 mM $MgCl_2$

Method

All manipulations to be carried out at 1–4 °C

1　The gradient chambers (A and B) must be dry before use. Remove air trapped in the connecting tap by displacing with a small amount of LDSS. Close tap. Remove excess LDSS with paper tissue.

2　Connect input tube of peristaltic pump to output tube of gradient mixer (chamber B).

3　Dispense 15 ml of LDSS into the left chamber (A) and 15 ml of HDSS into the right chamber (B).

4　Place small magnetic stirrer into each chamber and place chamber B centrally over electric stirrer. The magnetic stirrer in chamber B should spin in the centre of the chamber while the magnetic stirrer in chamber A spins against the output of chamber A.

5　With the tap closed, switch on the peristaltic pump and withdraw HDSS until it reaches the output tube of the pump. Stop pumping.

Protocol 7 continued

6 Position end of pump output tube in the bottom of centrifuge tube. Switch on pump, whilst opening the gradient mixer tap slowly clock-wise. When a double helix of LDSS appears in chamber B stop turning the tap.

7 Position the pump output tube so that its end rests on the meniscus of the fluid collecting in the centrifuge tube. Do not drop the solution in, as this will mix the gradient. It will be necessary to gradually raise the output tube as the gradient forms. To dispense all the chamber contents it may be necessary to tilt the mixer very slightly when the chambers are nearly empty.

8 When the solutions have been delivered, carefully place the centrifuge tube in ice to avoid disrupting the gradient.

9 The microsomal fraction, suspended in 1–2 ml of suspension buffer, should be layered carefully onto the top of the gradient. Avoid overloading the gradient and use about 10–15 mg of protein.

10 Centrifuge at 80 000 g in a swing-out rotor in 30–50 ml centrifuge tubes using the slow start and stop facility on the ultra-centrifuge to avoid disruption of the gradient.

11 Remove the centrifuge tube from the rotor and collect fractions. Thirty-millilitre gradients are usually divided into 20 fractions of about 1.45 ml each, which allows for microsomal volume. Fractions can be collected in plastic cuvettes with the 1.45 ml level previously marked on the side. The input tube of the peristaltic pump is held on the meniscus of the gradient whilst the output tube dispenses the fractions into the cuvettes.

12 Mix each fraction and take a small sample from each to determine the density using a refractometer.

13 To recover the membranes dilute at least 10-fold in suspension buffer, mix thoroughly, and centrifuge at 100 000 g for 45 min in 30–50 ml centrifuge tubes. Suspend the pellets in suspension buffer and either use immediately or freeze in liquid nitrogen and store at −80 °C.

A similar procedure to that described in *Protocol 7* is used to produce a discontinuous sucrose gradient, but in this case a range of sucrose solutions are prepared [e.g. 22, 26, 30, 34, 38, and 45% (w/v) sucrose]. These are then carefully layered into a centrifuge tube, starting with the highest density at the bottom, and the microsomal fraction is layered on top as described in *Protocol 7*. Following centrifugation, usually for 2–3 h, the membranes are collected from the interfaces and washed also as described in *Protocol 7*.

5 Biochemical analysis of soluble proteins and cytoplasmic components

5.1 Extraction of protein for biochemical analysis

Many biochemical analyses simply require measuring the activity of a soluble protein. In this case, the tissue can be frozen and ground in liquid nitrogen and the protein extracted by gentle stirring in extraction buffer. This buffer can include many of the protectants used for the isolation of membrane fractions (see *Table 1*), but the osmoticum can be omitted and replaced with a cryoprotectant such as ethylene glycol (25% v/v is recommended). For soluble proteins, 100 mM ammonium sulphate can be included as a protease inhibitor. An example of such an extraction, the isolation of apophytochrome for chromophore assembly *in vitro*, is given in Terry and Lagarias (19). The homogenate should be centrifuged at the highest possible speed to remove all membranous material (up to 200 000 g for 30 min). The protein in this supernatant fraction can then be concentrated by adding finely powdered ammonium sulphate followed by further centrifugation. For apophytochrome 0.23 g/ml is sufficient (19), but this step represents an opportunity for a partial purification of the protein of interest and the concentration of ammonium sulphate required should be determined empirically (see ref. 40).

There can be problems with this procedure for analysing cytoplasmic proteins, including the dilution of cytoplasmic proteins with proteins from other cellular compartments and the damage that can result from lysis of the vacuole as discussed before. One rather ingenuous way to counteract these problems was recently described by Harter *et al.* (41). They prepared cytosolic fractions directly by gently lysing evacuolated protoplasts and removing organelles by centrifuging through a silicon oil gradient. The final cytosolic fraction had a protein concentration of 2–4 µg/ml (41).

5.2 Extraction of protein for immunological analysis

Plant tissue can easily be extracted directly for immunological analysis by grinding in liquid nitrogen followed by the addition of 2× strength SDS-sample buffer [80 mM Tris–HCl pH 8.8, 10% (w/v) glycerol, 10% (w/v) SDS, 5% (v/v) β-mercaptoethanol and 0.002% (w/v) bromophenol blue]. Samples are then heated at 65 °C for 20 min and centrifuged at 13 000 g for 10 min at 4 °C. One problem using this protocol is that it is difficult to determine the protein concentration of these samples. Recently, Martinez-Garcia *et al.* (42) have described a rapid quantitative method for preparing protein extracts for immunoblot analysis, which is compatible with commercial systems used to determine protein concentration. It is known as the EZ procedure and was developed for *Arabidopsis*, but it may be suitable for other plant material. Plant material is initially ground in a buffer (buffer E) containing 125 mM Tris–HCl pH 8.8, 1% (w/v) SDS, 10% (v/v) glycerol, and 50 mM $Na_2S_2O_5$. Following centrifugation in a microcentrifuge, protein is quantified in aliquots using a detergent compatible protein assay (see section 6.1 below), and the remaining sample is diluted in buffer Z containing 125 mM Tris–HCl pH 6.8, 12% (w/v) SDS, 10% (v/v) glycerol, 22% (v/v) β-mercaptoethanol, and 0.001% (w/v) bromophenol blue, and analysed using SDS–PAGE and western blotting techniques in the usual way (see *Chapters 7* and *10*, also *Volume 1, Chapter 10*).

6 Assessment of cellular fractions and the use of marker enzymes

6.1 Estimating protein concentration

No protein assay is entirely satisfactory and will depend on the amino acid composition of the protein(s) in question. The Bradford method (43) is a simple and popular method for protein estimation, and is less subject to interference than some methods. In *Protocol 8* we describe a modified Bradford assay, which includes low levels of SDS to solubilize membrane-bound proteins. This method can be modified to a microtitre assay when large numbers of protein estimations are necessary, and is commercially available as the Bio-Rad Protein Assay (Bio-Rad). One problem is that this method is not compatible with high concentrations of detergent. If detergent is present (e.g. in the EZ protein extraction procedure, see section 5.2 above), then the Lowry method (44) or the bicinchonic acid method (45) will be more suitable. Protocols for these methods are commercially available with the Bio-Rad *DC* Protein Assay (Bio-Rad) and the BCA Protein Assay (Pierce, Rockford, IL, USA), respectively.

Protocol 8

Modified Bradford assay for protein estimation

Reagents

- 0.3% (w/v) SDS
- Coomassie stock solution: dissolve 200 mg of Coomassie Brilliant Blue G-250 in 50 ml of

95% ethanol. Add 100 ml of 85% (v/v) orthophosphoric acid
- 100 µg/ml BSA (protein standard)

Method

1 Prepare protein assay reagent by taking 15 ml of Coomassie stock solution, adding 1 ml of 0.3% SDS, and diluting to 100 ml. Mix thoroughly and filter (this can take over 1 h).

2 Prepare a calibration curve using 1–10 µg protein. Make up to 200 µl with distilled H_2O, add 1 ml of protein assay reagent, and mix thoroughly.

3 Take 200 µl of membrane sample, add 1 ml of protein assay reagent, and mix thoroughly. You may need to dilute your sample.

4 Incubate the sample for 3 min at room temperature and read absorbance at 595 nm (use either glass or plastic cuvettes; glass cuvettes can be cleaned in SDS solution or acid).

6.2 Marker analysis

The separation of different cell components can be monitored by following the distribution of molecular markers known to be associated with specific membrane or organelle fractions. Traditionally, these markers have taken the form of enzyme assays, but it has become increasingly common to use markers that can be detected immunologically. An ideal marker is one that is homogeneously distributed throughout one particular subcellular fraction. However, although most markers have a primary site, they are also found in other cellular locations. For this reason, it is best to use several markers to ascertain the distribution of a particular subcellular component. Here, we describe several enzyme assays, which can be used as markers to determine the presence of particular membranes. In addition, we describe a sensitive immunological approach, whereby antibodies are used to detect proteins associated with particular cell components and thus identify their cellular distribution. This method is particularly useful when material is in low abundance. *Table 2* gives a list of markers that can be used in combination to determine the enrichment of a particular subcellular component during purification.

6.2.1 Guidelines for conducting enzyme assays

Many of the biochemical assays to be performed, including most of the marker assays described below, will involve the measurement of enzyme activity. There are a few basic points to remember when conducting an enzyme assay:

1. The samples should be stored correctly for maximal activity. In most cases storage on ice is sufficient with the enzyme activity being lost at higher temperatures.

2. The enzyme reaction should be linear within the assay time and with the particular protein concentration used. This ensures that the substrates or co-factors are not limiting for the reaction. Always perform a time course to test the linearity of the reaction.

3. The temperature and pH should be optimized and carefully controlled throughout the assay.

163

Table 2 Biochemical and immunological markers used in the analysis of membrane distribution

Membrane	Biochemical marker	Immunological marker
Plasma membrane	Vanadate-sensitive, Mg/K-dependent ATPase (*Protocol 9*; ref. 46), glucan synthetase II (47)	Plasma membrane H^+-ATPase (48, 49)
Endoplasmic reticulum	NAD(P)H cytchrome c reductase (*Protocol 12*; ref. 34), stereospecificity of NADH-ferricyanide reductase (50)	BiP (51) calreticulin (52, 51) calnexin (53)
Tonoplast	Azide-insensitive, nitrate-sensitive ATPase (*Protocol 9*; ref. 54), bafilomycin-sensitive ATPase (*Protocol 9*; ref. 55), Mg/K-dependent PPase (Section 6.2.5; ref. 56)	vacuolar H^+-ATPase (57), vacuolar H^+-PPase (57), and tonoplast intrinsic protein (e.g. VM23; 58)
Golgi	latent NDPase, (Section 6.2.5; ref. 59) glucan synthetase I (60)	JIM 84 (61)
Mitochondrial	Azide-sensitive ATPase (*Protocol 9*; ref. 46), cytochrome c oxidase (*Protocol 11*; ref. 34)	
Chloroplast	chlorophyll (*Protocol 10*; ref. 62),	Triose phosphate translocator (28) chlorophyll *a/b*-binding protein (29)

4. Always run appropriate controls. For example, assays without membranes or substrate should be included to ensure that you are measuring the correct activity. If testing the effect of a particular reagent (e.g. a new inhibitor) on the enzyme activity, make sure this does not interfere with the assay. This can be tested on the standard curve.

5. For colorimetric assays ensure that the readings are taken within the correct time frame as the colour reaction may not be stable for extended periods.

[For further information on the principles and practice of enzyme assays see Eisenthal and Danson (63).]

6.2.2 General marker assays

6.2.2.1 ATPase assay.

Several marker enzyme assays rely on ATP hydrolysis for activity. ATPases associated with various membrane fractions have different enzymatic properties and it is therefore possible, by choosing appropriate conditions of pH and including various ions and/or inhibitors, to find out in one experiment the separation of these membranes following density gradient centrifugation or phase partitioning (see *Table 2*). For example, vanadate-sensitive MgK-ATPase activity assayed at pH 6.5 can be used to locate plasma membrane, azide-sensitive MgK-ATPase activity assayed at pH 8 can be used to detect mitochondrial membranes and azide-insensitive, nitrate-sensitive MgK-ATPases activity assayed at pH 8 can be used to detect tonoplast.

In *Protocol 9*, the Ohnishi assay (64), which measures ATPase activity by determining the free phosphate (Pi) release from ATP has been modified for use as a microtitre plate assay. It is a rapid and sensitive assay, which can be used for determining the kinetic parameters of a range of ATPases. It has the advantage that only small amounts of membrane protein are required.

Protocol 9

Microtitre assay for the determination of ATPase activity in membrane fractions

Equipment and reagents

- Microtitre plate reader
- 20 μl and 200 μl multipipettes
- Ohnishi components:

  (A) molybdate preservative, stored at room temperature and made fresh if precipitation occurs; 10 g of ammonium molybdate and 1.16 g EDTA (free acid) in 250 ml H_2O

  (B) reductant: 7.058 g hydroxylamine sulphate, 10 g PVP (mol. wt. 40 000), and 1.23 ml conc. H_2SO_4 in 250 ml H_2O; store refrigerated

  (C) Ohnishi colour developer: 1.32 g Na_2CO_3, 67.4 g NaOH in 250 ml H_2O. Store refrigerated

- Prepare Ohnishi A/B reagent fresh by mixing solutions A and B with water in a ratio of 2A:3B:1H_2O
- Stopping reagent: 10% (w/v) SDS (Sigma 99% GC grade)

- Standard assay reagents:
- 3 mM KH_2PO_4 for standard curve; pH 6.5 buffer (100 mM Tris-MES pH 6.5, 250 μM ammonium molybdate); pH 8.0 buffer (100 mM Tris-MES pH 8.0, 250 μM ammonium molybdate); 20 mM $MgSO_4$; 500 mM KCl; 20 mM ATP; membrane vesicles (100–140 μg protein/ml)
- Optional assay reagents:

  Detergent: 0.1% (v/v) Triton X-100 or 0.1% (w/v) Brij 58; 10 mM sodium azide; 500 mM KNO_3; 1 μM bafilomycin (in DMSO); 1 mM sodium orthovanadate. The orthovanadate must be freshly prepared, boiled, and the pH adjusted to 6.5 with HCl. For 100 ml put 0.0184 g sodium orthovanadate in a beaker and add 90 ml of H_2O. Adjust pH to 7.0. Bring to the boil on a hot plate with stirring. Allow the solution to cool and adjust pH to 6.5. Make up to 100 ml.

Method

1. Prepare a standard curve by setting up the following phosphate dilutions in duplicate or triplicate:

Tube no.	1	2	3	4	5	6
3 mM KH_2PO_4 (ml)	0	0.02	0.04	0.06	0.08	0.10
H_2O (ml)	0.50	0.48	0.46	0.44	0.42	0.40
nmol PO_4 in final assay (70 μl)	0	8.4	16.8	25.2	33.6	42

The standard curve may include reagents in the same concentrations as in the assay reaction mixes. The phase diluting medium (from Protocol 6) in the presence of the pH 6.5 and 8 buffers causes a slight overestimation of phosphate concentration, so it is advisable to include both in the standard curve by replacing 0.25 ml of the H_2O with 0.2 ml pH 6.5/8 buffer and 0.05 ml phase diluting buffer.

2. Prepare the assay reaction mixes in microcentrifuge tubes as indicated below. The membranes will be added later to start the reactions.

Tube no.	pH 6.5 or 8.0 buffer	ATP (2 mM final conc.)	$MgSO_4$ (2 mM final conc.)	KCl (50 mM final conc.)	H_2O
1(- ATP)	0.2 ml	–	0.05 ml	0.05 ml	0.15 ml
2(-memb)	0.2 ml	0.05 ml	0.05 ml	0.05 ml	0.15 ml
3(- Mg-K)	0.2 ml	0.05 ml	–	–	0.20 ml
4(- K)	0.2 ml	0.05 ml	0.05 ml	–	0.15 ml
5(+ Mg + K)	0.2 ml	0.05 ml	0.05 ml	0.05 ml	0.10 ml

The assay system may include other constituents, but the amount of water in the reaction mix must be adjusted to maintain a final volume of 0.50 ml. Examples are shown below:

(a) When assaying preparations of tightly sealed plasma membrane vesicles, include 0.01% Triton X-100 in the final assay mix (to allow access of the substrate to the active site of the enzyme) by substituting 0.05 ml of 0.1% (v/v) Triton X-100 for 0.05 ml of H_2O.

(b) Add 0.05 ml of 10 mM sodium azide (at pH 8.0) to inhibit the mitochondrial ATPase/synthase.

(c) Add 0.05 ml of 500 mM KNO_3 (replacing the KCl to maintain the potassium concentration) to preferentially inhibit the tonoplast H^+-ATPase (assayed in the presence of azide at pH 8.0).

(d) Add 0.05 ml of 1 mM sodium orthovanadate to preferentially inhibit the plasma membrane H^+-ATPase (assayed at pH 6.5).

(e) Add 0.05 ml of 1 μM bafilomycin to preferentially inhibit the tonoplast H^+-ATPase (assayed at pH 8.0).

When using inhibitors prepared in solvents you should check the effect of the solvent alone.

3 Vortex tubes to mix.

4 The assay is performed in a 96-well microtitre plate and each assay should be performed in triplicate. Add 70 μl from each of the standard mixes and 63 μl from each of the assay reaction mixes to the wells.

5 To start the reactions, add 7 μl of membranes (approximately 1 μg protein) to the appropriate wells and mix using a multipipette rinsing in water between rows to avoid cross-contamination.

6 Incubate for 1 h at 37 °C (the actual incubation time used will depend on activity). Mix at 20 min intervals during the incubation. If a time course is required, stagger the addition of membranes to the wells accordingly.

7 Stop the reactions by adding 8 μl of stopping reagent to each well, then add 180 μl Ohnishi A/B reagent and 18 μl Ohnishi developer. Mix thoroughly by drawing the mixes into a multipipette two or three times. Leave for 20 min to allow the colour to develop.

8 Read the microtitre plate in a microtitre plate reader at 550–720 nm. Plate readers can usually be programmed to take averages of the replicates, calculate a linear regression for the standard curve, and give the results in nmol phosphate. Divide the results by the amount of protein added to each well to give the results as nmol phosphate generated per μg protein.

6.2.2.2 Immunoblot analysis as a membrane marker.

There is now an increasing number of antibodies available to proteins that can be used as markers for subcellular components. Immunological marker analysis has the advantage that very small quantities of protein may be all that is required and immunoblots can be stripped and reprobed several times with different antibodies. Membrane samples should be prepared by diluting the sample 1:1 with 2× strength SDS-sample buffer [62.5 mM Tris–HCl pH 6.5, 5% (v/v) β-mercaptoethanol, 2% (w/v) SDS, 10% (v/v) glycerol, 0.002% (w/v) bromophenol blue] and heating at 70 °C for 20 min. Some membrane proteins will aggregate under these conditions and it may be better to incubate at 22 °C for 20 min. Samples can then be analysed using SDS–PAGE and western blotting techniques in the usual way (see *Chapters 7* and *10*). Examples of suitable immunological markers are given in *Table 2*.

6.2.3 Marker assays for organelles

There are a number of suitable markers for chloroplasts. A method for one of the simplest, total chlorophyll (Chl) determination, is shown in *Protocol 10*. This calculation gives the approximate concentration of Chl *a* and Chl *b* combined (62). To determine the relative

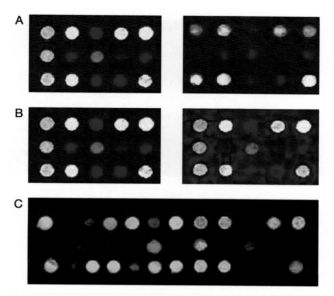

Plate 1 Example of DNA micro-array images. (A) Comparison of clones purified by ethanol precipitation (left) versus no purification (right). (B) Comparison of Cy3 (left) and Cy5 (right) labelled probes generated from the same RNA sample using the Genisphere 3DNA Expression Kit. C). False colour overlay image comparing expression levels in immature fruit (green) and ripe fruit (red) using Quantarray software. Yellow spots indicate similar expression levels.

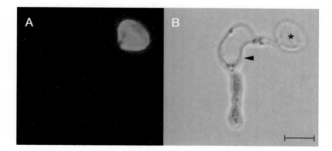

Plate 2 Mapping of a surface epitope of Phytophthora infestans germlings with a monoclonal phage displayed antibody. Germlings were immobilized on an ELISA plate and detected using 10 l of monoclonal phage displayed scFv purified by polyethylene glycol precipitation. Binding was detected using a rabbit anti-fd antibody and a goat anti-rabbit immunoglobulin fluorescein conjugate. Panel A shows the fluorescent image and panel B the corresponding image taken under normal light conditions. The zoospore body (asterisk) and appressorium (arrow head) are labelled and a scale bar of 10 m shown.

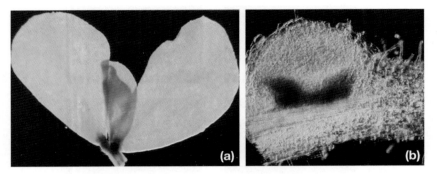

Plate 3 Examples of GUS staining. The images are derived from different lines of Lotus japonicus tagged with a promoterless GUS construct. (a) Whole mount staining showing GUS activity at the leaf bases of young leaves. (b) Hand-cut section showing GUS activity within the nodule vascular bundles. Images courtesy of L. Martirani and M. Chiurazzi (IIGB, Napoli, Italy).

Plate 4 A 3–4-week culture of wild-type Physcomitrella, growing on a 50 mm agar plate, showing leafy shoots and protonema (arrow). Scale bar 2 mm.

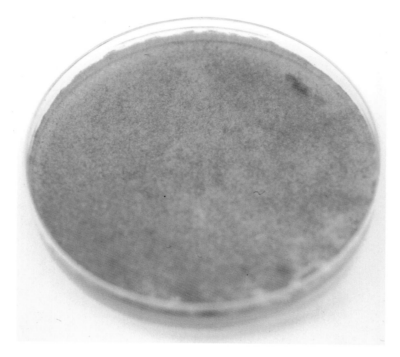

Plate 5 A 7-day homogenate culture of Physcomitrella, growing on a 90-mm agar plate, overlaid with cellophane, and showing protonemata only.

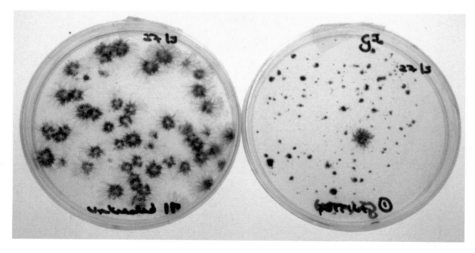

Plate 6 Appearance of transformants on 9-cm agar plates 14 days after transfer of cellophane overlay carrying regenerating colonies to medium containing antibiotic (right hand plate) and to medium without antibiotic (left hand plate).

concentrations of Chl *a* and Chl *b*, it is necessary to measure the absorbance maxima of both species separately [663 and 647 nm, respectively, in 80% (v/v) acetone] and use more recently derived equations (65, 66). This method can also be used for estimating Chl levels extracted directly from intact plant tissue, but in this case, methods utilizing other solvents such as *N,N*-dimethylformamide can have practical advantages (67). Care should be taken to use the appropriate molar absorption coefficients as these vary depending on the solvent used (66).

Protocol 10

Chlorophyll determination as a marker for chloroplasts

Method

1 Extract an aliquot of chloroplasts in a final concentration of 80% (v/v) acetone keeping the sample under low light conditions if possible.
2 Centrifuge sample in a bench top microcentrifuge at 12 000 g for 1 min.
3 Measure absorbance at 652 nm and calculate total chlorophyll concentration (in μg/ml) as A_{652} × 1000/34.5 (ref 62).

In addition to Chl determination, various other chloroplast markers can be used. For example, there are antibodies available to proteins such as the triose-phosphate translocator (28) or Chl *a/b*-binding proteins (29). Alternatively, $NADP^+$:triose phosphate dehydrogenase can be measured quite simply by following the oxidation of NADPH (18). For etioplasts that have been prepared under safelight conditions, the best marker to use is the Chl precursor protochlorophyllide. In practice, it is easier to measure protochlorophyll, which includes both the esterified and non-esterified forms of protochlorophyllide. These pigments can be extracted under the same conditions as given in *Protocol 11*, although for whole plant tissue acetone 0.1M NH_4OH (90/10, v/v) is recommended (68). Following a centrifugation step, protochlorophyll can be quantified by absorption spectroscopy using a molar absorption coefficient of 31 100 M^{-1} cm^{-1} at 626 nm (69).

The standard marker assay for mitochondria is cytochrome c oxidase (34). The assay is performed by monitoring the oxidation of cytochrome c spectrophotometrically and a protocol for this assay is shown below (*Protocol 11*). Remember that the presence of reducing agents in the suspension medium can interfere with both this assay and the assay for NAD(P)H cytochrome c reductase (see *Protocol 12*), so always run a control containing only the suspension medium with no membrane present. Azide-sensitive, MgK-ATPase activity assayed at pH 8 can also be used to detect mitochondrial membranes (46) and an assay for this enzyme is shown in *Protocol 9*.

Protocol 11

Assay for cytochrome c oxidase as a marker for mitochondria

Equipment and reagents

- Spectrophotometer (connected to chart recorder if no kinetic mode screen display)
- 50 mM K phosphate buffer pH 7.5
- 0.6 mM cytochrome c (reduced form) in K phosphate buffer (add a few crystals of
- sodium dithionite to reduced cytochrome c and remove excess dithionite by aeration for a few min)
- 0.3% (w/v) digitonin in K phosphate buffer

Protocol 11 continued

Method

1 In the cuvette, mix 33 μl of membrane vesicles (approximately 10 μg protein) with 33 μl digitonin. Wait 20–30 s and add 901 μl K phosphate buffer.

2 Start the reaction by adding 33 μl cytochrome c to give a final assay volume of 1 ml, and mix well.

3 Follow the oxidation of cytochrome c by monitoring the absorbance decrease at 550 nm at 25 °C.

4 Determine the rate of cytochrome c oxidation by calculating the rate of cytochrome c utilized using the molar absorption (extinction) coefficient for cytochrome c of 18.5 mM^{-1}cm^{-1}.

6.2.4 Marker assays for endo- and plasma membranes

Vanadate-sensitive, MgK-ATPase activity assayed at pH 6.5 is the most widely used marker for the plasma membrane (46) and measures the presence of the H^{+}-ATPase at this membrane. This enzyme can be assayed using the modified Ohnishi method (64) described in section 6.2.2.1 (*Protocol 9*). In some circumstances, contaminating ATPase activities may be a problem and in these cases glucan synthetase II activity could also be used (47). More recently, antibodies to the plasma membrane H^{+}-ATPase have proved to be useful for marker analysis (48, 49).

Antimycin A-insensitive NAD(P)H cytochrome c reductase is the most commonly used marker enzyme for the ER membrane. However, caution is warranted as this activity has been reported on other membranes (e.g. plasma membrane; 70). Thus, it should be used in conjunction with other markers. However, the assay is useful, particularly in the absence of detergents, for detecting the location of the bulk of the ER following gradient centrifugation. The assay is similar to that described for cytochrome c oxidase (*Protocol 11*) except that the reduction of cytochrome c is monitored, rather than its oxidation (34). A method for assaying NADH (NADPH) cytochrome c reductase is shown in *Protocol 12*. Another more specific marker for the ER, the stereospecificity of NADH-ferricyanide reductase activity, has been described (50), but this is not a routine technique and the preparation of the necessary reagents is time-consuming. Antibodies against the lumenal ER proteins BiP and calreticulin, and against the membrane-bound ER protein calnexin can also be used (51–53). However, care should be taken when using lumenal markers as these can be released during homogenization.

Protocol 12

Assay for NADH (NADPH) cytochrome c reductase as a marker for ER

Equipment and reagents

- Spectrophotometer (connected to chart recorder if no kinetic mode screen display)
- 50 mM K phosphate buffer pH 7.5
- 50 mM potassium cyanide in K phosphate buffer (be extremely careful when preparing this as cyanide is highly toxic)
- 0.6 mM cytochrome c (oxidized form) in K phosphate buffer
- 3 mM NADH (or NADPH) in K phosphate buffer
- 100 μM antimycin A in ethanol

Protocol 12 continued

Method

1 Add the following to a cuvette and mix: 833 μl of 50 mM K phosphate buffer, 66 μl of potassium cyanide, 33 μl of cytochrome c (oxidized form), and 10 μl of antimycin A.

2 Add 25 μl of membrane vesicles (containing ~10 μg protein).

3 Place parafilm over the top of the cuvette, invert 10 times, and place in spectrophotometer. Monitor non-specific reduction of cytochrome c at 550 nm for 3 min at 25 °C.

4 Add 33 μl of NADH, mix well, and monitor absorbance increase for 3 min.

5 Determine the rate of NADH-specific reduction by calculating the rate of cytochrome c utilized using a molar absorption (extinction) coefficient for cytochrome c of 18.5 mM^{-1}cm^{-1}.

Latent NDPase activity can be used as a marker for the Golgi apparatus although again this activity has been found in other locations (59). The modified Ohnishi assay described in *Protocol 9* can be adapted to measure this activity by replacing ATP with inosine diphosphate or uridine diphosphate, and assaying at pH 7.0 in the presence or absence of 0.01% (v/v) Triton X-100. The latent nucleotide diphosphatase (NDPase) activity is obtained by subtracting the MgK-dependent activity in the absence of detergent from that in the presence of detergent [for other enzymatic Golgi markers, including glucan synthetase I, see Green (60)]. Antibodies to JIM 84 can be used to indicate the presence of Golgi membranes although this may also label plasma membrane (61).

Bafilomycin-sensitive ATPase and azide-insensitive, nitrate-sensitive ATPase are the most commonly used tonoplast markers (see *Protocol 9*). An alternative marker is the K$^+$-stimulated H$^+$-pyrophosphatase (H$^+$-PPase), although this enzyme has been reported on other membranes (57, 71). PPase activity can also be assayed by modifying the ATPase microtitre assay (*Protocol 9*). It is assayed in the presence of 40 mM Tris-Mes pH 7.5–8.0, 0.1 mM ammonium molybdate, 2 mM MgSO$_4$, 0.2 mM pyrophosphate, and 50 mM KCl. Antibodies to the tonoplast H$^+$-ATPase and H$^+$-PPase can also be used to detect tonoplast membranes (56, 57) as can antibodies to certain tonoplast intrinsic proteins (58).

Acknowledgments

We would like to thank R. F. Mills for advice with protocols and critical reading of this chapter and to J. K. Pittman, J. L. Hall, and H. Okamoto for general comments. Thanks also to B. J. Whyte for introducing one of us (M.J.T.) to plastid isolation. The authors were both supported by Royal Society University Research Fellows.

References

1. Hall, J. L. and Moore, A. L. (ed.) (1983). *Isolation of membranes and organelles from plant cells*. Academic Press, London.

2. Linskens, H-F. and Jackson, J. F. (ed.) (1985). *Cell components: modern methods of plant analysis*. Springer-Verlag, Berlin.

3. Morré, D. J., Brightman, A. O. and Sandelius, A. S. (1987). In *Biological membranes: a practical approach* (ed. J. B. C. Findlay, and W. H. Evans), p. 37. IRL Press, Oxford.

4. Fowke, L. C. and Cutler, A. J. (1998). In *Plant cell biology: a practical approach* (ed. N. Harris, and K. J. Oparka), p. 177. IRL Press, Oxford.

5. Sandelius, A. S. and Morré, D. J. (1990). In *The plant plasma membrane* (ed. C. Larsson, and I. M. Møller), p. 44. Springer Verlag, Berlin.

6. Robinson, D. G., Hinz, G. and Oberdeck, K. (1998). In *Plant cell biology; a practical approach* (ed. N. Harris, and K. J. Oparka), p. 245. IRL Press, Oxford.

7. Williams, L. E., Nelson, S. J. and Hall, J. L. (1990). *Planta*, **182**, 532.
8. Thomson, L. J., Xing, T., Hall, J. L. and Williams, L. E. (1993). *Plant Physiol.*, **102**, 553.
9. Gianinni, J. L., Gildensoph, L. H., Reynolds-Niesman, I. and Briskin, D. P. (1987). *Plant Physiol.*, **85**, 1129.
10. Miernyk, J. A. (1985). In *Cell components: modern methods of plant analysis* (ed. H-F. Linskens, and J. F. Jackson), New Series Vol. 1, p. 259. Springer-Verlag, Berlin.
11. Berkowitz, G. A. and Gibbs, M. (1985). In *Cell components: modern methods of plant analysis* (ed. H-F. Linskens and J. F. Jackson), New Series Vol. 1, p. 152. Springer-Verlag, Berlin.
12. Walker, D. A., Cerovic, Z. G. and Robinson, S. P. (1987). In *Methods in enzymology* (ed. L. Packer, and R. Douce), Vol. 148, p. 145. Academic Press, London.
13. Pardo, A. D., Chereskin, B. M., Castelfranco, P. A., Franceschi, V. R. and Wezelman, B. E. (1980). *Plant Physiol.,* **65**, 956.
14. Fuesler, T. P., Castelfranco, P. A. and Wong, Y.-S. (1984). *Plant Physiol.*, **74**, 928.
15. Terry, M. J. and Kendrick, R. E. (1996). *J. Biol. Chem.*, **271**, 21681.
16. Joy, K. W. and Mills, W. R. (1987). In *Methods in enzymology* (ed. L. Packer and R. Douce), Vol. 148, p. 179. Academic Press, London.
17. Journet, E-T. and Douce, R. (1985). *Plant Physiol.*, **79**, 458.
18. Lee, H. J., Ball, M. D. and Rebeiz, C. A. (1991). *Plant Physiol.*, **96**, 910.
19. Terry, M. J. and Lagarias, J. C. (1991). *J. Biol. Chem.*, **266**, 22215.
20. Cline, K. (1985). In *Cell components: modern methods of plant analysis* (ed. H-F. Linskens, and J. F. Jackson), New Series Vol. 1, p. 182. Springer-Verlag, Berlin.
21. Douce, R., Holtz, R. B. and Benson, A. A. (1973). *J. Biol. Chem.*, **248**, 7215.
22. Dorne, A-J., Joyard, J. and Douce, R. (1990). *Proc. Natl Acad. Sci. USA*, **87**, 71.
23. Sato, N., Albrieux C., Joyard, J., Douce, R. and Kuroiwa, T. (1993). *EMBO J.*, **12**, 555.
24. Keegstra, K. and Yousif, A. E. (1986). In *Methods in enzymology* (ed. A. Weissbach, and H. Weissbach), Vol. 118, p. 316. Academic Press, London.
25. Smeekens, S., van Steeg, H., Bauerle, C., Bettenbroek, H., Keegstra, K. and Weisbeek, P. (1987). *Plant Mol. Biol.*, **9**, 377.
26. Roper, J. M. and Smith A. G. (1997). *Eur. J. Biochem.*, **246**, 32.
27. Douce, R. and Joyard, J. (1980). In *Methods in enzymology* (ed. A. S. Pietro), Vol. 69, p. 290. Academic Press, London.
28. Schunemann, D., Schott, K., Borchert, S. and Heldt, H. W. (1996). *Plant Mol. Biol.*, **31**, 101.
29. Høyer-Hansen, G., Bassi, R., Hønberg, L. S. and Simpson, D. J. (1988). *Planta*, **173**, 12.
30. Douce, R., Bourguignon, J., Brouquisse, R. and Neuburger, M. (1987). In *Methods in enzymology* (ed. L. Packer and R. Douce), Vol. 148, p. 403. Academic Press, London.
31. Moore, A. L. and Proudlove, M. O. (1987). In *Methods in enzymology* (ed. L. Packer and R. Douce), Vol. 148, p. 415. Academic Press, London.
32. Jackson, J. F. (1985). In *Cell components: modern methods of plant analysis* (ed. H-F. Linskens and J. F. Jackson), New Series Vol. 1, p. 353. Springer-Verlag, Berlin.
33. Datta, N., Chen, Y-R. and Roux, S. J. (1985). *Biochem. Biophys. Res. Comm.*, **128**, 1403.
34. Hodges, T. K. and Leonard, R. T. (1974). In *Methods in enzymology* (ed. S. Fleischer and L. Packer), Vol. 32, p. 392. Academic Press, London.
35. Larsson, C., Sommarin, M. and Widell, S. (1994). In *Methods in enzymology* (ed. H. Walter and G. Johansson), Vol. 228, p. 451. Academic Press, London.
36. Palmgren, M. G., Askerlund, P., Fredrikson, K., Widell, S., Sommarin, M. and Larsson, C. (1990). *Plant Physiol.,* **92**, 871.
37. Johansson, F., Olbe, M., Sommarin, M. and Larsson, C. (1995). *Plant J.*, **7**, 165.
38. Price, C. A. (1983). In *Isolation of membranes and organelles from plant cells* (ed. J. L. Hall and A. L. Moore), p. 1. Academic Press, London.
39. Lord, M. (1987). In *Methods in enzymology* (ed. L. Packer and R. Douce), Vol. 148, p. 576. Academic Press, London.
40. Scopes, R. K. (1993) *Protein purification—principles and practice*, 3rd edn. Springer-Verlag, New York.
41. Harter, K., Frohnmeyer, H., Kircher, S., Kunkel, T., Mühlbauer, S., and Schäfer, E. (1994). *Proc. Natl. Acad. Sci. USA*, **91**, 5038.
42. Martinez-Garcia, J. F., Monte, E. and Quail, P. H. (1999). *Plant J.*, **20**, 251.

43. Bradford, M. M. (1976). *Anal. Biochem.*, **72**, 248.
44. Lowry, O. H., Rosebrough, N. J., Farr, A. L. and Randall, R. J. (1951). *J. Biol. Chem.*, **193**, 265.
45. Smith, P. K., Krohn, R. I., Hermanson, G. T., Mallia, A. K., Gartner, F. H., Provenzano, M. D., Fujimoto, E. K., Goeke, N. M., Olson, B. J. and Klenk, D. C. (1985). *Anal. Biochem.*, **150**, 265.
46. Gallagher, S. R. and Leonard, R. T. (1982). *Plant Physiol.*, **70**, 1335.
47. Ray, P. M. (1979). In *Plant Organelles* (ed. E. Reid), p. 135. Ellis Horwood, Chichester.
48. Villalba, J. M., Lutzelschwab, M. and Serrano, R. (1991). *Planta*, **185**, 458.
49. Morsomme, P., d'Exaerde, A. D., DeMeester, S., Thines, D., Goffeau, A. and Boutry, M. (1996). *EMBO J.*, **15**, 5513.
50. Fredlund, K. M., Widell, S. and Møller, I. M. (1996). *Plant J.*, **10**, 925.
51. Napier, R. M., Trueman, S., Henderson, J., Boyce, J. M., Hawes, C., Fricker, M. D. and Venis, M. A. (1995). *J. Exp. Bot.*, **46**, 1603.
52. Nelson, D. E., Glaunsinger, B. and Bohnert, H. J. (1997). *Plant Physiol.*, **114**, 29.
53. Huang, L. Q., Franklin, A. E. and Hoffman, N. E. (1993). *J. Biol. Chem.*, **268**, 6560.
54. Wang, Y. and Sze, H. (1985). *J. Biol. Chem.*, **260**, 10434.
55. Bowman, E. J., Siebers, A. and Altendorf, K. (1988). *Proc. Natl Acad. Sci. USA*, **85**, 7972.
56. Rea, P. A., Britten, C. J. and Sarafian, V. (1992). *Plant Physiol.*, **100**, 723.
57. Oberbeck, K., Drucker, M. and Robinson, D. G. (1994). *J. Ex. Bot.*, **45**, 235.
58. Maeshima, M. (1992). *Plant Physiol.*, **98**, 1248.
59. Mitsui, T., Honma, M., Kondo, T., Hashimoto, N., Kimura, S. and Igaue, I. (1994). *Plant Physiol.*, **106**, 119.
60. Green, J. R. (1983). In *Isolation of membranes and organelles from plant cells* (ed. J. L. Hall, and A. L. Moore), p. 135. Academic Press, London.
61. Horsley, D., Coleman, J., Evans, D., Crooks, K., Peart, J., Satiatjeunemaitre, B. and Hawes, C. (1993). *J. Exp. Bot.*, **44**, 223.
62. Arnon, D. I. (1949). *Plant Physiol.*, **24**, 1.
63. Eisenthal, R. and Danson, M. J. (ed.) (1992). *Enzyme assays: a practical approach*. IRL Press, Oxford.
64. Ohnishi, T., Gall, R. S. and Meyer, M. L. (1975). *Anal. Biochem.*, **69**, 261.
65. Lichtenthaler, H. K. (1987). In *Methods in enzymology* (ed. L. Packer, and R. Douce), Vol. 148, p. 350. Academic Press, London.
66. Porra, R. J., Thompson, W. A. and Kriedmann, P. E. (1989). *Biochim. Biophys. Acta,* **975**, 384.
67. Inskeep, W. P. and Bloom, P. R. (1985). *Plant Physiol.*, **77**, 483.
68. Rebeiz, C. A., Mattheis, J. R., Smith, B. B., Rebeiz, C. C. and Dayton, D. F. (1975). *Arch. Biochem. Biophys.*, **171**, 549.
69. Kahn, A. (1983). *Physiol. Plant.*, **59**, 99.
70. Widell, S. and Larsson, C. (1990). In *The plant plasma membrane* (ed. C. Larsson and I. M. Møller), p. 16. Springer Verlag, Berlin.
71. Long, A. R., Hall, J. L. and Williams, L. E. (1997). *J. Plant Physiol.*, **151**, 16.

Chapter 9

Studying protein–protein interactions with the yeast two-hybrid system

Claus Schwechheimer* and Xing-Wang Deng

Department of Molecular Cellular and Developmental Biology, Osborn Memorial Laboratories, Yale University, 165 Prospect Street, New Haven CT 06520–8104, USA

*Current address: Centre for Plant Molecular Biology, Auf der Morgenstelle 5, 72076 Tübingen, Germany

1 Introduction

In recent years, the yeast two-hybrid system has become a popular technique to study the physical interactions between two (or more) known proteins or to screen cDNA expression libraries for proteins that interact directly with a protein under study (1–10). The yeast two-hybrid system uses molecular biology techniques to address problems that were, until recently, reserved to those who could master the more challenging biochemical approaches.

The two-hybrid system makes use of the fact that many transcriptional activators consist of a DNA-binding domain and an activation domain. These domains can be physically separated and when together are sufficient to activate transcription from a promoter which is recognized by the DNA-binding domain (*Figure 1a*). When two proteins of interest (X and Y in *Figure 1*) are fused to a DNA binding and an activation domain, respectively, then a functional transcriptional activator is reconstituted if the two proteins interact (*Figure 1*). Through the activation of reporter genes in the two-hybrid system, this interaction is phenotypically detectable. In the two-hybrid system, the vectors containing the DNA-binding domain are usually referred to as bait vectors, which encode bait proteins and their activation domain counterparts as prey vectors encoding prey proteins.

The concept of the two-hybrid system has been exploited in many different ways. Modified versions of the two-hybrid system are available that were designed to study the interactions between more than two proteins, between RNA and proteins, or between small ligands and their receptors as well as between DNA and protein in the form of the one-hybrid system (11).

The yeast two-hybrid system allows identification of relatively weak and transient interactions. This can be attributed to the fact that both the bait and the prey protein are abundantly expressed in the yeast cell. An increased degree of sensitivity also stems from the high affinity of the respective DNA-binding domain to the promoters driving the reporter genes and the accumulation over time of these reporter proteins.

This chapter provides the protocols for interaction screens using the two most common versions of the yeast two-hybrid system. More ample descriptions of yeast molecular biology and the yeast two-hybrid system in particular have been published elsewhere (11–14).

2 The LexA- and the GAL4-based two-hybrid systems—an overview

There are two common versions of the two-hybrid system: one version, which uses the DNA-binding domain of the bacterial repressor protein LexA and one version using the yeast

(a)

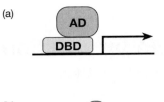

(b)

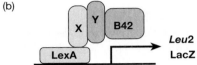

(c)

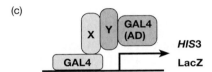

Figure 1 (a) Many transcription factors comprise a DNA–binding domain (DBD) and an activation domain (AD). These domains can be physically separated and when together are sufficient to activate transcription from a cognate promoter. (b) In the LexA two-hybrid system, a protein or a protein domain (X) under study is fused to the DBD of the bacterial LexA protein. The interaction between the protein under study and a second protein (Y), which is fused to the B42 activation domain is phenotypically detectable through the activation of the reporter genes LEU2 and LacZ. (c) Similarly, the GAL4 two-hybrid system uses the DBD and the AD of the yeast activator GAL4 to activate the reporter genes HIS3 and LacZ. Perhaps the most intriguing feature of the two-hybrid system is the possibility to identify new interaction partners for a given bait protein. This is achieved by selecting large numbers of yeast transformed with a specially constructed activation domain cDNA fusion library. By selecting the transformed yeast on media that requires the cells to express an amino acid marker gene, only cDNA prey constructs, which encode a polypeptide that interacts with the bait protein, are recovered. The cDNA construct encoding the putative interacting protein can then be directly isolated from the positive yeast clones.

activator GAL4 (see *Figure 1b* and *c*). In the context of this chapter, these two systems are referred to as LexA and GAL4 (two-hybrid) systems. The LexA and the GAL4 system use different sets of marker genes and are therefore not compatible.

2.1 The LexA-system

In the LexA-system, the candidate protein is fused to the DNA-binding domain of LexA, while the cDNA library clones or other candidate cDNAs are fused to the B42 activation domain (3). The acidic activation domain of B42 was originally identified in a yeast screen for bacterial peptides, which could activate transcription in yeast when fused to the LexA DNA-binding domain (15). When bait and prey proteins interact, they reconstitute a LexA/B42 transcriptional activator that activates *LEU2* and *LacZ* reporter genes (*Figure 1b*). *Table 1a* summarizes the features of some frequently used bait and prey vectors of the LexA-system.

In the LexA-system, only the *LEU2* reporter gene is stably integrated into the yeast genome, while the *LACZ*-reporter resides on a multicopy plasmid (see *Table 1b*), which like the bait and the prey vectors is transformed into cells and maintained as a plasmid. The presence of the *LacZ* reporter gene on a multicopy plasmid enhances the sensitivity of the system and compensates for the comparatively low level activation from the B42 activation domain. Alternative *LacZ* reporter plasmids are available with decreased sensitivity to the LexA DNA-

Table 1a Vectors for the LexA two-hybrid system

	Bacterial selection	Yeast Marker	Promoter	Epitope tag	Yeast ori	Reference
Bait vectors						
pEG202 (pLexA)[a]	Ampicillin	HIS3	ADH1	none	2 μ	(3, Clontech)
pHybLex/Zeo	Zeocin	Zeocin	ADH1	none	2 μ	(27, Invitrogen)
pGilda[ab]	Ampicillin	HIS3	GAL1	none	CEN/ARS	(28, Clontech)
Prey vectors						
pJG4-5 (pB42AD)	Ampicillin	TRP1	GAL1	V5	2 μ	(3, Invitrogen)
pYESTrp2	Ampicillin	TRP1	GAL1	HA	2 μ	(Invitrogen)

pEG202 (pLexA)[a]

EcoRI   BamHI   NcoI*   NotI   XhoI   SalI*

gaa ttc ccg ggg atc cgt cga cca tgg ccg ctc gag tcg acc tgc agc

pHybLex/Zeo

EcoRI   SacI   PvuII   ApaI   KpnI   NotI   XhoI   SalI   PstI

gaa ttc aag ctt gag ctc aga tct cag ctg ggc ccg gta ccg ctc gag tcg acc tgc agc caa

pGilda[ab]

EcoRI   BamHI   NcoI*   NotI   XhoI   SalI*

gaa ttc ccg ggg atc cgt cga cca tgg ccg ctc gag tcg acc tgc agc

pJG4-5 (pB42AD)

EcoRI   XhoI

gaa ttc ggc cga ctc gag

HindIII   KpnI   SacI   BamHI   ApaI   BstXI   EcoRI

aag ctt ggt acc gag ctc gga tcc act agt aac ggc cgc cag tgt gct gga att ctg cag …

pYESTrp2

BstXI   NotI   XhoI   ApaI

… ata tcc atc aca ctg gcg gcc gct cga ggc atg cat ctagag ggc

[a] There are two Sal I sites in the multiple cloning site of these vectors.

[b] This vector has a CEN/ARS replication origin and will be propagated in yeast as a single copy plasmid. All other two-hybrid vectors in this table have a 2 μ origin of replication. A yeast cell can contain up to 50 copies of a 2 μ plasmid.

Table 1b LacZ-reporter plasmids for the LexA-system[a]

pSH18-34	LacZ gene under control of 8 LexA operator sites
pJK 103	LacZ gene under control of 2 LexA operator sites
pRB1840	LacZ gene under control of 1 LexA operator site

[a] For reference see ref. 3.

binding domain. This sensitivity is determined by the number of LexA-binding sites in the promoter of the *LacZ* gene (see *Table 1b*). This variability is a convenient tool in cases where the bait protein activates the reporter genes in the absence of a prey construct (autoactivation; 16).

In the LexA-system, expression of the prey protein is under control of the galactose-inducible *GAL1* promoter, and the expression of the reporter genes in the presence and absence of prey protein can therefore be directly monitored. Similar to the blue/white selection in bacteria, the yeast strains used in the LexA-system can be stained for β-galactosidase activity when grown on solid medium containing the XGAL substrate.

2.2 The GAL4-system

In the GAL4-system, the bait protein for the interaction test is fused to the DNA binding domain of GAL4, while the cDNA library or the second candidate protein is fused to the GAL4 activation domain. When both gene products interact, a functional reconstituted GAL4 transcription factor activates expression from two reporter genes, *HIS3* and *LacZ*. Both reporter genes are stably integrated into the yeast genome. The properties of a number of GAL4-system vectors are shown in *Table 2*.

In most cases, the expression from both interaction plasmids is driven by the constitutive *ADH1* promoter. Expression from the *ADH1* promoter varies between different vectors depending on the size of the promoter, e.g. pACT2, pAS2–1, and pGAD GH express at higher levels than other plasmids. In the case of pACT2, the promoter activity appears to be increased by a cryptic enhancer sequence within the plasmid rather than by the promoter length.

2.3 Constructing two-hybrid cDNA libraries

Activation domain fusion expression libraries are generated by insertion of cDNA fragments into activation domain vectors. The cDNAs can be generated using any standard protocol (12, 17–19). When using a directional cloning protocol, approximately one third of the clones represent functional fusions. Fusion libraries should represent at least 10^6 independent cDNA clones. The cDNA inserts are either cloned into the plasmid vectors or into a phagemid vector from where the two hybrid vectors are excised using the *Cre-Lox* recombination excision system (19; Stratagene).

Random peptide libraries are an alternative to cDNA libraries. Rather than representing the mRNA from a particular organism or tissue, these libraries have DNA fragments, encoding short random peptides, fused to the DNA-binding domain (20). One peptide library will match all the two-hybrid screening needs, independent of the origin of the bait. After screening the two-hybrid library, the peptides are used as query sequences in database searches to identify proteins that contain the interacting peptide. The bait/prey interaction surface can then be further defined by mutating individual amino acids in the peptide.

When this text was being prepared, the Arabidopsis Biological Resource Center at Ohio State University (http://aims.cps.msu.edu) provided two *Arabidopsis* libraries for the two-hybrid system, one of which had been successfully used in a number of two-hybrid screens (9,

Table 2 Vectors for the GAL4 two-hybrid system

	Multiple cloning site	Bacterial selection	Yeast Marker	Promoter	Epitope tag	Yeast ori	Reference
Bait vectors							
pAS2-1	*Nde* I, *Nco* I, *Sfi* I, *EcoR* I, *Sma* I, *BamH* I, *Sal* I, *Pst* I — cat atg gcc atg gag gcc gaa ttc ccg ggg atc cgt cga cct gca gcc aag	Ampicillin	TRP1	ADH1	none	2 μ	(1, 12, Clontech)
pBD-GAL4-2.1	*EcoR* I, *Sma* I, *Xho* I, *Sma* I*, *Sal* I, *Pst* I, *Xba* I — gaa ttc gcc cgg gcc tcg agc ccg ggt cga ctc tag agc cct ata gtg agt cgt att act gca gcc	Chloramphenicol	TRP1	ADH1	none	2 μ	(Stratagene)
pGBT9[c]	*EcoR* I, *Sma* I, *BamH* I, *Sal* I, *Pst* I — gaa ttc cgg ggg atc cgt cga cct gca gcc aag	Ampicillin	TRP1	ADH1	none	2 μ	(14, 29, Clontech)
pGBKT7[a]	*Nde* I, *Nco* I, *Sfi* I, *EcoR* I, *Sma* I, *BamH* I, *Sal* I, *Pst* I — cat atg gcc atg gag gcc gaa ttc ccg ggg atc cgt cga cct gca gcc aag	Ampicillin	TRP1	ADH1	c-Myc	2 μ	(30, Clontech)
pHybLex/Zeo	*EcoR* I, *Sac* I, *Pvu* II, *Apa* I, *Kpn* I, *Not* I, *Xho* I, *Sal* I, *Pst* I — gaa ttc aag ctt gag ctc aga tct cag ctg ggc ccg gta ccg cgg ccg ctc gag tcg acc tgc agc caa	Zeocin	Zeocin	ADH1	none	2 μ	(Invitrogen)
Prey vectors							
pGAD10	*Nco* I, *Sfi* I, *Sma* I, *BamH* I, *EcoR* I, *Sac* I, *Xho* I — atg gcc atg gag gcc ccg ggg atc cga att cga gct cga gag	Ampicillin	LEU2	ADH1	HA	2 μ	(12, 19, Clontech)
pAD-GAL4-2.1	*BamH* I, *Nhe* I, *EcoR* I, *Xho* I, *Sal* I, *Xba* I, *Pst* I, *Bgl* II — gga tcc tct gct agc aga gaa ttc aat tct cta atg ctt ctc gag agt att cga ctc tag agc ... act gca gag atc tat	Ampicillin	LEU2	ADH1	none	2 μ	(Stratagene)
pGAD10	*Xho* I, *BamH* I, *EcoR* I — tct cga gga tcc gaa ttc cag	Ampicillin	LEU2	ADH1	none	2 μ	(14, 31, Clontech)
pGAD424[c]	*EcoR* I, *Sma* I, *BamH* I, *Sal* I, *Pst* I, *Bgl* II — gaa ttc ccg ggg atc cgt cga cct gca gag atc tat	Ampicillin	LEU2	ADH1	none	2 μ	(14, 31, Clontech)
pGAD GL	*Spa* I, *BamH* I, *Sma* I, *EcoR* I, *Sal* I, *Xho* I, *Apa* I — gaa cta gta gat ccc ccg ggc tgc agg aat tcg ata tca agc tta tcg ata tca agc ggg ggc ccg	Ampicillin	LEU2	ADH1	none	2 μ	(14, 31, Clontech)
pGAD GH	*Spe* I, *BamH* I, *Sma* I, *EcoR* I, *Cla* I, *BamH* I, *Sac* I, *Xho* I, *Apa* I — gaa cta gtg gat ccc ccg ggc tgc agg aat tcg ata tca agc tta tcg ata tca agc ggg ggc ccg	Ampicillin	LEU2	ADH1	none	2 μ	(14, 31, Clontech)
pGADT7[a]	*Not* I, *Sfi* I, *EcoR* I, *Sma* I, *BamH* I, *Sal* I, *Xho* I — cat atg gcc atg gag gcc agt gaa ttc cac ccg ggt ggg atc cat cga gct cga gct gca gat	Kanamycin	LEU2	ADH1	HA	2 μ	(31, Clontech)

[a] These vectors contain a T7 promoter upstream of the insert and the epitope tag for coupled *in vitro* transcription translation reactions.
[b] There are two *Sma* I sites in the multiple cloning site of these vectors.
[c] See also reference 29.

177

10). An *Arabidopsis* library from mature *Arabidopsis* plants and a peptide two-hybrid library can be purchased from Clontech.

2.4 Practical considerations

The choice between the two-hybrid systems described above is commonly determined by a number of factors.

1. Some proteins when fused to a DNA-binding domain function as transcriptional activators in the two-hybrid system (autoactivators). These autoactivators cannot be used for two-hybrid screens. If the bait construct is an autonomous activator, one should consider using the other two-hybrid system since the bait protein can behave differently in a different protein fusion context (see also *Protocol 6*, step 3). Autoactivation in yeast does not automatically imply that the protein under study is a transcriptional activator in the organism it is derived from.

2. In the LexA-system, expression of the prey protein is induced by galactose and repressed by glucose from the *GAL1* promoter. This configuration facilitates work with proteins that are toxic to yeast. Before the induction of expression of a potential toxic prey protein, the transformants can be grown as colonies that allow assessment of reporter gene activation. The inducibility of the prey protein also facilitates discrimination between reporter gene activation in the presence and absence of prey protein expression.

3. Constructing and characterizing a cDNA library for the two-hybrid system is technically challenging and time-consuming. Therefore, many researchers prefer to screen available libraries before they venture into making their own library. The type of library available will often determine which vector system is to be used for the interaction screening.

3 Working with yeast—the techniques for the yeast two-hybrid system

The strategy for two-hybrid library screens with the LexA-system differs slightly from screens with the GAL4-system. However, most methods are common to both systems. These methods are described here, while the next section provides a detailed outline for the specific screening strategies for each system.

3.1 Growth and maintenance of yeast

Saccharomyces cerevisiae yeast strains are commonly grown on yeast extract and peptone supplemented with dextrose or glucose as a carbon source (YEPD or YPD; see *Protocol 1*). Under optimal growth conditions, 28–30 °C and good aeration, yeast has a doubling time of 90 min. The vectors used in the two-hybrid system (*Tables 1* and *2*) are shuttle vectors that can be maintained in both bacteria and yeast. For cloning and DNA amplification, they are used like any other bacterial vector. For maintenance in yeast, most of the vectors carry an amino acid marker gene and the 2 μ replication origin. The yeast strains used in the two-hybrid system (*Table 3*) carry several mutations that confer amino acid auxotrophy. After transformation of yeast with a plasmid that complements an amino acid auxotrophy, the transformed yeast is grown on synthetic dropout (SD) medium, which lacks that specific amino acid. Transformed yeast has to be maintained on SD media to select for the transformed plasmid. However, yeast loses plasmids at a lower rate than bacteria and it is possible to grow yeast for short periods in full medium without selection (e.g. for mating, see *Protocol 9*).

Yeast can be stored for several months at 4 °C on solid medium. For long-term storage, glycerol stocks are prepared by mixing an overnight yeast culture with an equal volume of sterile 50% (v/v) glycerol. The glycerol stocks can then be stored at −80 °C for several years.

Protocol 1

Preparation of yeast growth media

Reagents

- Bacto-tryptone
- Yeast extract
- Yeast nitrogen base without amino acids (Difco)
- Bacto-agar
- 40% (w/v) glucose (autoclaved)

- 40% (w/v) raffinose (autoclaved)
- 40% (w/v) galactose (autoclaved)
- Autoclaved amino acid drop out solutions (see *Protocol 1C*)
- 1 M 3-amino-1,2,4-triazole (3-AT), filter sterilized (for use with the GAL4-system only)

Methods

(A) YPD medium

1 Weigh out 20 g Bacto-tryptone, 10 g yeast extract, and 20 g Bacto-agar (for plates only).

2 Add to 950 ml and adjust the pH to 5.8.

3 After autoclaving, allow the medium to cool to 50 °C before adding 50 ml/l 40% (w/v) glucose to a final concentration of 2%.

(B) Synthetic drop-out (SD) glucose medium

1 Weigh out 6.7 g of yeast nitrogen base without amino acids and 20 g Bacto-agar (for plates only).

2 Add to 850 ml H$_2$O and adjust the pH to 5.8.

3 After autoclaving, cool the medium to 50 °C and add 100 ml of the appropriate 10× amino acid drop out solution, 10 ml of the appropriate 100× amino acid to supplement the dropout solutions as determined by the selection criteria of the specific plasmid vector (see *Protocol 1D*), and 50 ml 40% glucose solution.

4 The resulting SD media lacking for example leucine, tryptophan, and histidine would be referred to as Glu SD/-Leu/-Trp.

In the GAL4-system, 3-AT is included in the medium when testing for the activity of the HIS3 reporter gene (see *Protocol 11*).

(C) Synthetic drop-out (SD) galactose induction medium

1 Weigh out 6.7 g yeast nitrogen base without amino acids and 20 g Bacto-agar (for plates only).

2 Add to 825 ml H$_2$O and adjust the pH to 5.8.

3 After autoclaving, cool the medium to 50 °C and add 100 ml of the appropriate 10× amino acid drop out solution, 10 ml of the appropriate 100× amino acid to supplement the dropout solutions as determined by the selection criteria of the specific plasmid vector (see *Protocol 1D*), and 50 ml 40% galactose and 25 ml of a 20% raffinose stock solution.

4 The resulting SD media lacking for example uracil, leucine, and tryptophan would be referred to as Gal/Raff SD/-Ura/-Leu/-Trp

(D) Amino acid drop out solutions

Prepare a 10× amino acid drop out mix that contains all amino acids that are not required for the selection of marker genes in the respective two-hybrid system. For example, a mix omitting tryptophan, histidine, uracil, and leucine for the LexA-system, or a mix lacking tryptophan, leucine, and histidine for the GAL4 system.[b] In addition, a 100× mix for each of the omitted amino acids should be prepared to supplement the 10× drop out media when required. The amino acid mixes are autoclaved and can be stored at 4 °C for several months up to 1 year.

Protocol 1 continued

From the following list, omit those amino acids that will be used for plasmid selection from the 10× stock and make separate 100× stocks of these amino acids. We obtain these reagents from Sigma:

10× amino acid stock: 200 mg/l adenine hemisulfate salt, 200 mg/l L-arginine HCl, 200 mg/l L-histidine HCl monohydrate, 300 mg/l L-isoleucine, 1000 mg/l L-leucine, 300 mg/l L-lysine HCl, 200 mg/l L-methionine, 500 mg/l L-phenylalanine, 2000 mg/l L-threonine, 200 mg/l L-tryptophan, 300 mg/l L-tyrosine, 200 mg/l uracil, 1500 mg/l valine.

100× histidine stock: 200 mg/100 ml L-histidine HCl monohydrate

100× leucine stock: 1000 mg/100 ml L-leucine

100× tryptophan stock: 200 mg/100 ml L-tryptophan

100× uracil stock: 200 mg/100 ml uracil

[a] Solid SD medium cannot be re-autoclaved or be kept for long times at high temperatures as the agar will become mushy. Yeast grows best on dry solid medium plates. Therefore, solid media should always be prepared well in advance and the plates should be dried in a flow hood.

[b] Some commercial suppliers provide premixed amino acid powders for the two-hybrid system.

3.2 Yeast transformation

Below are two protocols for small and large scale yeast transformation (21). The large-scale transformation procedure is only required for cDNA library transformation, while the small-scale transformation procedure is sufficient for all other applications.

Protocol 2

Small scale yeast transformation

Reagents

- Yeast strain and transforming plasmids
- SD plates (see *Protocol 1*)
- YPD liquid medium (see *Protocol 1*)
- 10× LiAc: 1 M lithium acetate, pH 7.5 (dilute acetic acid)
- 10× TE: 100 mM Tris–HCl, pH 8.0; 10 mM EDTA
- 50% PEG: 50% (w/v) polyethylene glycol, MW = 3350
- 10 mg/ml sheared salmon sperm DNA[a]
- 100% DMSO

Method

1   Confirm that the yeast strain to be used shows the expected auxotrophy phenotype by growing it on plates containing SD media lacking His, Leu, Trp, or Ura, as appropriate at 30 °C for 3–5 days.

2   Grow a 50 ml yeast culture in a conical flask overnight in YEPD (or appropriate SD medium when transforming a yeast strain already containing a plasmid) at 30 °C and 200–250 rpm to stationary phase (OD_{600} = 1.5–2).

3   The following day, dilute the culture 1 in 4 with either YPD or appropriate SD medium, and grow 50 ml of this culture at 30 °C and 200–250 rpm for 2–3 h. A 50 ml yeast culture is sufficient for approximately 10 transformations.

Protocol 2 continued

4 Prepare the following solutions from the stock solutions:

15 ml $1\times$ TE/$1\times$ LiAc;

5 ml 40%PEG/$1\times$ LiAc/$1\times$ TE (make fresh every time);

15 ml $1\times$ TE.

5 Pellet the cells by centrifugation for 5 min at 1000 g resuspend in sterile H_2O to wash the pellet.

6 Recentrifuge the culture for 5 min at 1000 g and resuspend the pellet in 500 μl H_2O.

7 Make 50 μl aliquots of yeast cells in 1.7-ml microcentrifuge tubes and add 1 ml $1\times$ TE/$1\times$ LiAc to each tube.

8 Pellet the cells for 1 min at 1000 g in a microcentrifuge, discard the supernatant and resuspend the pellet in 50 μl $1\times$ TE/$1\times$ LiAc solution.

9 Add 5 μl of 10 mg/ml salmon sperm DNA and 300 μl of the 40% PEG/$1\times$ LiAc/TE solution to the cells.

10 Add approximately 1 μg of plasmid DNA (5 μl of a DNA miniprep can be used for transformation of a single construct). For simultaneous transformation of several plasmids, double the amount of DNA for both constructs. The transformation efficiency depends on the yeast strain used.

11 Mix the cells and DNA well by pipetting up and down and incubate the mixture for 30 min at 30 °C with shaking at 200–250 rpm.

12 (Optional: before the heat shock, carefully add 35 μl DMSO to the transformation mix.) Heat shock the cells for 15 min at 42 °C.

13 Pellet the cells at 1000 g in a microcentrifuge for 1 min and resuspend the pellet in 1 ml $1\times$ TE.

14 Pellet the cells again by centrifugation at 1000 g in a microcentrifuge for 1 min and resuspend the pellet in 200 μl $1\times$ TE.

15 Plate 180 and 20 μl of the cells on the appropriate SD medium (see *Protocol 1*). Colonies will be visible 2–3 days after transformation.

[a] The salmon sperm DNA for yeast transformation can be prepared either from lyophilized DNA

Protocol 3

Library scale yeast transformation[a]

Reagents

- Yeast strain and transforming plasmids
- SD plates (see *Protocol 1*)
- YPD liquid medium (see *Protocol 1*)
- $10\times$ LiAc: 1 M lithium acetate, pH 7.5 (dilute acetic acid)
- $10\times$ TE: 100 mM Tris-HCl, pH 8.0; 10 mM EDTA

- 50% PEG: 50% (w/v) polyethylene glycol, MW = 3350
- 10 mg/ml sheared salmon sperm DNA[b]
- 100% DMSO

Method

1 Confirm that the yeast strain to be used shows the expected auxotrophy phenotype by growing it on plates containing SD media lacking His, Leu, Trp, or Ura, as appropriate, at 30 °C for 3–5 days.

Protocol 3 continued

2 Grow a 150 ml yeast culture in YEPD or SD medium[c] in a conical flask at 30 °C and 200–250 rpm overnight to stationary phase (OD 600 = 1.5–2).

3 Pellet the cells at 1000 g for 5 min and resuspend the pellet in fresh YEPD. Use the entire culture to inoculate one litre of YEPD in a large conical flask.

4 Grow the 1 l culture at 30 °C and 200–250 rpm for 2–3 h until OD 600 = 0.5–1.

5 Prepare the following solutions from the stock solutions:

10 ml 1× TE/1× LiAc

100 ml 40%PEG/1× LiAc/1× TE

10 ml 1× TE

6 Pellet the cells by centrifugation at 1000 g for 5 min and resuspend the pellet in 500 ml H$_2$O.

7 Pellet the cells again by centrifugation at 1000 g for 5 min and resuspend the pellet in 8 ml 1× TE/1× LiAc.

8 Add between 100 and 300 μg of transforming DNA,[d] 2 ml 10 mg/ml salmon sperm DNA and 60 ml 40% PEG/1× LiAc/1× TE to the 8 ml of resuspended yeast cells. Mix the contents well with a pipette.

9 Incubate the cells for 30 min at 30 °C with shaking at 200–250 rpm.

10 Slowly add 7 ml DMSO to the cells and mix gently.

11 Heat shock the cells for 15 min at 42 °C. Invert the tubes occasionally during the heat shock to mix the culture and ensure heat exchange.

12 Pellet the cells by centrifugation at 1000 g for 5 min. Resuspend the pellet in up to 5 ml 1× TE.

13 Plate the cells on appropriate selective SD medium agar plates.[e]

14 In order to calculate the transformation efficiency, plate serial dilutions of the transformed cells onto appropriate selective SD medium plates.

15 Colonies will become visible after 2–3 days.

[a] All procedures are carried out at room temperature unless otherwise stated.

[b] The salmon sperm DNA for yeast transformation can be prepared either from lyophilized DNA using published protocols (22) or purchased ready for transformation.

[c] For sequential transformation as recommended for the LexA-system use a culture grown in SD medium.

[d] For cotransformation of the bait plasmid and prey library in the GAL4-system, use up to 1 mg of bait DNA. As a rule of thumb, use twice as much bait as prey DNA. This ensures that all cells receiving a prey plasmid will also have a bait plasmid.

[e] One transformations will require between 20 and 30 15cm diameter agar plates.

3.3 Measurement of *LacZ* reporter gene activity

While the expression of amino acid reporter genes in yeast is directly assessed by the ability of the yeast to grow on a medium lacking the respective amino acid, a number of qualitative and quantitative assays are available to measure *LacZ* reporter gene activity. A fast indication of whether a particular combination of bait and prey constructs can mediate expression of the *LacZ* gene, can be obtained through colony staining on the plate with XGAL as a substrate (*Protocol 4*). However, only the yeast strains used in the LexA system take up XGAL directly from the medium. In the more sensitive colony lift assay, which is suitable for both versions of the two-hybrid system, filter lifts of the yeast colonies are prepared (*Protocol 5*). First, the cells are permeabilized by crack-freezing in liquid nitrogen and then stained for β-galactosidase activity. Both assays provide a first semi-quantitative measurement of the *LacZ* expression levels. Quantitative β-galactosidase assays using fluorescent or luminescent substrates like ONPG or GalactonStar can be undertaken (*Protocols 6 and 7*). When performing a *LacZ* expression assay, it is good practice to include an established bait/prey interactor pair as positive control. Lamin C is often used as a non-interacting negative control protein (1–3, 23, 24). The four different β-galactosidase assays for monitoring or *LacZ* gene expression are presented in *Protocols 4–7*.

Protocol 4

Qualitative plate assay for β-galactosidase activity

This protocol is only suitable for the LexA system and cannot be used for the GAL4 system as the cells used for the GAL4 system do not directly take up XGAL.[a]

Reagents

- SD agar plates (see *Protocol 1*)
- 40 mg/ml XGAL (5-bromo-4-chloro-3-indoyl-β-D-galactopyranoside in dimethylformamide; prepare this solution fresh every time.
- 10× BU salts: 70 g/l $Na_2HPO_4.7H_2O$, 30 g/l NaH_2PO_4, pH 7.0; filter sterilized.

Method

1 Prepare solid Gal/Raff SD/-Ura and Glu SD/-Ura as described (see *Protocol 1*), but add 750 ml H_2O instead of 850 ml for glucose medium or 725 ml H_2O instead of 825 ml for galactose/raffinose medium.

2 When the medium has cooled to 50 °C, add 100 ml 10× BU salts and 2 ml XGAL solution per litre. Mix the contents well and pour the plates.

3 Replicate or streak colonies to be analysed on both the Gal/Raff SD/-Ura and Glu SD/-Ura XGAL medium plates, and incubate at 30 °C. When staining becomes visible, compare the blue colour formation on the two different media. The development of the blue coloration may take some time and the staining is usually most informative after 2–3 days.

[a] Including XGAL in the medium slows down growth of yeast colonies.

Protocol 5

Qualitative colony-lift assay for β-galactosidase activity

This protocol can be used with either the GAL4 system or the LexA system

Reagents

- SD agar plates (see *Protocol 1*)
- 3MM Whatman paper No. 1
- LacZ buffer: 16.1 g/l $Na_2HPO_4.7H_2O$, 5.5 g/l $NaH_2PO_4.H_2O$, 0.75 g/l KCl, 0.246 g/l $MgSO_4.7H_2O$, pH 7.0, and sterilize by autoclaving
- 20 mg/ml XGAL (5-bromo-4-chloro-3-indoyl-β-D-galactopyranoside in dimethylformamide (DMF); make this solution fresh every time
- β-mercaptoethanol
- Liquid nitrogen

Protocol 5 continued

Method

1 Replicate or double streak the clones on solid Glu SD agar plates lacking appropriate amino acids when using the Gal4 system, and both Glu SD and Gal/Raff SD agar plates lacking appropriate amino acids when using the LexA system.

2 After 2–3 days, when the cells form dense colonies, make a lift from both plates using Whatman 3MM filter paper. Make sure to obtain a good and even transfer by pushing the filter paper down with a dry sponge or any other suitable item. If necessary, mark the filter position on the plate to be able to identify the colonies after the staining.

3 Prepare the LacZ staining solution:

100 ml LacZ buffer

0.27 ml β-mercaptoethanol

1.67 ml 20 mg/ml XGAL solution

4 For each plate of colonies to be analysed, soak a Whatman filter paper in LacZ staining solution and place the filter paper in a Petri dish. Remove any excess remaining staining solution so that the filter papers are not too wet and all filter papers are incubated in a similar amount of staining solution.

5 Lift the filter paper from the yeast plate and freeze it in liquid nitrogen for 2–3 min. Let the filter paper thaw and then place it carefully—colony-side up—on the pre-soaked filter from step 4. Avoid trapping air bubbles between the two filter papers.

6 Incubate the filters at 30°C. Strong positive clones stain blue after 30 min, but weak positives may only be visible after 8–12 h of staining. Use the staining of the positive and negative controls as an indication for when to stop the incubation.

7 Air dry the filters in a fume hood as β-mercaptoethanol is present in the staining solution.

Protocol 6

Fluorescent ONPG assay for β-galactosidase activity

Equipment and reagents

- Spectrophotometer
- Liquid SD medium (see *Protocol 1*)
- Liquid YEPD medium (see *Protocol 1*)
- LacZ buffer (see *Protocol 5*)
- LacZ buffer/β-mercaptoethanol buffer: 270 μl /β-mercaptoethanol/100 ml LacZ buffer
- 1 M Na_2CO_3

- 4 mg/ml o-nitrophenyl β-D-galactopyranoside in LacZ buffer: prepare the solution fresh before each use but ensure sufficient time for the o-nitrophenyl β-D-galactopyranoside to dissolve .
- Liquid nitrogen

Method

1 Inoculate 5 ml of the appropriate SD liquid medium with a yeast colony. Use Glu SD and Gal/Raff SD medium lacking the appropriate amino acids when using the LexA-system and Glu SD medium lacking the appropriate amino acids when using the Gal4 system. Use 4–8 replicates for each clone to be analysed.

2 Incubate the cultures overnight at 30°C with shaking at 200–250 rpm.

Protocol 6 continued

3 The next day, inoculate 2 ml of the overnight culture into 8 ml of YEPD medium and grow for 3–5 h until mid-log phase ($OD_{600} = 0.5$–0.8).

4 Vortex the culture to dissolve cell clumps and measure the OD_{600}.

5 Split the cultures into three 1.5 ml aliquots in microcentrifuge tubes for three replicate measurements and pellet the cells by centrifugation for 1 min at 1000 g.

6 Remove the supernatant, resuspend the cells and wash them in 1.5 ml LacZ buffer.

7 Centrifuge the cells 1 min at 1000 g, remove the supernatant and resuspend the pellet in 300 μl LacZ buffer.

8 Transfer 100 μl of the cells into a fresh 1.5 ml microcentrifuge tube. Set up a blank with 100 μl LacZ buffer only.

9 Rupture the cells by freezing the tubes in liquid nitrogen for 1 min and subsequently thaw them for 1 min in a 37 °C water bath.

10 Add 0.7 ml of LacZ/β-mercaptoethanol buffer to the cell suspension.

11 Add 0.16 ml of the ONPG solution to both the cell suspension and the blank. Start a timer and place the tubes in a 30 °C water bath.

12 When yellow colour develops,[a] add 0.4 ml 1 M Na_2CO_3 to stop the reaction and record the elapsed time.

13 Centrifuge the cells at 1000 g for 10 min in a microcentrifuge to pellet the cell debris.

14 Carefully transfer the supernatant to a cuvette and measure the OD_{420}.[b]

15 Calculate the β-galactosidase activity in Miller units using the following equation:

Miller units = $1000 \times OD_{420}/(t \times V \times OD_{600})$

where t = elapsed time in min, V = 0.1 ml × 5 (concentration factor from steps 6 and 7 of this protocol), and $OD_{600} = A_{600}$ of 1 ml of culture.

[a] The time taken for the yellow colour to develop will be different for each construct tested.

[b] To be within the linear range of the assay, the OD_{420} of the samples has to be between 0.2–1.

Protocol 7

Quantitative chemiluminescent GalactonStar Plus™ assay for β-galactosidase activity

Equipment and reagents

- Luminometer or scintillation counter
- Spectrophotometer
- Galacto-Light Plus™ kit (Tropix Inc.)
- LacZ buffer (see *Protocol 3*)
- Bradford assay reagent (BioRad)
- 425–600 μm diameter acid washed glass beads (Sigma)

Method

1 Inoculate 5 ml of the appropriate SD medium lacking amino acids with a yeast colony. Use Glu SD lacking appropriate amino acids for the Gal4 system, and both Glu SD and Gal/Raff SD medium lacking appropriate amino acids for the LexA system. Use 4–8 replicates for each clone to be analysed.

Protocol 7 continued

2 Incubate the culture overnight at 30 °C with shaking at 200–250 rpm

3 The following day, add 1 ml of culture to 4 ml of fresh medium and incubate 5 ml of the diluted culture for an additional 2 h.

4 Prepare the reaction buffer by diluting the Galacton-Plus™ substrate 100-fold with Galacto-Light™ reaction buffer diluent. Make sure all reagents are at room temperature. Only prepare the amounts necessary for the experiment.

5 Pellet 1 ml of the yeast cultures from step 3 by centrifugation at 1000 g for 1 min.

6 Resuspend the pellet of cells in 300 μl LacZ buffer and add approximately 100 μl glass beads.

7 Vortex the cells with the glass beads vigorously five times for 1 min each time, cooling the tubes on ice for 1 min after vortexing.

8 Pellet the cell debris and the glass beads by centrifugation in a microcentrifuge at 10 000 g for 5 min.

9 Carefully remove the supernatant containing the protein extract from the pellet and place it in a separate microcentrifuge tube.

10 Mix 20 μl of the protein extract with 70 μl of Galacton-Star™ reaction buffer and incubate at room temperature for 1 h.[a]

11 Add 100 μl of Light Emission Accelerator to the reactions.[b]

12 Measure relative light units (RLU) over a 5-s interval in a luminometer or microplate luminometer.

13 Perform a Bradford protein assay with the total cell extract and calculate the amount of protein in the 20 μl volume of cell extract used in the assay (see *Chapter 8*).

14 Calculate the RLU per mg protein.

[a] For large numbers of assays, and if a microplate luminometer is available, this can be set up in 96-well microtitre plates.

[b] With the Galacto-Light Plus™ kit, light emission will be stable for 1 h.

3.4 Recovering plasmid DNA from yeast

After a two-hybrid cDNA library screen it is important to isolate the prey cDNA construct from the yeast cells. However, at this stage, the yeast cells also contain the bait plasmid and, with the LexA system, also the reporter vector. One of the following strategies can be chosen to facilitate selection of the prey plasmid:

1. Some bait/prey vector combinations have different antibiotic resistance genes (see *Table 2*) and after isolation of the plasmid DNA from yeast, bacterial transformants can be separated based on antibiotic resistance.

2. The positive yeast clone can be grown for several days in liquid SD medium without selection for the bait and the reporter plasmids. Plating the yeast on solid SD medium with the appropriate amino acids for selection of the prey plasmid, and subsequent re-streaking of colonies on appropriate SD plates, will allow the identification of yeast clones that have lost the bait and where relevant the reporter plasmid. Plasmid preparations from these colonies should then contain only the prey construct. Furthermore, yeast clones selected in this way can be used for further testing of the putative positive clone by mating them with other yeast clones containing control constructs (see *Protocol 5*). The presence of the *Cyh2* susceptibility gene on the bait vector will render the yeast strain

susceptible to cycloheximide (see *Table 3*). When plated on cycloheximide medium (1 μg/μl for CG-1945, 10 μg/μl for Y190), growth of cells that have lost the plasmid will be favoured (13).

Table 3 Genotypes of yeast strains commonly used in the two-hybrid system

A. *LexA-based system*

EGY48	*MATα, ura3-52, trp1-901, his3-200, 6lex A$_{UAS}$-LEU2.*	(16)
YM4271[a]	*MATa, ura3-52, his3-200, lys2-801, ade2-101, ade5, trp1-901, leu2-3,112, try1-501, gal4- Δ512, gal80-d538, ade::hisG.*	(32)

B. *GAL4-based system*

AH109[abc]	*MATa, ura3-52, his3-200, lys2-801, ade2-101, trp1-901, leu2-3,112, gal4Δ gal80Δ, LYS2::GAL1$_{UAS}$-GAL1$_{TATA}$-HIS3, GAL2$_{UAS}$-GAL2$_{TATA}$-ADE2, URA3::MEL1$_{UAS}$-MEL1$_{TATA}$-LACZ.*	(25)
CG-1945[ab]	*MATa, ura3-52, his3-200, lys2-801, ade2-101, trp1-901, leu2-3,112, gal4-542, gal80-538, cyh^{r}2, LYS2::GAL1$_{UAS}$-HIS3$_{TATA}$-HIS3, URA3::GAL1$_{UAS}$-GAL1$_{TATA}$-LACZ.*	(33)
Y187[ab]	*MATα, ura3-52, his3-200, ade2-101, trp1-901, leu2-3,112, gal4Δ, gal80Δ, met,⁻ URA3::GAL1$_{UAS}$-GAL1$_{TATA}$-LACZ.*	(1)
Y190[ab]	*MATa, ura3-52, his3-200, lys2-801, ade2-101, trp1-901, leu2-3,112, gal4Δ, gal80Δ, cyh^{r}2, LYS2::GAL1$_{UAS}$-HIS3$_{TATA}$-HIS3, URA3::GAL1$_{UAS}$-GAL1$_{TATA}$-LACZ.*	(1)

[a] Yeast strains carrying the *ade2-101* mutation form redish colonies. White revertants occur at a low frequency due to spontaneous mutations and should not be used.

[b] The GAL4-system makes use of the GAL4 DNA-binding and activation domains. Therefore, the endogenous *GAL4* and *GAL80* genes of the GAL4-system yeast strain are deleted. For this reason the *GAL1* promoter cannot be used for the expression of the bait and the prey protein in the GAL4-system.

[c] This strain has a third reporter gene, *ADE2*, in addition to the LacZ and HIS3 reporter genes.

The *TRP1* and *LEU2* marker genes on the two-hybrid system vectors are functional in bacteria and can be used to rescue tryptophan and leucine auxotroph bacteria such as strains KC8 and HB101 (see *Table 4*).

Table 4 Genotypes of bacterial strains which can be used for the rescue of specific two-hybrid vectors.

KC8	*hsdR leuB600 trpC9830 pyrF::Tn5 hisB463 lacΔ74 strA galU galK*
HB101	*supE44 ara14 galK2 lacY1 Δ(gpt-proA)62 rpsL20 (Str^r) xyl-5 mtl-1 recA13 Δ(mcrC-mrr) HsdS⁻ (r-m⁻)*

Protocol 8

Rescuing prey plasmid DNA from yeast

Equipment and reagents

- Electropulser (e.g. Gene Pulser® II, BioRad) and electrotransformation cuvettes
- Glu SD/-Trp liquid medium when using the LexA-system (see *Protocol 1*)
- Glu SD/-Leu liquid medium when using the GAL4-system (see *Protocol 1*)
- Phenol/chloroform (1:1)
- 425–600 μm diameter acid washed glass beads (Sigma)
- Yeast lysis solution: 2% Triton X-100, 1% SDS, 100 mM NaCl, 10 mM Tris pH8.0, 1 mM EDTA
- TE buffer: 10 mM Tris–HCl pH 8.0, 1 mM EDTA

Protocol 8 continued

- Selection medium for complementation in KC8 or HB101: 15 g Bacto-agar, 1 g $(NH_4)_2SO_4$, 4.5 g, KH_2PO_4, 10.5 g K_2HPO_4, 0.5 g Na-citrate.H_2O per litre of H_2O. Sterilize by autoclaving and when the medium has cooled to 50 °C, add 1 ml 1 M $MgSO_4.7H_2O$, 10 ml of 20 % glucose, 1 ml 1% thiamine-HCl, and 10 ml 10 mg/ml ampicillin. Also add 40 mg/ml each of histidine, leucine, and uracil when using the LexA-system, or 40 mg/ml each of histidine, tryptophan and uracil when using the GAL4-system. For HB101, add also 1 ml of 40 mg/ml proline.
- Electrocompetent *E. coli* KC8 or HB101 cells
- LB ampicillin agar plates: LB plates and top agarose—10 g bactotryptone, 5 g bactoyeast extract, 10 g NaCl make up to 1 l with deionized water. Adjust to pH 7.0 with NaOH. add 15 g agar and autoclave.

Method

1 Grow a 2-ml overnight culture of the yeast clones in the appropriate medium to select only for the presence of the prey construct.

2 Pellet the cells from 1 ml of the culture by centrifugation at 1000 g for 1 min in a microcentrifuge and resuspend the pellet in 300 µl yeast lysis solution.

3 Add 100 µl of glass beads and break the cells by vigorous vortexing for 5 min.

4 Add 300 µl phenol/chloroform and vortex for another min.[a]

5 Separate the phases by centrifugation at 10 000 g in a microcentrifuge for 5 min.

6 Remove the aqueous phase to a new microcentrifuge tube and repeat the phenol/chloroform extraction from step 4.

7 Separate the phases by centrifugation at 10 000 g in a microcentrifuge for 5 min.

8 Remove the aqueous phase to a new microcentrifuge tube and add 2 vols of 100% ethanol and mix.

9 Pellet the DNA by centrifugation at 10 000 g in a microcentrifuge for 15 min at 4 °C.

10 Discard the supernatant and wash the DNA pellet in 70% ethanol.

11 Pellet the DNA by centrifugation at 10 000 g in a microcentrifuge for 5 min and carefully remove the supernatant.

12 Dry the DNA pellet in a rotary vacuum dessicator.

13 Resuspend the pellet in 50 µl TE.[b]

14 Prepare electrocompetent *E coli* KC8 or HB101 cells as previously described (22) for the complementation strategy outlined above.

15 Transform 10 µl of the plasmid DNA from step 13 into the electrocompetent cells following a previously published protocol (e.g. 22).

16 Plate the transformation on LB agar plates containing the appropriate antibiotic for plasmid selection and isolate plasmid DNA from colonies as previously described (22).

17 For the complementation of tryptophan or leucine auxotrophy, re-streak colonies on the appropriate medium.

18 The colonies that grow on selection medium contain the prey plasmid. Prepare plasmid DNA from these colonies. Use the plasmid DNA for sequencing (for sequencing primers see *Table 5*), restriction digestions and control transformation.

[a] Be aware that some glass beads might be trapped between the lid and the tube, which may cause phenol to splash during vortexing. Take the necessary precautions.

[b] The plasmid DNA can also be used for retransformation of yeast to verify the authenticity of the clones or to separate prey plasmids if multiple cDNA prey clones are present within one positive yeast clone.

Table 5 Sequencing and PCR primers for two-hybrid system vectors.

LexA- system	
LexA-DBD forward[a]	5′-CGT CAG CAG AGC TTC ACC ATT G-3′
LexA-AD forward[a]	5′-GAT GTT AAC GAT ACC AGC C-3′
LexA-AD reverse[ab]	5′-GGA GAC TTG ACC AAA CCT CTG GCG-3′
LexA-AD reverse 2[ab]	5′-GCG TGA ATG TAA GCG TGA C-3′
GAL4- system	
GAL4-DBD forward[a]	5′-TCA TCG GAA GAG AGT A-3′
GAL4-AD forward[a]	5′-CTA TTC GAT GAT GAA GAT ACC CCA CCA AAC CC-3′
GAL4-AD reverse[a]	5′-ACA GTT GAA GTG AAC TTG CG-3′

[a] DBD primers are for sequencing of DNA-binding domain vectors while AD-primers are for sequencing and insert amplification of activation domain vectors.

[b] The LexA-AD reverse primer is required for insert amplification from the vectors pJG4-5 (pB42AD). The LexA-AD reverse 2 primer can be used for insert amplification from pYESTrp2.

3.5 Yeast mating

When the interaction between a large number of bait and prey constructs is being tested, it is most convenient to initially transform the bait and the prey vectors separately into yeast strains of differing mating types. For example EGY48 and YM4271, when using the LexA-system, and Y190 and Y187, when using the GAL4-system. In an easy procedure these constructs are then brought together through mating of an a and an α mating type transformant. Transformants containing a single plasmid can be kept as glycerol stocks (see Section 3.1) or on plates and used at a later stage to test other putative interactions. Several laboratories are currently working on approaches that in the near future will allow testing of a bait protein construct under study against a well-defined or catalogued set of prey constructs. For these procedures, yeast mating will almost certainly be the method of choice for fast and reliable screening.

Protocol 9

Yeast mating

Reagents

- Yeast strains with a and α mating type (see *Table 3*)
- Transforming DNA
- YEPD agar plates (see *Protocol 1*)
- SD medium Glu SD/-Ura/-His/-Trp agar plates for the LexA system (see *Protocol 1*)
- SD medium Glu SD/-Trp/-Leu agar plates for the Gal4 system (see *Protocol 1*)

Method

1 Transform the bait and the prey constructs separately into an a and an α mating type strain (see *Protocol 2*).

2 Grow transformants for 2–3 days on the required SD medium to select for the plasmid used. Re-streak individual transformants onto appropriate SD agar plates containing the appropriate amino acids for plasmid selection.

3 When colonies have grown, streak the desired combinations of bait and prey transformants in the form of a cross on a YEPD plate[a] and incubate the plates 30 °C.

4 When the cells have grown sufficiently, usually overnight, streak cells from the centre of the cross onto the appropriate selective SD agar plates[b] and incubate at 30 °C.

Protocol 9 continued

5   After 2 days of incubation, colonies should be visible. These colonies represent diploid yeast cells containing both plasmids. Restreak colonies from these plates onto appropriate SD plates and assay for the expression of the amino acid marker or the *LacZ* reporter gene.

^a It is essential to perform the mating on full medium (YEPD).

^b Use Glu SD/-Ura/-His/-Trp selection for the LexA system and Glu SD/-Trp/-Leu selection for the GAL4-system to select for the presence of both plasmids.

4 Performing two-hybrid screens

4.1 Performing a two-hybrid library screen with the LexA-system

An outline of the protocol is given in *Figure 2*. First, a two-hybrid construct is generated and tested for its suitability for the cDNA library screening (see Section 2.4). After transformation of the two-hybrid cDNA library and the bait construct into yeast, it is advisable to select only for co-transformants and not directly for the activation of reporter genes. Expression of the prey fusion proteins is subsequently induced on galactose containing medium and putative interacting clones are selected for the activation of the LEU2 reporter gene. To further test whether the putative positive clones resulting from this screen are, indeed, from a genuine bait/prey interaction, the clones are re-streaked on medium containing XGAL and either glucose or galactose. Clones are selected which show stronger staining on galactose than on glucose medium. The prey plasmid is then isolated from the positive clones for further tests.

Protocol 10

Performing a yeast two-hybrid screen with the LexA-system

Equipment and reagents

- Glu SD/-Ura/-His agar plates (see *Protocol 1*)
- Glu SD/-Ura/-His/-Trp agar plates(see *Protocol 1*)
- Gal/Raff SD/-Ura/-His XGAL agar plates (see *Protocol 1*)
- Gal/Raff SD/-Ura/-His/-Leu agar plates (see *Protocol 1*).
- Gal/Raff SD/-Ura/-His/-Trp/-Leu agar plates (see *Protocol 1*).
- liquid Gal/Raff SD/-His/-Trp/-Ura medium
- Gal/Raff SD/-Ura XGAL agar plates

- Glu SD/-Ura XGAL agar plates.
- Liquid YEPD medium (see *Protocol 1*)
- Sterile 50% glycerol
- Yeast strains
- Plasmid contructs
- 2× Cracking buffer (optional)
- 100 mM Tris-Cl (pH 6.8), 4% SDS, 0.2% bromophenol blue, 20% glycerol, 200 mM DTT (or β mercaptoethanol)

Methods

(A) Design and characterization of the bait construct

1   Make an in frame translational fusion with the LexA bait vector (see *Table 1*) and the gene of interest. Confirm the construct by DNA sequencing (for sequencing primers see *Table 5*).

2   Co-transform the LexA bait construct together with the LacZ reporter plasmid pSH18–34 into the appropriate yeast cells using the small-scale yeast transformation protocol (see *Protocol 2*).

3   Transform the empty bait vector and pSH18–34 into the appropriate yeast cells using the small-scale yeast transformation protocol (see *Protocol 2*) as a negative control.

Protocol 10 continued

4   Select for co-transformants on Glu SD/-Ura/-His medium.[a]

5   Re-streak or replicate the co-transformants onto inductive Gal/Raff SD/-Ura/-His XGAL and Gal/Raff SD/-Ura/-His/-Leu plates. If the yeast colonies do not grow on the selective plates lacking leucine and do not stain blue on the XGAL medium, then the bait construct is suitable for library screening (continue with *Protocol 10B*). If the colonies grow on medium lacking leucine and/or stain blue on XGAL plates, then the construct is autoactivating. In this case, several options are available:

(a) try a less sensitive reporter plasmid e.g. pJK103 (see *Table 1b*).

(b) try to delete the region of the clone that encodes the domain that confers autoactivation— there are no rules how to do this but it is worthwhile testing a number of constructs.

(c) if it is not convenient to modify the protein by deletion, consider switching to the GAL4-two-hybrid system—fusion proteins can behave differently in different fusion contexts.

(d) if the autoactivation is weak, it is in some cases possible to identify strong interactors despite the background.

(B) Analysis of bait construct by Western blotting (optional)

1   Grow a 2-ml culture of yeast overnight in appropriate selective media to an $OD_{600} = 2$.

2   Pellet the cells by centrifugation at 1000 g in a microcentrifuge for 10 min.

3   Resuspend the pellet in 200 μl 2× cracking buffer.

4   Boil cells for 10 min and perform a Western blot as previously described (22).[b]

(C) Library transformation

1   If the cDNA library is available as a single-library plasmid stock, perform a library scale transformation to introduce the library into cells carrying the bait plasmid (see *Protocol 3*). If your cDNA is aliquoted in small batches after a library amplification, perform several (approximately 30 small scale transformations) and plate each transformation on large Petri dishes.

2   Plate the transformation on Glu SD/-Ura/-His/-Trp medium and incubate for several days at 30 °C.[c] Also plate several serial dilutions to calculate the transformation efficiency.

3   When the transformants are visible (usually after 2–3 days), calculate the transformation efficiency from the serial dilutions. It is desirable to screen at least 10^6 independent cotransformants.

4   Using a sterile glass scraper, scrape the cells from one plate into a 15-ml tube.

5   Pellet the cells by centrifugation at 1000 g for 5 min and wash the cells once in YEPD.

6   Resuspend the cells in 5 ml YEPD and add an equal volume of sterile 50% glycerol. Make 1 ml aliquots and store the glycerol stocks at −80 °C (see Section 3.1).[d]

7   Dilute an aliquot of the glycerol stock to an $OD_{600} = 0.5$ in liquid Gal/Raff SD/-His/-Trp/-Ura medium to induce expression of the prey fusion protein.

8   Grow the liquid culture at 30 °C with 200–250 rpm for 4–6 h.

9   Plate approximately 10 times the number of cells than the number of calculated co-transformants to make sure to represent all co-transformants, and to account for losses due to freezing. Plate 10^6 cells/100 mm plate on Gal/Raff SD/-Ura/-His/-Trp/-Leu medium and incubate at 30 °C.

10   Check plates daily for growth. Putative positive clones will start growing after 3 days (but incubate the plates for a total of 7 days). Positive clones will form solid colonies that turn pink or red after prolonged incubation (provided the yeast strain carries the *ade2–101* mutation).[e]

11   Pick the putative positive clones and re-streak onto Gal/Raff SD/-Ura/-His/-Trp/-Leu agar plates and grow for 2–5 days.

Protocol 10 continued

12 Re-streak or replicate the putative positive clones onto Gal/Raff SD/-Ura XGAL and Glu SD/-Ura XGAL plates. Also include a positive and a negative control to evaluate the XGAL staining.[f]

13 Over a period of 1–3 days check regularly for blue staining. Screen for clones that accumulate blue stain faster on galactose than on glucose medium.

14 PCR amplify the cDNA inserts of the prey vectors directly from the positive yeast clones. For this purpose resuspend a small amount of yeast in 50 μl H$_2$O and boil the yeast for 5 min. Use 5 μl of this mix for the PCR. Classify the prey clones by restriction digestion with frequent cutting enzymes, e.g. *Alu* I and *Hae* III. The PCR products can be used directly for the restriction digestion.

15 Isolate the bait plasmid DNA from the positive yeast clones (see *Protocol 8*).

16 Retransform the crude plasmid preparation into a yeast strain carrying the reporter and the bait construct.[g]

17 Select for co-transformants using appropriate selective media and assay again for *LacZ* expression as in step 12.[h] The corresponding empty vector construct and an unrelated bait construct should serve as negative controls.[f] A true interactor will show specific interaction with the bait construct and not with the empty vector construct alone or an unrelated protein.

18 Sequence the plasmids encoding the true interactors (for primers see *Table 5*).

[a] If available, test a known interaction partner in the two-hybrid system to see whether interaction can be detected with this particular fusion construct.

[b] Antibodies are available from several commercial suppliers.

[c] Using non-inductive conditions during this pre-amplification step maximizes the plasmid copy number in each cell and permits higher fusion protein expression under inductive conditions.

[d] Keep co-transformants from different library batches separate.

[e] Depending on the yeast strain and the bait construct, some background growth can be observed at this stage, but the size of these colonies will never exceed 1–2 mm.

[f] See references 1–3, 23, 24 for positive and negative controls.

[g] This step can also be done by mating (see also *Protocol 9*).

[h] This step is important as primary transformants may contain more than one prey plasmid and the contaminating prey plasmids can be singled out in this step. This step also confirms the authenticity of the interactions.

4.2 Performing a two-hybrid screen with the GAL4-system

Screening a GAL4-system two-hybrid library is similar to the procedure described for the LexA-system. A major difference between the LexA-system is that the *LacZ* and *HIS3* reporter genes are both stably integrated into the genome (25). Therefore, the bait and the prey cDNA library can be co-transformed as an alternative to the sequential transformation of the bait and the prey construct in the LexA-system. The resulting co-transformants are directly screened for putative interactors without the amplification step. For these reasons, performing a GAL4-system cDNA library screening is less lengthy than screening with the LexA-system. For an outline of the GAL4-system screening procedure, refer to *Figure 3*.

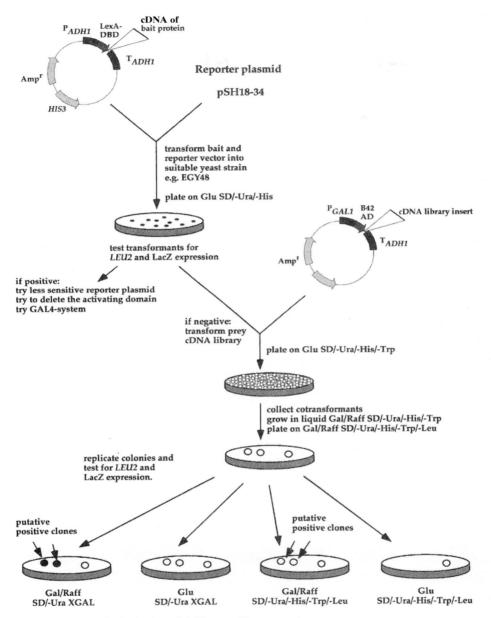

Figure 2 Flow chart for the LexA–two-hybrid system library screening.

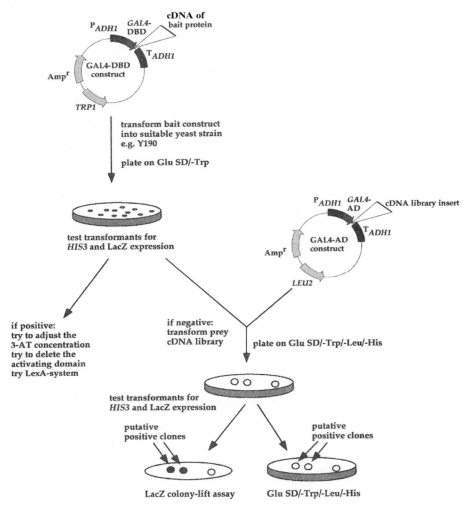

Figure 3 Flow chart for the GAL4–two hybrid system library screening.

Protocol 11

Performing a yeast two-hybrid screen with the Gal4-system

Equipment and reagents

- Glu SD/-Trp/-Leu/-His agar plates containing 3-AT[b] (see *Protocol 1*)

- Glu SD/-Trp/-Leu agar plates (see *Protocol 1*)

Method

Design and characterization of the GAL4-fusion protein

1   Construct an in-frame translational fusion with the GAL4-DNA binding domain using one of the vectors described in *Table 2*.

Protocol 11 continued

2 Verify the construct by DNA sequencing.

3 Transform yeast as outlined in *Protocol 2* with the bait construct and the empty prey vector or a negative control prey construct. Use the empty bait vector as an additional negative control.

4 Select for co-transformants on Glu SD/-Trp/-Leu medium.

5 Restreak the cotransformants on Glu SD/-Trp/-Leu/-His (+ 3-AT).[b] Adjust the 3-AT concentration if necessary to optimize the selection in the absence of prey library. If the bait construct does not auto-activate and the optimal 3-AT concentration has been determined, perform the library transformation (continue with step 6). If the bait construct is auto-activating, see *Protocol 10A*, Step 5.

6 Optional: Check the expression of the bait fusion protein on a Western blot (see *Protocol 10B*).

7 Transform the cDNA library following the sequential or simultaneous library transformation protocol (see *Protocol 3*).

8 Select for expression of the *HIS3* reporter gene by plating on Glu SD/-Trp/-Leu/-His (+ 3-AT)[b] and grow cells for up to 8 days at 30 °C.

9 Putative positive clones will start appearing after 3–4 days.[c]

10 After 8 days, pick the putative positive clones and restreak on Glu SD/-Trp/-Leu/-His (+ 3-AT) to give single colonies.

11 Restreak the putative positive clones on Glu SD/-Trp/-Leu and grow to high density. Then perform filter lift LacZ assay as described in *Protocol 5*.

12 Continue with control experiments and the isolation of the prey plasmid as outlined in *Protocol 10*, steps 14–18.

[a] If available, test a known interaction partner in the two-hybrid system to see whether interaction can be detected with this particular fusion construct.

[b] 3-AT is added as yeast can grow to some extent on medium lacking histidine. This leakiness can be suppressed by 3-AT. The required concentrations of 3-AT vary between different strains and should be titrated in 5 mM steps for each strain. To some extent, increasing the 3-AT concentration also suppresses background activation conferred by the bait construct. As a general guideline, CG1945, Y190, and HF7c require around 5 mM, 15 mM, and 25 mM 3-AT, respectively, for suppression of the *HIS3* gene.

[c] Background growth may be observed depending on the bait, the yeast strain, and 3-AT levels. Background colonies never exceed 1–2 mm in size.

4.3 Studying interactions between two known proteins

The two-hybrid system can also be employed to study the interaction between a known set of proteins or their individual domains (7–10). The LexA as well as the GAL4 system can be employed for this approach and the general strategy is identical to the one described for two-hybrid screening. All steps can be carried out using the small scale transformation procedures. When large numbers of bait and prey constructs are to be assayed, the mating strategy (see *Protocol 5*) should be given preference over the cotransformation approach outlined below.

Protocol 12

Studying interactions between two known proteins

Equipment and reagents

- Gal/Raff SD/-His/-Ura agar plates when using the LexA-system (see *Protocol 1*).[a]
- Glu SD/-Leu agar plates when using the GAL4-system (see *Protocol 1*).[a]
- Gal/Raff SD/-His/-Leu/-Ura agar plates when using the LexA-system (see *Protocol 1*).
- Glu SD/-Leu/-His (+ 3AT) agar plates when using the GAL4-system (see *Protocols 1 and 11*)
- Glu SD/-His/-Trp agar plates when using the LexA-system (see *Protocol 1*).
- Glu SD/-Trp/-Leu agar plates when using the GAL4-system (see *Protocol 1*)
- Glu SD/-His/-Trp/-Leu agar plates when using the LexA-system (see *Protocol 1*).
- Gal/Raff SD/-His/-Trp/-Leu agar plates when using the LexA-system (see *Protocol 1*).
- Glu SD/-Trp/-Leu/-His (+ 3AT) agar plates when using the GAL4-system (see *Protocol 1*)

Method

1 Generate fusion constructs by fusing one of the interaction partners to the DNA-binding domain vector and the other partner to a compatible activation domain vector (see *Tables 1* and *2*). The correct open reading frame must be maintained between the fused domains. Check the construct by DNA sequencing (see *Table 4* for sequencing primers).

2 Transform the DNA-binding domain construct(s) (together with the pSH18–34 reporter plasmid in the LexA-system) into the desired yeast strain and grow on Gal/Raff SD/-His/-Ura (LexA-system) or Glu SD/-Leu (GAL4-system).[a]

3 Select for autonomous activation by re-streaking the colonies on Gal/Raff SD/-His/-Leu/-Ura (LexA-system) or Glu SD/-Leu/-His (+ 3AT) (GAL4-system).

4 Perform a qualitative *LacZ* expression assay (see *Protocol 5*). If no reporter gene activity is detected, the DNA-binding construct is suitable for the interaction screen. If reporter gene activity is detected the second protein under investigation can be fused to the DNA-binding domain vector and subjected to the same test.[b]

5 If the auto-activation test is negative or if the autoactivation is weak, transform the bait and the prey construct into the desired yeast strain by sequential or simultaneous transformation (*Protocol 2*), and select for co-transformants on the respective media (Glu SD/-His/-Trp for the LexA-system and Glu SD/-Trp/-Leu for the GAL4-system). Include also a number of positive and negative control constructs.[d]

6 Streak the resulting colonies on media selecting for amino acid reporter gene activation (Glu and Gal/Raff SD/-His/-Trp/-Leu for the LexA-system and Glu SD/-Trp/-Leu/-His (+ 3AT) for the GAL4-system) and perform a qualitative or quantitative *LacZ* expression assay (see *Protocol 5*).

[a] Include positive and negative control plasmids to monitor the activity of the constructs.

[b] See also Section 3.5.

[c] See also the recommendations in *Protocol 6*, step 3.

[d] For control constructs see references 1–3, 23, 24.

4.4 Substantiating the findings from the two-hybrid assay

As the yeast two-hybrid system—despite its simplicity—makes use of a very complex assay system, a yeast cell, interactions detected using the two-hybrid system cannot automatically be considered as genuine interactions. For publication, the detected interactions usually have to be substantiated by an independent interaction assay (26). As the majority of interactions detected in the two-hybrid system are transient interactions, a proof of *in vivo* interaction by co-immunoprecipitation is difficult. Therefore, most laboratories prefer *in vitro* interaction assays for this purpose. Some two-hybrid system vectors now permit to use the two-hybrid constructs directly for coupled *in vitro* transcription translation reactions. Overlapping expression data as obtained from *in situ* hybridization, existing genetic interaction data or results obtained from the analysis of sense and antisense expression lines can substantiate the interaction findings even further. In addition, the numerous reverse genetic approaches, especially in *Arabidopsis*, will soon allow the direct identification of mutants for most given genes.

References

1. Harper, J. W., Adami, G. R., Wei, N., Keyomarsi, K. and Elledge, S. J. (1993). *Cell*, **75**, 805.
2. Fields, S. and Song, O-K. (1989). *Nature*, **340**, 245.
3. Gyuris, J., Golemis, E., Chertkov, H. and Brent, R. (1993). *Cell*, **75**, 791.
4. Chen, C-T., Bartel, P. L., Sternglanz, R. and Fields, S. (1991). *Proc. Natl Acad. Sci.*, **88**, 9578.
5. Chevray, P. M. and Nathans, D. (1992). *Proc. Natl Acad. Sci.*, **89**, 5789.
6. Zhou, J., Tang, X. and Martin, G. B. (1997). *EMBO J.*, **16**, 3207.
7. Ang, L-H., Chattopadhyay, S., Wei, N., *et al.* (1998). *Mol. Cell*, **1**, 213.
8. Goff, S. A., Cone, K. C. and Chandler, V. L. (1992). *Genes Dev.*, **6**, 864.
9. Kim, J., Harter, K. and Theologis, A. (1997). *Proc. Natl Acad. Sci.*, **94**, 11786.
10. Ni, M., Tepperman, J. M. and Quail, P. H. (1998). *Cell*, **95**, 657.
11. Brachmann, R. K. and Boeke, J. D. (1997). *Curr. Opin. Biotech.*, **8**, 561.
12. Bartel, P. L. and Fields, S. F. (ed.) (1997). *The yeast two-hybrid system*, Oxford University Press, New York.
13. Guthrie, C. and Fink, G. R. (ed.) (1991). *Guide to yeast genetics and molecular biology. Methods in Enzymology*, 194th edn. Academic Press, San Diego.
14. Bartel, P. L., Chien, C-T., Sternglanz, R. and Fields, S. (1993) In: *Cellular interactions in development: a practical approach* (ed. D. A. Hartley), p. 153 Oxford University Press, Oxford.
15. Ma, J. and Ptashne, M. (1987). *Cell*, **51**, 113.
16. Estojak, J., Brent, R. and Golemis, E. A. (1995). *Mol. Cell. Biol.*, **15**, 5820.
17. Ausubel, F. M., Brent, R., Kingston, R., *et al.* (Ed.) (1995). *Current Protocols in Molecular Biology*, John Wiley & Sons, London.
18. Gubler, U. and Hoffman, B. J. (1983). *Gene*, **25**, 263.
19. Durfee, T., Becherer, K., Chen, P-L., *et al.* (1993). *Genes Dev.*, **7**, 555.
20. Yang, E. W., Zha, J., Jockel, J., Boise, L. H., Thompson, C. B. and Korsmyer, S. J.(1995). *Cell*, **80**, 265.
21. Gietz, D., Jean, A. S., Woods, R. A. and Schiestl, R. H. (1992). *Nucl. Acids Res.*, **20**, 425.
22. Sambrook, J., Fritsch, E. F. and Maniatis, T. (ed.) (1989). *Molecular cloning. A laboratory manual*, 2nd edn. Cold Spring Harbor Laboratory Press, New York.
23. Bartel, P., Chien, C.-T., Sternglanz, R. and Fields, S. (1993). *BioTechniques*, **14**, 920.
24. Li, B. and Fields, S. (1993). *FASEB J.*, **7**, 957.
25. James, P., Halliday, J. and Craig, E. A. (1996). *Genetics*, **144**, 1425.
26. Phizicky, E. M. and Fields, S. (1995). *Microbiol. Rev.*, **59**,94.
27. Serebriiski, I., Khazak, V. and Golemis, E. A. (1999). *J. Biol. Chem.*, **274**, 17080.
28. Gimeno, R. E., Espenshade, P. and Kaiser, C. A. (1996). *Mol. Biol. Cell*, **7**, 1815.
29. Roder, K. H., Wolf, S. S. and Schweizer, M. (1996). *Anal. Biochem*, **241**, 260.
30. Louvet, O., Doignon, F. and Crouzet, M. (1997). *BioTechniques*, **23**, 816.

31. Chien, C. T., Bartel, P. L., Sternglanz, R. and Fields, S. (1991). *Proc. Natl Acad. Sci.*, **88**, 9578.
32. Liu, J., Wilson, T. E., Milbrandt, J. and Johnston, M. (1993). *Methods: a companion to methods in enzymology*, **5**, 125.
33. Feilotter, H. E., Hannon, G. J., Ruddel, C. J. and Beach, D. (1994). *Nucl. Acids Res.*, **22**, 1502.

Chapter 10
Antibody techniques

William G. T. Willats, Clare G. Steele-King, Susan E. Marcus and J. Paul Knox

Centre for Plant Sciences, University of Leeds, Leeds, LS2 9JT, UK

1 Introduction

Exploiting the mammalian immune system to generate molecular probes for the detection and location of biomolecules has resulted in a well-established series of techniques. This chapter has two parts: making antibodies and using antibodies. The technologies directed at making antibodies have evolved alongside the development of cellular and molecular techniques in general. They range from the animal-based production of antisera, through cell-based generation of hybridomas and monoclonal antibodies, to gene-based phage display and related technologies. The later technologies are complementary to the earlier technologies and both are now widely used. The protocols presented here are based on the generation and use in our laboratory of antibodies to cell wall antigens, but in all cases the protocols are applicable to a wide range of antigens. Protocols for the use of antibodies in immunolocalization studies, and in qualitative and quantitative immunochemical assays are described.

2 Making antibodies

2.1 Antisera

The easiest way to make an antibody is by immunization of an animal and preparation of an antiserum from a blood sample. If a relatively pure or defined antigen is available for immunization, specific antisera with high titres, and good specificity can be readily achieved and will be useful for most purposes. Proteins are generally sufficiently immunogenic to be used directly as immunogens and very often the protein is generated in recombinant form using one of the procedures described in *Chapter 6*.

Small peptides or small molecular weight compounds (plant hormones, secondary metabolites, etc.) need to be coupled to a non-glycosylated protein such as bovine serum albumin (BSA) or keyhole limpet haemocyanin (KLH) to elicit an appropriate response (1). Immunization with polysaccharide antigens may result in poor responses, but the coupling of oligosaccharide haptens or polysaccharides to proteins such as BSA will produce immunogenic neoglycoproteins. Problems inherent in the generation of antibodies to complex plant heteropolysaccharides are discussed elsewhere (2).

Although larger mammals, such as rabbits or sheep, are often used to generate antisera, a useful strategy is to immunize mice or rats to generate antisera for initial studies that can then, if appropriate, be taken further to hybridoma preparation and monoclonal antibodies. A single tail bleed of a rat can produce up to 0.2 ml of antiserum that, when of high specificity and titre (i.e. the extent to which that it can be diluted and still give a strong signal in an appropriate assay) can be put to wide use. *Protocol 1* describes the immunization of rats for the generation of antisera as used in our laboratory. In general, the larger the animal immunized, the higher the dose of immunogen required, the longer the gap between subsequent injections and the larger volume of antiserum obtained from each bleed (for details of immunization procedures and handling of animals see ref. 1). Good antibody responses are obtained when immunogens are mixed with adjuvants, the most commonly used being Freund's adjuvants. Complete Freund's adjuvant, used for the first injection, consists of mineral oil with a suspension of killed mycobacteria. The emulsification of the immunogen with the mineral oil results in sustained release of the immunogen and the mycobacteria promote a general stimulation of the immune response. The preparation of an antiserum requires several injections of immunogen. For subsequent (booster) injections Freund's incomplete adjuvant (without the mycobacteria) is used.

Depending on the animal, a dose of 10–100 μg of immunogen is used at each injection. As a guide, an antiserum should have a titre of at least 1000 to be of general use, i.e. when diluted 1000-fold it produces a significant signal/response in an appropriate assay when compared with a pre-immune serum (a serum prepared from a blood sample taken prior to immunization).

Protocol 1

Generation of an antiserum

Equipment and reagents

- Immunogen, e.g. recombinant protein (see *Chapter 6*)
- Disposable syringes and needles (Terumo).
- Freund's adjuvants (Sigma)
- Phosphate-buffered saline pH 7.2 (PBS, 8 g NaCl, 2.86 g $Na_2HPO_4.12H_2O$, 0.2 g KCl, 0.2 g KH_2PO_4 in 1 l H_2O)

Method

1 Antigen is dissolved in PBS to give ideally 2 mg/ml (if in short supply 50 μg/ml or less can be effective).

2 0.5 ml of antigen in PBS is emulsified with 0.5 ml Complete Freund's adjuvant by repeated passage through a needle until a drop of the emulsion does not disperse when placed on a water surface.

3 100–200 μl of the emulsified mixture is injected subcutaneously[a] into each rat. It is useful to have two to four rats per antigen as the response can vary.

4 Thirty days later the same antigen, this time prepared and mixed with Incomplete Freund's adjuvant, is administered in the same way.

5 Ten days after the second injection (first booster) the rat can be tail bled[a] to provide approximately 0.5 ml blood, collected in a microcentrifuge tube.

6 To prepare the antiserum from the blood sample: place tube at 37°C for 1 h to promote clotting, then at 4°C for 1 h to promote shrinkage of clot. Spin at maximum speed in a bench microcentrifuge for 5 min. Carefully collect the antiserum (a semi-transparent liquid) from above the clot.

7 Antisera should be stored in aliquots at −20°C or below.[b]

8 Subsequent booster injections should be made at 30-day intervals, with tail bleeds and preparation of antisera 10 days after each boost.

[a] For details of animal handling and procedures see ref. 1.

[b] Repeated thawing and freezing of all antibodies should be avoided. It may be useful to store a 25-fold dilution of antisera in PBS containing 0.05% sodium azide at 4°C for frequent use.

2.2 Hybridoma technology

Hybridoma technology involves the isolation, selection, and immortalization of specific lymphocytes that secrete antibodies with appropriate specificities. This is done using mice or rats, and the techniques are essentially the same for both species. In-depth discussion of procedures can be found in Liddell and Cryer (3), and Peters and Baumgarten (4). The spleen is generally selected as a source of lymphocytes and they are fused with an immortal non-secreting myeloma cell line to produce hybridomas. Several myeloma lines are available. We use IR983F (5), but other rat lines include YB2/0 (6) and a suitable mice line is NSO/1 (7, 8). Hybridomas are selected by the use of media containing hypoxanthine, aminopterin and thymidine (HAT media). Aminopterin blocks the *de novo* synthesis of nucleotides promoting the use of the salvage pathway requiring hypoxanthine-guanine phosophoribosyl transferase (HPRT). Only successful fusions of lymphocytes with myeloma cells (hybridomas) survive due to the combination of HPRT, provided by the lymphocyte, and the proliferative capacity provided by the myeloma. All unfused spleen cells and myeloma cells (without HPRT) die. Rats are immunized as described in *Protocol 1* and when an appropriate immune response has been achieved, as indicated by antisera with good titre and specificity, the spleen is isolated for the generation of hybridomas as described in *Protocol 2*. During early stages of hybridoma production and during cloning by limiting dilution (*Protocol 3*), a fibroblast cell line (MRC5) is used as a feeder cell layer to promote cell proliferation at low cell densities.

Protocol 2

Generation of hybridomas and selection of monoclonal antibodies

Equipment and reagents

- All reagents (except the rat) must be sterile
- Immunized rat (*Protocol 1*)
- Dulbecco's modified Eagles medium (DMEM, Gibco BRL, Life Technologies)
- HAT supplements (Life Technologies)
- Foetal calf serum (FCS, Life Technologies)
- MRC5 fibroblast feeder cell line (European Collection of Animal Cell Cultures).
- 50 ml screw-cap Falcon tubes (Becton Dickinson)
- Nunc™ Cell culture plastics (Life technologies)
- 50% PEG-1500 (Boehringer Mannheim)
- Cell counting device (Haemocytometer, Scientific Laboratory Supplies)
- Multi-channel pipettes (Jencons)
- Automatic pipette filler (Pipet Boy, Integra Biosciences)
- Laminar flow hood (Inter Med, MDH)
- CO_2 incubator for animal cell culture (Gallenkamp or Integra Biosciences)

Method

1 Three days before a lymphocyte-myeloma fusion is to take place, give the selected rat an intra-peritoneal injection of 100 µg of antigen in 1 ml PBS (no adjuvant).

2 Three days before the fusion takes place prepare ten 96-well cell culture microtitre plates with the MRC5 feeder cells growing in DMEM with 20% FCS (20 FCS-DMEM).

3 Grow IR983F myeloma cells in 20 FCS-DMEM and make sure that the cells are actively growing, but not overgrown.

Protocol 2 continued

4 On the day of the fusion, take 200 ml of actively growing myeloma cells for each half spleen to be fused and pellet by centrifugation at 180 g for 10 min. Resuspend in 20 ml of serum-free DMEM medium. Dilute a small aliquot (e.g. 100 μl) 10-fold and count myeloma cells.

5 Dissect out the spleen from the immunized rat and place into 10 ml of 2.5% FCS-DMEM. Rinse twice more in the same medium. Remove all fat from the spleen with a pair of scissors, cut into six or seven small pieces and push through a sieve with a plunger from a plastic syringe into 2.5 FCS-DMEM. Suspend in 20 ml of 2.5 FCS-DMEM. Leave for 5 min, and transfer suspended cells to another tube and spin at 180 g for 10 min. Resuspend in 20 ml serum-free medium. Dilute a small aliquot 10-fold and count lymphocytes.

6 Mix spleen cells and myeloma cells at a ratio of 2:1.

7 Add serum-free medium to cells up to 40 ml (at this point, remove 1 ml of mixed cells prior to the fusion as a control for the fusion plates). Spin remaining cells at 180 g for 10 min.

8 Remove supernatant thoroughly and loosen cells in the pellet by flicking the outside of the tube.

9 Add 0.8 ml of 50% PEG-1500, while gently stirring at 37°C over 1 min. Continue stirring for a further 1.5 min. Add 1 ml serum-free medium over 1 min. Repeat with another 1 ml. Add 1 ml serum-free medium over 30 s and repeat again. Add 1 ml serum-free medium over 15 s and repeat. Continue additions up to 10 ml. Make up to 25 ml with serum-free medium.

10 Spin at 50 g for 10 min.

11 Re-suspend pellet to give a cell density of 2×10^6 cells/ml in 20 FCS-DMEM. Add 100 μl of cell suspension to each well of the MRC5 plates.[a]

12 Next day remove 100 μl of medium and replace with 20 FCS-DMEM medium containing double HAT concentration (20 FCS-DMEM-HAT).

13 Continue to feed the cells with 20 FCS-DMEM-HAT when necessary. Hybridomas should be visible by day 7 after fusion.

14 Hybridoma supernatants are taken to screen for antibody specificities when the base of the microtitre well is at least half covered with cells. The supernatants are screened in an appropriate rapid and sensitive assay. The most commonly used (and easiest to integrate with hybridoma production) are screens using ELISAs (see *Protocol 13*).

15 The proliferation of selected hybridoma cells is maintained and they should be transferred to larger wells and flasks, and frozen (*Protocol 3*) for storage as soon as possible (see *Protocol 3*, steps 6–8).

[a] Up to 10 plates can be used for a fusion involving half a spleen.

To ensure that a selected hybridoma cell line, and hence the secreted antibody, is monoclonal and not a mixture of cell lines secreting different antibodies, selected hybridomas must be cloned to ensure that a population has been derived from a single cell. This is achieved by culturing cells at very low cell densities and selecting single colonies. The limiting dilution procedure for cloning hybridoma cells in 96-well microtitre plates is described in *Protocol 3*. The procedures required to freeze hybridoma cells for storage and the preparation of the hybridoma supernatant for use are also described in *Protocol 3*.

Protocol 3

Limiting-dilution cloning of hybridomas

Equipment and reagents

- All reagents must be sterile
- Dulbecco's modified Eagles medium (DMEM, Life Technologies)
- Foetal calf serum (FCS, Life Technologies)
- HEPES/azide: 5 g sodium azide, 29.8 g HEPES in 100 ml PBS. Store in a dark bottle for up to 3 months
- 50 ml screw-cap Falcon tubes (Becton Dickinson)
- Nunc™ Cell culture plastics (Life Technologies)
- Cell counting device (Haemocytometer, Scientific Laboratory Supplies)
- Cryotubes (Nunc™, Life Technologies)
- DMSO (Sigma)
- Multi-channel pipettes (Jencons)
- Pipet Boy Automatic pipette filler (Integra Biosciences)
- Laminar Flow hood (Inter Med, MDH)
- CO_2 incubator for animal cell culture (Gallenkamp or Integra Biosciences)
- 96-well microtitre plates prepared with MRC5 feeder cell layer (*Protocol 2*)

Method

1 Count an aliquot of actively growing selected hybridoma cells as obtained in *Protocol 2*.

2 Plate out into MRC5 microtitre plates with 20 FCS-DMEM (see *Protocol 2*) at a range of dilutions in the range of 5 cells/well to 0.5 cells per well.

3 After approximately 7 days, select colonies that are judged to be single colonies (monoclonal) from the highest dilutions of cells.

4 Check supernatants of selected cells for appropriate specificity using ELISA (*Protocol 13*) or other suitable immunoassay.

5 Proliferate selected cells through 24- and 6-well culture plates into culture flasks (80 cm^2 culture area).

6 When cells are confluent in culture flasks, cell samples should be frozen as soon as possible. To do so, centrifuge cells at 180 g and resuspend cells from one flask in 0.5–1 ml of a mixture of 90% FCS and 10% DMSO. Place 0.5 ml aliquots (around 10^5–10^7 cells/ml) in cryotubes.

7 The cells are slowly cooled to −80 °C by enclosure within polystyrene in −80 °C freezer for a few days prior to being placed in a liquid nitrogen cryostore.[a]

8 Animal cells are frozen slowly and thawed quickly. Cells are thawed by immersing closed cryotubes to just below the stopper in a 37 °C water bath. Cells are removed and washed in DMEM medium and placed in the wells of culture plates or in a culture flask to promote proliferation.

9 Established cell lines are gradually weaned from 20 FCS-DMEM-HAT through 20 FCS-DMEM-HT (medium without the aminopterin inhibitor) to 20 FCS-DMEM and eventually the level of FCS supplement may also be reduced.

10 Hybridoma supernatants are isolated by removing cells by centrifugation (180 g) and a 50-fold dilution of HEPES/Azide is added. Supernatants can be stored at 4 °C for many years.

11 Repeat cloning procedure at least two times.

[a] Several cryotubes of cells should be frozen for each line of potential interest.

2.3 Isolation of phage display antibodies from naive libraries

Antibody phage display technology is a more recently developed strategy for exploiting the diversity of the mammalian immune system. The technique involves the expression of antibody binding fragments at the surface of bacteriophage. Whole phage particles or soluble antibody fragments may be used as antibody probes in the same range of applications as conventional antibodies. Antibody phage libraries are constructed by the fusion of rearranged antibody variable region genes (*V*-genes) to genes encoding phage coat proteins, for example, the M13 coat protein p111 (9, 10, see *Chapter 11*). Expression of the fusion product results in the antibody fragment being presented on the surface of phage and, since each phage particle also carries the genetic material encoding the displayed antibody fragment, the nucleotide sequence encoding selected antibody fragments may be readily determined and manipulated.

One approach for selecting phage antibodies is to construct a library for each antigen of interest using antibody gene sequences derived from lymphocytes of immunized animals. The construction and use of such immune libraries is described in *Chapter 11*. Alternatively, a single, extremely large naive combinatorial library may be constructed from antibody *V*-genes derived from an un-immunized donor and used to screen for phage antibodies against all antigens. This 'single pot' approach is the focus of this chapter and, for some applications, has advantages over the use of immune libraries because animal immunization is completely by-passed and there is no requirement for the target antigen to be immunogenic. *Protocols 4–8* are based on the use of the Synthetic scFv Library (#1), obtained from the Medical Research Council (MRC) Centre for Protein Engineering at Cambridge (11). The protocols are adapted from those produced for use with the Synthetic scFv Library (#1) with some modifications based on our experience in using this library to select antibodies to plant cell wall antigens (12). Further information is available at the following web site: (http://www.mrc-cpe.cam.ac.uk/).

2.3.1 Screening of naive phage display libraries.

Selection of phage display antibodies from phage libraries is based on the panning of diverse phage populations against antigens of interest. The protocols detailed here are based on antigens immobilized on tubes known as immunotubes. A schematic overview of the procedures is shown in *Figure 1*. A diverse phage population from the library is amplified (stages 1 and 2) and panned against the antigen of interest (stages 3a and b). Non-binding phage are removed by washing (stage 3c). Phage with binding specificity for the immobilized antigen are retained, eluted from the antigen (stage 4) and infected into host bacteria (stage 5). A selected library is thereby produced that should be greatly enriched in clones that are infected with phage with antigen-binding potential. This library is then amplified (stages 6a and 7) and the panning process repeated (usually at least four times) starting at stage 2. These panning rounds result in the production of a polyclonal phage population (stage 8), from which monoclonal phage antibodies may be isolated and screened (stages 9 and 10).

Protocol 4 describes the growth of the phage libraries and secretion of phage particles into the growth medium and applies to both the original library stock and selected libraries produced after panning (stages 1 and 7 in *Figure 1*). In the Synthetic scFv Library (#1) the genes encoding scFvs are present on the pHEN1 plasmid grown in *Esherichia coli* strain TG1.

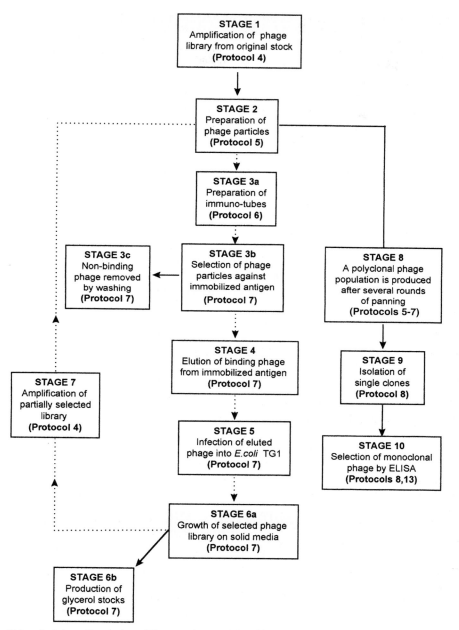

Figure 1 Schematic overview of the procedures required for generating monoclonal antibodies by phage display technology. The repetitive bio-panning stages (2–7) are indicated with dotted lines.

Protocol 4

Growth of phage libraries

Stages 1 and 7 (*Figure 1*)

Equipment and reagents

- Synthetic scFv Library (#1)
- Orbital incubator (Gallenkamp)
- 50 ml screw-cap polypropylene tubes (Falcon)
- 2TY broth (16 g Tryptone, 10 g yeast extract, 5 g NaCl in 1 l water)
- Ampicillin stock solution (100 mg/ml in 70% ethanol)
- Kanamycin stock solution (25 mg/ml in filter-sterilized de-ionized water)
- Glucose stock solution (25% w/v) in de-ionized water
- Helper phage: VCS-M13 or M13-KO7 (New England BioLabs)
- *E. coli* TG1[a]

Method

1 Inoculate 50 μl of library stock into 50 ml 2TY containing 100 μg/ml ampicillin and 1% glucose.

2 Grow with shaking at 37 °C until an OD of 0.5 at 600 nm is reached (this corresponds to approximately 4×10^8 bacteria/ml), which takes 1–2 h.

3 Infect 10 ml of the bacterial culture from step 2 with helper phage in the ratio of 1:20 (bacterial cells: helper phage particles).

4 Incubate at 37 °C without shaking for 30 min.

5 Centrifuge the infected cells at 3300 g for 10 min.

6 Remove the supernatant and resuspend the pellet in 10 ml of fresh 2TY. Add to 300 ml of fresh 2TY containing 100 μg/ml ampicillin and 25 μg/ml kanamycin.

7 Incubate with shaking at 30 °C overnight.

[a] Uninfected *E. coli* strain TG1 is supplied with the Synthetic scFv Library (#1).

The phage particles produced in *Protocol 4* must be purified from the bacteria and concentrated before they can be screened by panning against immobilized antigen. Protocol 5 describes the preparation of phage particles.

Protocol 5

Preparation of phage particles

Stage 2 (*Figure 1*)

Equipment and reagents

- Incubators (Gallenkamp)
- 300 ml of overnight culture of *E. coli* TG1 (step 7, *Protocol 4*)
- Sterile PBS (*Protocol 1*)
- PEG/NaCl (20% (w/v) polyethylene glycol 6000, 2.5 M NaCl in de-ionized water)
- Sterile de-ionized water.
- 80% glycerol in de-ionized water

Method

1 Spin the overnight culture from step 7 in *Protocol 4* at 10 000 g for 10 min or 3300 g for 30 min.

2 Add 1/5 volume of PEG/NaCl to the supernatant, mix well, and incubate on ice for 1 h.

Protocol 5 continued

3 Centrifuge at 10 000 g for 30 min, remove the supernatant and discard.

4 Resuspend the pellet in 40 ml de-ionized water and add 8 ml PEG/NaCl. Mix well and incubate on ice for 30 min.

5 Centrifuge at 3300 g for 30 min, remove the supernatant, and discard.

6 Resuspend the pellet in 2 ml of sterile PBS and centrifuge at 3300 g for 10 min. Remove the supernatant and centrifuge it again at 3300 g for 10 min.

7 The supernatant contains the phage particles. For short-term storage (1 or 2 weeks) add 0.05% sodium azide to the supernatant and store at 4 °C. For long-term storage add 7% DMSO to the supernatant and store at −80 °C.

Phage particles from step 7, *Protocol 5* are selected for binding to antigen immobilized on immunotubes in a procedure known as panning. *Protocol 6* describes the preparation of immunotubes.

Protocol 6

Preparation of immunotubes

Stage 3a (*Figure 1*)

Equipment and reagents

- Antigen
- PBS containing 3% fat-free milk protein (MP/PBS)
- Immunotubes (Nunc-Immuno™ Tubes with MaxiSorp™ surface, Life Technologies)
- PBS (*Protocol 1*)

Method

1 Prepare 4 ml of a solution of antigen in PBS.[a]

2 Fill immunotube with 4 ml of antigen solution and incubate overnight at 4 °C.

3 If proceeding directly with selection pour out antigen solution and fill immunotube with 4 ml of blocking solution (MP/PBS) and incubate for at least 2 h at 37 °C or overnight at 4 °C.

4 Pour out blocking solution from immunotube.

5 Wash immunotube three times with PBS.[b]

[a] The density of immobilized antigen influences the efficiency of selection and the relative proportions of phage retained with the same specificity, but different avidities. As a guide, 10–100 µg/ml is usually satisfactory.

[b] All washing steps are performed by squirting solutions from a wash bottle into immunotube and pouring out immediately.

Purified and concentrated phage particles (prepared as described in *Protocol 5*) derived from the original library stock or from partially selected libraries are screened against antigen immobilized on immunotubes (prepared as described in *Protocol 6*).

Protocol 7

Selection of phage by screening against immobilized antigen

Stages 3b to 6b (*Figure 1*)

Equipment and reagents

- Coated immunotube (*Protocol 7*)
- Phage particles (*Protocol 5*)
- PBS (*Protocol 1*)
- PBS containing 3% fat-free milk protein (MP/PBS)
- PBS containing 0.1% Tween-20
- 2TY (*Protocol 4*)
- TYE (15 g BACTO-agar, 8 g NaCl, 10 g Tryptone, 5 g yeast extract in 1 l de-ionized water)
- An exponentially growing culture of *E. coli* strain TG1.[a]

- A freshly made solution of 7.18 M triethylamine (700 μl triethylamine in 50 ml de-ionized water).
- 1 M Tris pH 7.4
- Nunc[TM] Bio-assay dish (245 × 245 × 25 mm, Life Technologies) containing 400 ml TYE with 100 μg/ml ampicillin and 1% glucose
- Cell scraper III (Costar)
- 15ml and 50 ml screw cap polypropylene tubes (Falcon)
- Sterile 80% glycerol in de-ionized water

Method

1 Combine the 2 ml of phage in PBS from step 7 in *Protocol 5* with 2 ml of MP/PBS. This mixture should contain approximately 10^{11}–10^{13} phage particles/ml.

2 Add the 4 ml of phage solution to an antigen-coated and blocked immunotube (*Protocol 6*). Seal the tube with parafilm and incubate with continuous end-over-end rotation for 30 min, and then standing for 90 min.

3 Pour off the phage/blocking PBS solution (containing unbound phage), wash the tube 20 times with PBS containing 0.1% Tween-20, and then a further 20 times with PBS.

4 After the final wash add 1 ml of 7.18 M triethylamine to the immunotube and incubate with end-over-end rotation for 10 min.

5 Remove the triethylamine (now containing eluted phage) and add to 0.5 ml 1 M Tris in a 15 ml polypropylene screw cap tube.

6 Immediately add 200 μl of 1 M Tris to the now empty immunotube.

7 Add 9 ml of an exponentially growing culture of TG1 to the 1.5 ml of eluted phage solution in the 15 ml screw cap tube and mix briefly end-over-end. Then add 4 ml of the TG1 culture to the 200 μl 1 M Tris in the immunotube. Seal with parafilm and mix briefly end-over-end.

8 Incubate both tubes for 30 min at 37 °C without shaking.

9 Add the 4.2 ml of solution from the immunotube to the 10.5 ml in the screw cap tube and centrifuge at 3300 g for 10 min. Resuspend the pelleted bacteria in 1 ml of 2TY and spread out onto a Bio-assay dish containing 300 μl TYE containing 100 μg/ml ampicillin and 1% glucose.

10 Grow overnight at 30 °C.

11 After overnight growth, remove the bacteria from the surface of the TYE with a cell scraper and put into a 50 ml screw cap centrifuge tube. Add 1.7 ml 2TY and mix well by vortexing.

12 Glycerol stocks may be produced by adding 80% sterile glycerol to the bacteria/2TY solution from step 11 to a final concentration of 15%. Glycerol stocks may be stored indefinitely at −80 °C. Alternatively, 50 μl of the bacteria may be added to 50 ml of 2TY as described in step 1 in *Protocol 4* and a further round of panning begun without delay.

[a] Uninfected *E. coli* strain TG1 is supplied with the Synthetic scFv Library (#1).

After several rounds of panning against immobilized antigen (usually at least four) the selected phage library should be greatly enriched in clones that have binding specificity towards the immobilized antigen (see also *Chapter 11*). The binding of this polyclonal population may be rapidly assessed by immuno-dot-assays or ELISAs as described in *Protocols 12* and *13*. *Protocol 8* describes the isolation and screening of monoclonal phage antibodies (stages 9 and 10 in *Figure 1*).

Protocol 8

Selection of phage monoclonal antibodies

Stage 9 (*Figure 1*)

Equipment and reagents

- 2TY containing 100 μg/ml ampicillin
- Sterile 96-well microtitre culture plates (CellWells™, Corning)
- 96-well microtitre ELISA plates (Nunc™ Immunoplate with Maxisorp™ surface, Life Technologies)
- Six 9 cm Petri dishes containing TYE with 100 μg/ml ampicillin and 1% glucose
- Sterile wooden tooth picks
- Sterile 80% glycerol in de-ionized water
- Helper phage (*Protocol 4*)

Method

1 Prepare 10 ml of antigen solution as in step 1, *Protocol 6* and use it to coat (100 μl/well) a 96-well ELISA plate overnight at 4 °C.

2 Take approximately 20 μl of *E. coli* clone from the final round of panning (either fresh or from a glycerol stock (*Protocol 7*) and add to 20 ml of 2TY containing 100 μg/ml ampicillin. Mix well.

3 From the 1/1000 dilution in step 2 produce a further five 1/100 serial dilutions in 2TY containing 100 μg/ml ampicillin.

4 Take 100 μl of each of the dilutions and spread out onto TYE containing 1% glucose and 100 μg/ml ampicillin, and grow overnight at 37 °C.

5 Select a plate from step 4 that has at least one hundred well spaced individual colonies.

6 Prepare a 96-well culture plate containing 100 μl/well 2TY with 100 μg/ml ampicillin and 1% glucose (Plate 1)

7 Using sterile wooden tooth picks inoculate individual colonies from step 5 into each well of Plate 1 and grow with shaking overnight at 37 °C.

8 Use a 96-well transfer device or multi-channel pipette to transfer a small inoculum (e.g. 3 μl) from Plate 1 to a fresh 96-well culture plate (Plate 2) prepared as in step 6. Grow with shaking for 1 h. Add sterile glycerol to a final concentration of 15% to each of the wells of Plate 1, mix briefly and gently (avoiding cross-well contamination) on a vortexer and store Plate 1 at −80 °C.

9 To each well of Plate 2 add 25 μl 2TY containing 100 μg/ml ampicillin, 1% glucose, and 10^9 helper phage.

10 Leave Plate 2 to stand for 30 min (to allow for infection of the helper phage) then shake for 1 h at 37 °C.

11 Centrifuge Plate 2 at 1800 g for 10 min, then aspirate off the supernatant from each well.

12 Resuspend the pellet in each well of Plate 2 in 200 μl 2TY containing 100 μg/ml ampicillin and 50 μg/ml kanamycin, and grow overnight with shaking at 30 °C.

13 Centrifuge Plate 2 at 1800 g for 10 min.

Protocol 8 continued

14 Transfer the supernatants from each well of Plate 2 (without dilution) to a 96-well ELISA plate, coated with antigen as described in step 1, and perform a standard antibody capture ELISA as described in *Protocol 13*.

15 The results of the ELISA enable the monoclonal phage with the most desirable binding characteristics to be identified. Clones of interest may then be grown up from the glycerol stocks in each well of Plate 1.

3 Using antibodies

The use of antibodies to detect antigens of interest involves the use of marker molecules. Markers include enzymes, such as peroxidase and alkaline phosphatase, for ELISA and immuno-blotting techniques, fluorochromes, such as FITC for antigen detection using a fluorescent microscope and gold particles for electron microscopy. In the majority of cases an indirect immunolabelling protocol is used. This entails incubation with a (primary) antibody to the antigen of interest, followed by incubation with a secondary antibody (that recognizes the constant region of the primary antibody) attached to the marker. This system offers immense versatility and is supported by the commercial availability of a vast range of products.

Here, we cover the use of antibodies to detect antigens within plant tissues and in immunochemical assays. Most of these protocols are applicable for use with antisera, hybridoma antibodies, and phage display antibodies. The minor modifications required when phage display antibodies are used are indicated where appropriate. Dilutions of antibodies given are guidelines only and appropriate conditions should be determined in each case. Controls should always be included with any immunolabelling or any immunochemical protocol. A negative control could consist of a pre-immune serum if available, or an antibody that is not expected to bind to the material under study. A positive control primary antibody, that is known to bind to the antigen or section, is also useful in initial phases of immuno-chemical and immunolocalization studies.

3.1 Immunolocalization

A range of techniques is available for the preparation of plant material for the *in situ* detection of antigens within organs, tissues, cells, and organelles. Generally, the speed and ease of the preparative stages (before antibodies are used) is inversely proportional to the level of resolution obtained. An important consideration with some techniques is the loss of antigenicity. This can be due to fixation, the heat used to polymerize an embedding medium or the loss of soluble antigens during preparation. In certain cases, it may be useful to start with a technique that preserves maximum antigenicity and then extend to other procedures for more detailed work. For example, initial techniques, such as tissue printing and labelling of sections prepared by cryo-sectioning or hand-sectioning of agarose-embedded material, will give a good indication of the levels and distribution of an antigen. Where higher resolution is required, we routinely embed material in a medium such as LR White resin that may be sectioned for immunolabelling and examination with both light and electron microscopes.

Tissue printing, as described in *Protocol 9*, provides a very rapid indication of antibody binding to cells and tissues in plant materials without the need for fixation and with minimum preparation (12, 13). It also has an added application for the immunolocalization of soluble antigens that may not be cross-linked to plant material by aldehyde fixatives and which therefore cannot be studied using conventional immunolocalization techniques because of their loss during processing.

Protocol 9

Tissue printing

Equipment and reagents

- Nitrocellulose membrane (200 × 200 mm, Schleicher and Schuell)
- PBS containing 3% fat-free milk protein (MP/PBS)
- Primary antibody
- Secondary antibody, e.g. anti-rat IgG coupled to alkaline phosphatase (AP; Sigma) for conventional antibodies or anti-M13 phage coupled to horseradish peroxidase (HRP; Pharmacia) for phage display antibodies)
- AP substrate buffer (100 mM NaCl, 5 mM $MgCl_2$, 100 mM Tris, pH 9.5)

- NBT stock for AP substrate (0.5 g nitro-blue tetrazolium in 10 ml 70 % dimethylformamide)
- BCIP stock for AP substrate (0.5 g bromochloroindolyl phosphate in 10 ml dimethylformamide)
- AP substrate (add 66 μl of NBT stock to 10 ml AP substrate buffer, mix well, and add 33 μl of BCIP stock).
- Scalpel or blade to cut plant material

Method

1 Firmly press the freshly cut surface of a plant organ onto a piece of nitrocellulose for up to 30 s. A second impression of the same surface may be useful in cases where the material is highly pigmented or where the antigen is very abundant.

2 Block all protein binding sites on the nitrocellulose sheet by incubation with irrelevant protein in PBS such as MP/PBS for 1 h (if using peroxidase as a marker, the addition of 0.1 % sodium azide to this buffer will inhibit endogenous peroxidases).

3 Incubate in primary antibody for 1.5 h.[a]

4 Wash extensively with water.[b]

5 Incubate in secondary antibody diluted in the region of 1000-fold in MP/PBS for 1.5 h.

6 Wash extensively in water.[b]

7 Determine antibody binding to the nitrocellulose by incubation in appropriate substrate system for the marker enzyme leading to an insoluble product (e.g. AP substrate, prepared immediately before use) and allow to develop.

8 Stop reaction, when appropriate, by washing extensively in tap water.

[a] Hybridoma supernatants are diluted at least 10-fold in MP/PBS. Phage display antibodies (whole phage particles) are used at a concentration of around 10^{12} phage/ml in MP/PBS.

[b] The use of phage display antibodies may result in higher backgrounds and require more stringent washing. We use two 10-min washes in PBS containing 0.3% Tween 20 prior to the water washes.

3.1.1 Immunolabelling for the light microscope

A routine method of immunolabelling plant material for the light microscope is indirect immunofluorescence labelling as described in *Protocol 10*. Plant material to be labelled can be presented in a variety of ways. These include the direct immunolabelling of suspension-cultured cells, protoplasts, or root squashes, and the immunolabelling of sections prepared by cryo-sectioning, or from agarose, wax, polyethylene glycol, or resin-embedded materials. Details of fixation, embedding and sectioning procedures and microscopy procedures can be found elsewhere (14–17).

The careful selection of a fixative for plant material is essential if optimal ultra-structural detail and antigenicity are to be preserved. Using a fixative with an osmoticum similar to that of the specimen, will help to ensure good fixation, and the addition of small amounts of calcium chloride or sucrose may also help. Too much aldehyde fixative or fixing for too long may reduce antigenicity. Usually fixation for 1–2 h at room temperature is sufficient. It is essential that the fixative is able to enter in to the plant material to promote cross-linkage as quickly as possible and, therefore, specimens should be as small as possible; blocks of material much larger than $2 \times 2 \times 2$ mm will show a reduced quality of fixation when viewed with an electron microscope. If specimens do not sink in fixative (a common problem in tissues such as leaf mesophyll, where large air spaces occur), fixing under vacuum will help to expel air and allow the fixative in. Similarly, the action of a surfactant such as Brij 35 may help in the penetration of fixative into plant materials. We use glutaraldehyde (EM grade, 2.5% w/v, Agar Scientific) at 1–2.5 % in a microtubule stabilizing buffer (MTSB, 50 mM PIPES, 5 mM $MgSO_4$, 5 mM EGTA, pH 6.9) prior to embedding in resin for EM microscopy. Glutaraldehyde may cause a problem of autofluorescence during immunofluorescent microscopy, in which case 2–4% (w/v) formaldehyde should be used instead.

After fixation, the dehydration of plant material is essential before infiltration with a water immiscible resin. Dehydration of the tissue can potentially cause ultra-structural damage to the tissue, and the application of a graded ethanol series consisting of more stages and smaller steps between ethanol concentrations is recommended. At the end of the dehydration process the specimen should be submerged in dried absolute ethanol, which can then gradually be substituted with resin. LRWhite resin is routinely used for embedding plant material for EM and has the advantage that it can be polymerized by UV light in the presence of a catalyst (0.5% benzoin methyl ether) and absence of oxygen (by the use of sealed gelatin capsules). This eliminates the need for heat polymerization, which may reduce the antigenicity of the specimen.

It is important to use an adhesive reagent such as Vectabond or poly-L-lysine to ensure that material is adhered firmly to glass slides. The starting point for *Protocol 10* is fixed plant material that can be from a range of preparations (thin sections, whole mounts of cells, etc.) adhered to glass slides. Protoplasts or suspension-cultured cells can be immunolabelled in a variety of ways, for example, in a microcentrifuge tube with a centrifugation step to pellet the cells between protocol steps, or embedded in low melting agar and sectioned, or pipetted directly onto slides and left for a few minutes to adhere before proceeding. In the latter case, it should be noted that the majority of the protoplasts may be lost during the protocol, so apply protoplasts generously. The use of multi-well slides is recommended, so that small droplets of reagent (e.g. 50 μl) can be pipetted on and off each well, and immersion in buffers is avoided (which requires large volumes of reagents and can often lead to loss of sections). Permeabilization of cells is required if the starting point is not sectioned material and antigens other than those recognizing the cell wall or outer face of the plasma membrane are to be investigated. *Protocol 10* can be used to investigate a variety of antigen distributions, including studies of the ER, Golgi, and cytoskeleton.

Protocol 10 is a starting point. Primary antibody dilutions and blocking buffers may need to be optimized for specific labelling. Plant material should not be allowed to dry out at any point during the immunolabelling protocol. The use of a humid chamber is therefore recommended for longer incubations. For immunolabelling of sectioned material where permeabilization of cells is not required proceed directly to Step 4 of *Protocol 10*. Fluoroscein isothiocyante (FITC) is the most widely used fluorochrome and a wide range of secondary antibodies linked to FITC is available commercially. Other fluorochromes with absorption and emission spectra distinct from FITC are available and these can allow dual labelling of two antigens in the same section.

Protocol 10

Immunofluorescence labelling for light microscopy

Equipment and reagents

- Adhesive reagent (Vectabond, Vector Laboratories or poly-L-lysine, Sigma)
- Multi-well glass slides (8-well multi-test slides, ICN Biomedicals, Inc.)
- PBS
- PBS with 3% milk protein (MP/PBS)
- Primary antibody

- Secondary antibody (e.g., anti-rat-IgG coupled to FITC)
- Anti-fade reagent (Citifluor AF1, Agar Scientific)
- Microscope equipped with epifluorescence illumination

Method

1 If using intact cells or protoplasts allow to adhere to wells of multi-well glass slides.

2 Permeabilize cells or protoplasts with 0.5% Triton X-100 diluted in buffer. In some cases it may also be useful to use cell wall-degrading enzymes to increase antibody access to antigen.

3 Wash cells or protoplasts by incubation with three changes of PBS for 5 min each.

4 Block non-specific binding sites of prepared cells/protoplasts or section of plant material adhered to glass slide by incubation in MP/PBS for at least 30 min.

5 Incubate with primary antibody diluted in MP/PBS for at least 1 h at room temperature or overnight at 4 °C.

6 Wash three times with PBS with at least 5 min for each change.

7 Incubate with secondary antibody diluted in the region of 100-fold in MP/PBS for 1 h at room temperature.

8 Wash three times with PBS with at least 5 min for each change.

9 Mount slides using anti-fade reagent and examine in the microscope.

3.1.2 Immunolabelling for the electron microscope

Protocol 11 assumes that ultra-thin resin sections, supported on EM grids have already been prepared. Only a brief introduction to the essential preparative steps and some points of consideration are provided and details can be found elsewhere (16).

Ultra-thin sections cut on an ultra-microtome should be silvery gold in colour (approximately 80 nm in thickness) and supported on gold or nickel grids. If sections tear easily under the electron beam, it is advisable to coat the grids with a supportive medium, such as formvar (Agar Scientific, UK). However, this introduces an increased risk of non-specific antibody binding and should be used only when necessary. Grids should not be allowed to dry out at any stage during the protocol and the longer incubation steps should be done in a humid chamber to prevent this. Gently blotting the edges of grids with Whatman Grade 50 hardened filter paper (unlike Grade 1, it does not leave fibres) between steps is also recommended, although care should be taken to prevent the loss or damage of sections on the grid.

Protocol 11

Immunogold electron microscopy

Equipment and reagents

- EM sections on nickel or gold grids
- EM forceps, preferably anti-capillary (Agar Scientific)
- Parafilm
- Syringes and syringe filters
- PBS (*Protocol 1*)

- PBS with 3% BSA (BSA/PBS)
- Primary antibody
- Secondary antibody (e.g. anti-rat-IgG coupled to 10 nm gold particles, Sigma)
- Transmission electron microscope

Method

1 Block the section to prevent non-specific binding by floating the EM grid section side down on a droplet (at least 20 μl) of BSA/PBS on Parafilm[a] for 30 min.

2 Transfer grid to a droplet of primary antibody BSA/PBS. Monoclonal antibodies should be diluted to between 5- and 100-fold.

3 Wash grids by incubation in a minimum of three changes of PBS.[b]

4 Transfer grids to secondary antibody[c] diluted 1 in 20 with BSA/PBS.

5 Wash as in step 3 and then extensively in distilled water.[d]

6 Allow the grid to dry and then examine in an electron microscope.

[a] Parafilm provides a clean surface onto which droplets of reagent can be pipetted. If grids sink during incubation they should be turned section side up in the droplet.

[b] The addition of Tween 20 (1–3%) will increase the stringency of washes.

[c] Gold conjugated secondary antibodies are available in a variety of sizes. Generally, if only one antigen is being detected, 10 nm gold conjugate provides a good compromise between ease of viewing without causing steric hindrance. Five and 20 nm of gold can be used to identify two antigens on the same section.

[d] Grids can either be taken through a series of droplets of distilled water or washed by aiming a flow of distilled water from a wash bottle onto forceps so that it trickles over the grid. Do not aim water directly at the grid as this may wash away the sections.

3.2 Immunochemical assays

A wide range of assays are available in which to use antibodies to detect and/or quantify antigens. Here, we describe the most commonly used simple and rapid assays using nitrocellulose and microtitre plates as solid supports. These protocols are not only useful for antibody detection and quantification of antigens, but also invaluable assay systems for screening, selection, and characterization purposes during the preparation of antibody probes whether they be antisera, hybridoma, or phage antibodies. A feature of immuno-dot assays (IDAs, *Protocol 12*) and enzyme-linked immunosorbent assays (ELISAs, *Protocols 13* and *14*) is their immense flexibility. The protocols below are for a guide and starting point only—they can be adapted in many ways to suit many systems and antigens. In depth discussions and details of immunochemical assays and related techniques such as immunoaffinity chromatography are available elsewhere (1, 18).

3.2.1 Immunochemistry on nitrocellulose

Substrates such as nitrocellulose sheets are versatile supports for a range of immunochemical techniques. Their properties allow for ease of antibody application and rapid washing between antibody incubation steps. Antigens can be applied to sheets in a number of ways. Tissue printing onto nitrocellulose has already been covered in *Protocol 8*. Other applications of nitrocellulose for immunochemistry include immunoblotting (western blotting) and immuno-dot-assays (IDAs). Once antigens are applied to nitrocellulose, their detection with antibodies is essentially the same in all cases. *Protocol 12* below is for an IDA, which provides a very rapid qualitative and semi-quantitative analysis of antibody binding. In some cases mixtures of antigens can have differing mobilities during application to the nitrocellulose, resulting in dots of differing dimensions. This may have analytical value and has been discussed elsewhere (19).

Enzyme marker systems that produce an insoluble product are used with nitrocellulose. Most commonly, these are horseradish peroxidase (HRP) and alkaline phosphatase. As an example, we include details of an HRP substrate system in *Protocol 12*. Details of an alkaline phosphatase substrate system are included in *Protocol 9* (see also *Chapter 2*).

Protocol 12

Immuno-dot-assay (IDA)

Equipment and reagents

- Nitrocellulose membrane (200 × 200 mm, Schleicher and Schuell)
- PBS (*Protocol 1*)
- PBS containing 3% fat-free milk protein (MP/PBS)
- Secondary antibody [e.g. anti-rat IgG coupled to HRP (Sigma) or anti-M13 phage coupled to HRP (Pharmacia)]

- Primary antibody
- HRP substrate [25 ml de-ionized water, 5 ml methanol containing 10 mg/ml 4-chloro-1-naphthol, 30 μl 6% (v/v) H_2O_2]
- Pipette capable of accurately delivering 1 μl (e.g. Gilson)

Method

1 Mark out assay sites on an appropriately-sized piece of nitrocellulose by carefully ruled lines with a soft pencil. We use 5 × 5 mm squares.

2 1-μl aliquots of test compounds dissolved in water or appropriate buffer are applied to nitrocellulose and left to dry for 1 h. It is useful to use a dilution series.

3 Incubate sheet in MP/PBS for 1 h to block all binding sites.

4 Incubate in primary antibody for 1.5 h.[a]

5 Wash extensively in PBS.[b]

6 Incubate in secondary antibody (e.g. anti-rat IgG-HRP or anti-M13-HRP conjugate) diluted in the range of 1000-fold in MP/PBS for 1.5 h.

7 Wash extensively in PBS.[b]

8 Determine antibody binding to the nitrocellulose by incubation in enzyme marker substrate (e.g. HRP substrate, prepared immediately before use).

9 When dark dots have appeared at antigen sites, stop reaction by washing extensively in tap water.

[a] Hybridoma supernatants are diluted at least 10-fold in MP/PBS. For phage display antibodies use around 10^{12} phage/ml and see note in *Protocol 9* regarding more stringent washing appropriate to phage particles.

[b] In most cases we find that tap water is a suitable alternative to PBS.

3.2.2 Enzyme-linked immunosorbent assays (ELISAs)

Enzyme-linked immunosorbent assays (ELISAs) are complementary to IDAs in immuno-chemical analyses in that they are capable of providing a quantification of antigen and antibody levels. ELISAs require more expensive equipment than IDAs due to the need for multi-channel pipettes and a microtitre plate reader.

For ELISAs, an antigen is immobilized onto the wells of 96-well microtitre plates. We recommend the use of microtitre plates with absorbent surfaces specially prepared for ELISAs, such as Nunc[TM] immunoplates with Maxisorp[TM] surface. Antibody binding is detected through a secondary antibody using an enzyme-linked system producing a soluble coloured product that is quantified spectrophotometrically. Details for the use of HRP as an enzyme marker are provided here.

Titration of antibody or antigen components is a very important aspect of ELISAs and five- or ten-fold dilution series should be used within columns or rows of the microtitre plates. Two ELISA protocols are provided. *Protocol 13* is a standard antibody capture ELISA in which an antigen is immobilized and antibody binding to it is determined. *Protocol 14* is a competitive-inhibition ELISA in which antibody binding to soluble antigen is assessed.

Protocol 13

Enzyme-linked immunosorbent assay (ELISA)

Equipment and reagents

- 96-well microtitre plates (Nunc-Immuno[TM] plates with Maxisorp[TM] surface, Life Technologies)
- Microtitre plate reader
- Coating buffer (50 mM sodium carbonate, pH 9.6)
- PBS (*Protocol 1*)
- PBS with 3% bovine serum albumin (BSA/PBS)
- Primary antibody
- Secondary antibody (e.g. anti-rat IgG coupled to HRP)
- HRP substrate (18 ml of de-ionized water, 2 ml of 1 M sodium acetate buffer, pH 6.0, 200 μl of tetramethylbenzidene, and 20 μl of 6% H_2O_2)

Method

1. Adjust the sample that is to be used as an immobilized antigen to a concentration of approximately 50 μg/ml in sodium carbonate coating buffer. It is useful to make a dilution series.

2. Add 100 μl of antigen to appropriate wells of a microtitre plate and incubate overnight at 4 °C (or 2 h at room temperature). It is useful to have wells without antigen to determine background signal with no antibody binding.

3. Remove solutions containing antigen.

4. Block all binding sites on the plate with 200 μl/well of BSA/PBS for 2 h at room temperature or overnight at 4 °C.

5. Wash plates with PBS. This is most readily done by the repeated submerging of plates in trays full of PBS, shaking and then forcibly throwing the PBS out (into a sink). This is repeated at least 15 times.[a]

6. Add 100 μl/well of the primary antibody at appropriate dilution in BSA/PBS for at least 1.5 h.[b] Wash plates 15 times as in step 5.

7. Add 100 μl/well of secondary antibody (e.g. anti-rat IgG-HRP conjugate) diluted in the range of 1000-fold in BSA/PBS.

Protocol 13 continued

8 Incubate for at least 1.5 h.

9 Wash plates 15 times as in step 5.

10 Determine antibody binding to the plates by the addition of 150 μl/well of HRP substrate (prepared immediately before use) and watch for appearance of blue colour in wells.

11 Colour development is stopped by the addition of 35 μl/well of 2 N sulphuric acid.

12 Absorbances are determined at 450 nm in a microplate reader.

[a] In our hands, tap water is often a suitable alternative washing medium.

[b] As a guide, hybridoma supernatants are diluted 10-fold followed by 5- or 10-fold dilutions. Phage particles are diluted to around 10^{12} phage/ml followed by 5- or 10-fold dilutions.

In the second type of ELISA, known as competitive-inhibition ELISA, quantification of the antigen-antibody interaction is entirely in the soluble phase (see *Figure 2* and *Protocol 14*). Competitive-inhibition ELISA can provide a very sensitive quantification of all antigens and is favoured for the precise and sensitive estimation of antigen levels or the determination of antibody specificity by extent of its binding to a range of antigens. It is a particularly suitable technique for the assay of antibody binding to haptens or small molecules that do not stick efficiently to microtitre plates. Competitive-inhibition ELISA must be carried out in two steps as shown in *Figure 2*. First of all, the working dilution of the antibody is determined using an antibody capture ELISA (*Protocol 13*) and the titration of the antibody against constant levels of immobilized antigen. A typical plotted result is shown diagrammatically in Figure 2A. An antibody dilution that provides 90% of the maximal binding is used in the second step, because at this dilution the antibody binding is most sensitive to inhibition. Secondly, the inhibition of this level of antibody binding to the immobilized antigen by the presence of soluble antigens or haptens is determined as described in *Protocol 14* and is shown diagrammatically in *Figure 2B*.

Protocol 14

Competitive-inhibition ELISA

Equipment and reagents

- 96-well microtitre plates (Nunc-Immuno™ plates with Maxisorp™ surface, Life Technologies)
- Microtitre plate reader
- Coating buffer (50 mM sodium carbonate, pH 9.6)
- PBS (*Protocol 1*)
- PBS with 3% bovine serum albumin (BSA/PBS)

- Primary antibody
- Secondary antibody (e.g. anti-rat IgG coupled to HRP)
- HRP substrate (18 ml of de-ionized water, 2 ml of 1 M sodium acetate buffer, pH 6.0, 200 μl of tetramethylbenzidene and 20 μl of 6% H_2O_2)

Method

1 Immobilize antigen onto microtitre plates and block with BSA/PBS as in *Protocol 13*.

2 Determine a primary antibody dilution (x) that gives 90% of maximal binding with this antigen in antibody capture ELISA (*Protocol 13* and *Figure 2A*).

Protocol 14 continued

3 Add 50 μl/well of a serially-diluted soluble antigen or hapten in PBS to appropriate wells of a microtitre plate.[a]

4 Add to each well containing antigen from step 3, 50 μl/well of primary antibody in BSA/PBS at a concentration of 2x. This will give a final concentration of antigen of 1 mg/ml (with 10-fold dilutions) and an antibody concentration of x. Agitate the microtitre plate gently to ensure mixing.

5 Incubate for at least 1.5 h.

6 Wash the microtitre plate 15 times (see step 5, *Protocol 13*).

7 Add 100 μl/well of secondary antibody (e.g. anti-rat IgG-HRP conjugate) diluted 1/1000 in BSA/PBS.

8 Incubate for at least 1.5 h.

9 Carry out steps 9–12 in *Protocol 13*.

10 Determine the concentration of antigen required to inhibit antibody binding by 50% as shown in *Figure 2B*. This is known as the IC_{50}. Comparing the IC_{50} of a range of antigens gives a good indication of extent of their recognition by a particular antibody.

[a] As a guide, start at 2mg/ml and prepare 10-fold dilutions.

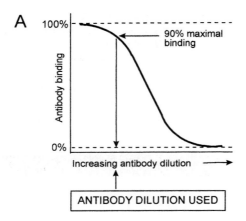

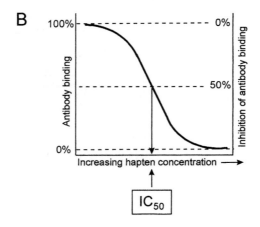

Figure 2 Schematic outline of antibody-antigen binding characteristics in ELISAs. (A) Schematic plot to indicate the determination of the dilution of an antibody required to provide 90% of maximal binding to an immobilized antigen using an antibody capture ELISA. (B) Schematic plot to indicate the determination of the capacity of soluble haptens or antigens to inhibit antibody binding to immobilized antigen by 50% (IC_{50}) in a competitive-inhibition ELISA.

References

1. Harlow, E. and Lane, D. (1988). *Antibodies, a laboratory manual.* Cold Spring Harbor Laboratory Press, New York.
2. Willats, W. G. T., Steele-King, C. G., McCartney, L., Orfila, C., Marcus S. E. and Knox, J. P. (2000) *Plant Physiol. Biochem.*, **38**, 27.
3. Liddell, J. E. and Cryer, A. (1991). *A practical guide to monoclonal antibodies.* John Wiley & Sons, Chichester.
4. Peters, J. H. and Baumgarten, H. (1992). *Monoclonal antibodies.* Springer-Verlag, Berlin.
5. Bazin, H. (1982). In *Protides of the biological fluids* (ed. H. Peters, and L. Pergamon), p. 615. Pergamon Press, New York.
6. Kilmartin, J. V., Wright, B. and Milstein, C. (1982). *J. Cell Biol.*, **93**, 576.
7. Galfrè, G. and Milstein, C. (1981). *Methods Enzymol.*, **73**, 3.
8. Taggart, R. T. and Samloff, I. M. (1982). *Science*, **219**, 1228.
9. Winter, G., Griffiths, A. D., Hawkins, R. E. and Hoogenboom, H. R. (1994). *Ann. Rev. Immunol.*, **12**, 433.
10. Nissim, A., Hoogenboom, H. R., Tomlinson, I. M., Flynn, G., Midgley, C., Lane, D. and Winter, G. (1994). *EMBO J.* **13**, 692.
11. Willats, W. G. T., Gilmartin P. M., Mikkelsen, J.D. and Knox, J. P. (1999). *Plant J.*, **18**, 57.
12. Varner, J. E. and Ye, Z. (1994). *FASEB J.*, **8**, 378.
13. Reid, P. D. and Pont-Lezica, R. F. (1992). *Tissue printing.* Academic Press, San Diego.
14. O'Brien, T. O. and McCully, M. E. (1981). *The study of plant structure. Principles and selected methods.* Termarcarphi Pty. Ltd., Melbourne.
15. Ruzin, S. E. (1999). *Plant microtechnique and microscopy.* Oxford University Press, New York.
16. Hall, J. L. and Hawes, C. (1991). *Electron microscopy of plant cells.* Academic Press, London.
17. Rawlins, D. J. (1992). *Light microscopy.* BIOS Scientific Publishers, Oxford.
18. Johnstone A. H. (1996). *Immunochemistry in practice*, 3rd edn. Blackwell Scientific Publications. Oxford.
19. Willats, W. G. T. and Knox J. P. (1999). *Anal. Biochem.*, **268**, 143.

Chapter 11
Antibody phage display libraries

Kevin C. Gough, Yi Li and Garry C. Whitelam
Department of Biology, Faculty of Medicine and Biological Sciences, University of Leicester, Adrian Building, University Road, Leicester, LE1 7RH, UK

1 Introduction

The display of proteins on the surface of filamentous bacteriophage has led to the development of powerful methods for selecting proteins with particular characteristics, together with the genes encoding them, from mixtures of closely related proteins. Bacteriophage fd is a filamentous, single-stranded DNA phage that infects *Escherichia coli* cells. Rather than lysing their host, filamentous phages are continuously secreted by infected bacterial cells. These bacteriophages can display peptides or proteins on some of their surface proteins with a minimal impairment of their function. Filamentous phages are easily separated from bacteria and so can be readily used as particles displaying proteins or peptides for selection. Furthermore, because more than 10^{13} bacteriophage can be accommodated in 1 ml of solution, huge numbers of phage-displayed proteins can be screened with a minimal investment of time and space.

Peptide phage display technology has proved useful for mapping epitopes of monoclonal antibodies (1, 2) and for the isolation of peptide ligands that bind specifically to a range of 'ligates' (3), including extracellular and intracellular proteins and DNA (4). There is a vast potential for isolating phage-displayed peptides that can interfere with protein–protein interactions (5, 6). For example, phage display peptides have been isolated that bind to the core antigen of hepatitis B virus and interfere with the assembly of virus particles (7). Phage display has also been used to isolate peptides that bind specifically to plant viruses (8).

Numerous phage libraries displaying random peptides have been produced and the size and presentation of the peptides vary considerably among the various libraries. Short peptides, comprising 6–8 amino acids, have been successfully displayed on the major capsid protein, pVIII (9), whereas libraries of peptides ranging from 6 (10) to 38 (11) amino acids have been displayed on the minor capsid protein, pIII. Libraries of longer peptides typically have an incomplete repertoire and can be considered as a library of smaller peptides embedded within variable scaffolds (12). Phage displayed peptides have been presented linearly (13) or 'constrained' between two cysteine residues (14).

Phage display technology has proved to be especially valuable for the selection of high affinity, antigen-specific antibody fragments. Diverse libraries of antibody fragments, derived from immune, naïve, and semi-synthetic sources have been generated and efficiently screened by repeated rounds of affinity selection (panning) followed by amplification via infection of *E. coli*. Antibodies have been displayed on the pIII capsid protein as functional binding molecules in the form of single-chain Fv (scFv) fragments, Fab fragments and *VH* fragments. Antibody phage display technology offers extremely powerful approaches for exploiting the immense diversity of the mammalian immune system, for bypassing immunization, and for the *in vitro* manipulation of the affinity and specificity of antibody binding fragments.

Increasingly, developments in recombinant antibody technology are impacting on plant biology, and the expression of recombinant antibodies and antibody fragments in plants has both a commercial and scientific potential. Developments include the use of crop plants for large-scale production of antibodies for use in therapy and the development of novel plant disease and pest resistance strategies based upon *in planta* expression of antibodies (15, 16). The use of antibody expression to confer disease resistance, an approach that has been called intracellular immunization, often requires production of the appropriate antibody fragment in a particular cell compartment.

Antibody fragments have been produced within the cytosol of transgenic plants, with the protein accumulating at levels up to 0.1% of total soluble protein (17). Targeting antibody fragments via the endoplasmic reticulum to the apoplast has been shown to result in much higher levels of accumulation of functional antibody proteins (18) and the addition of the endoplasmic reticulum retention tetrapeptide KDEL to the C-terminus may increase antibody accumulation even further (19). The addition of a C-terminal KDEL sequence also appears to be a simple strategy to stabilize antibody fragment proteins expressed in the cytosol (20).

Although most research to date on the expression of antibody fragments in plants has utilised the constitutive 35S promoter from cauliflower mosaic virus (21), seed-specific expression of antibody fragments, directed by the *Vicia faba* legumin B4 promoter has been performed (22, 23). Seed-expressed antibody fragments can accumulate to very high levels and are stable for long periods during dry storage of seeds.

2 Naïve, synthetic, and immune scFv phage display libraries

Single-chain Fvs are versatile, small antibody fragments that retain the full antigen-binding site of intact antibodies. These antibody fragments are frequently used in the generation of antibody phage display libraries (24). There are essentially three kinds of scFv phage display libraries: naïve, synthetic, and immune. Naïve and synthetic antibody phage display libraries are constructed in order to bypass immunization. Naïve libraries employ antibody variable (*V*) gene sequences (*VH* and *VL*) from a non-immunized donor. Synthetic libraries use *VH* and *VL* sequences containing synthetic complementarity determining regions (CDRs). These are the hyper-variable peptide loops, within the variable regions, that define the binding activity of the antibody molecule. There are three key advantages to bypassing immunization; first, a very diverse array of different binders can be selected from a single library; secondly, human antibodies can be readily isolated, and thirdly, antibodies directed against toxic molecules can be directly isolated. However, in order to ensure a high probability that an antibody fragment with the desired affinity/specificity will be present these libraries need to be very large (25). For example, recombinant antibody libraries produced at the Medical Research Council (MRC, Cambridge) contain 10^8–10^9 individual clones (26, 27) and that from Cambridge Antibody Technology contains 10^{10} clones (28). Although such libraries have proved to be a valuable source of new binding specificities, they are difficult and expensive to construct. In contrast, when libraries are constructed using *V* gene sequences from animals that have previously been immunized with the target antigen (immune phage display libraries), sequences encoding high affinity antibodies, generated during the immunization processes, should be well represented in the library. Thus, in the case of immune libraries the production of very large libraries should not be necessary to ensure a high likelihood of obtaining antibodies that bind the target antigen with high affinity. In addition, a single immune library can also be generated for the isolation of a range of binders against different targets by simply immunizing an animal with all of the required targets (29, 30). Thus, despite the many successful applications of naïve or synthetic antibody phage display libraries, the immune antibody phage display library approach remains a very attractive, rapid, and economical alternative to hybridomas for the production of high affinity antibodies.

2.1 Construction of scFv phage display libraries

Details of the methods used for the construction of naïve and synthetic antibody phage display libraries can be found in various references (26, 31–34). The following protocols (*Protocols 1–3*) have been developed in our laboratory for the construction of phage display libraries derived from immunized rabbits. Total RNA can be isolated from rabbit spleen or peripheral blood lymphocytes using commercially available kits, such as the Qiagen RNeasy Kit, and first strand-cDNA can be synthesized with a Pharmacia kit using random hexamer primers, according to the manufacturer's instructions. First strand cDNA is used in the amplification of rabbit *VH*, *V*κ and *V*λ genes by PCR. The required primers are listed in *Table 1* (see also 35). The products of *VH* and *VL* PCR are digested with *Sfi* I and *Pfl* MI and ligated into the phagemid vector pSD3 (36) to generate scFv-ΔgIII fusion genes (*Figure 1*).

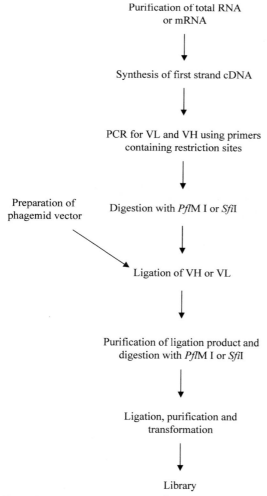

Figure 1 Flow diagram for the cloning of scFv sequences into the phagemid vector pSD3.

Table 1 Primers for amplification of rabbit *VH* and *VL* genes

vh reverse	
rbvh_1	aatgaataatgaaaaccaaggagtgggttctSAGSMGYTGRWGSAGTCCGG
rbvh_2	aatgaataatgaaaaccaaggagtgggttctCAGTCGBTGRAGGAGTCCGRS
rbvh_3	aatgaataatgaaaaccaaggagtgggttctCAGSAGCAGYTGVWGGAGTMCVG
vh forward	
rbch_γ	aagaaatgataaaaaccaccaactggctGACTGAYGGAGCCTTAGGTTGC
vκ reverse	
rbvκ_1	taagtcaatttcaatggcccaaccggccatggctGMBVNHGWKMTGACCCAGACTCCA
rbvκ_2	taagtcaatttcaatggcccaaccggccatggctGCCGAAGTAGTGATGACCCAGACTCCA
rbvκ_3	taagtcaatttcaatggcccaaccggccatggctGCTCAAGTGCTGACCCAGAC
vκ forward	
rbvκ_j1	tatatatataattatggcctccctggccTHBRAYBWCYASHWYGGTCCC
rbvκ_j2	tatatatataattatggcctccctggccTTTGATCTCCAGCTTGGTCTC
vλ reverse	
rbvλ	taagtcaatttcaatggcccaaccggccatggctCAGCCTGTGCTGACTCAGTCG
vλ forward	
rbvλ_j	tatatatataattatggcctccctggccACCTGTGACGGTCAGCTGGGTCC

Notes:
1 *V* gene sequences are in upper case and sequences for introducing restriction enzymes are in lowercase. *Sfi* I and *Pfl* MI sites are underlined
2 Redundant bases in the sequences are M = A/C, R = A/G, W = A/T, S = C/G, Y = C/T, K = G/T, V = A/C/G, H = A/C/T, D = A/G/T, B = C/G/T, N = A/C/G/T.

Protocol 1

Amplification of *VH* and *VL* genes

Equipment and reagents

- First strand cDNA prepared from rabbit RNA
- 50 pmol/μl primers in 10 mM TE pH 8.5
- dNTP mixture containing dATP, dGTP, dCTP, and dTTP, each at a concentration of 10 mM in water (Invitrogen)
- Expand high fidelity Taq and 10× PCR buffer (Boehringer)
- QIAquick DNA gel extraction kit (Qiagen)
- 0.2 ml microcentrifuge tubes
- Thermal cycler

Method

1 Combine the following in a 0.2 ml microcentrifuge tube: 36 μl of water, 5 μl of 10× buffer, 1 μl of dNTP mixture, 1 μl of reverse primer (50 pmol/μl), 1 μl of forward primer (50 pmol/μl), 5 μl of cDNA product, and 1 μl Expand high fidelity Taq.[a]

2 Carry out the PCR reaction as follows: 1 cycle of 95 °C for 2 min, five cycles of 94 °C for 30 s, 50–60 °C for 15–30 s, 72 °C for 30 s, 25 cycles of 94 °C for 30 s, 55–60 °C for 15–30 s, 72 °C for 30 s with a 5 s extension for each cycle, and the PCR is finished with one final cycle of 72 °C for 10 min.

Protocol 1 continued

3 Run a 5µl aliquot of the PCR reaction on a 2% (w/v) agarose gel containing ethidium bromide. The *VH* gene (approximately 400 bp, containing restriction sites for *Pfl* MI) and *Vκ/Vλ* genes (approximately 380 bp, containing the restriction sites for *Sfi* I) should be clearly visible.[b]

4 Four *VH*, or two *Vκ* and one *Vλ* gene products can be mixed and purified using a Qiagen gel extraction kit.

[a] For the amplification of *Vκ*, two forward primers can be mixed, although it is better to use reverse primers separately for each reaction.

[b] If *VH* or *VL* amplification is not satisfactory with the *Sfi* I or *Pfl* MI-containing primers, a primary PCR can be carried out using shorter primers containing only *V* gene sequences. In this case, the PCR products should be approximately 350 bp for *VH* and approximately 330 bp for *VL*. Restriction sites can be added by a second PCR using the primary product as template and the *Sfi* I and *Pfl* MI-containing primers. The PCR programme is essentially the same as that for the primary PCR reaction except that the annealing temperature is 55–60 °C for all of the reactions.

Protocol 2

Preparation of *VH* and *VL* genes for construction of scFvs

Equipment and reagents

- 10× restriction buffer for *Sfi* I and *Pfl* MI (New England Biolabs)
- Restriction enzymes *Sfi* I 20 U/µl, and *Pfl* MI 8 U/µl (New England Biolabs)
- 10 mg/ml BSA (New England Biolabs)
- QIAquick PCR purification kit (Qiagen)
- DNA molecular weight markers (Gibco BRL)
- Mineral oil (Sigma)
- EB buffer: 10 mM Tris–HCl pH 8.5

Method

1 The *Sfi* I and *Pfl* MI PCR products, purified with a Qiagen Kit, are digested as follows: 70 µl of water, 20 µl of 10× buffer, 100 µl of PCR products (approximately 30 ng/µl), 2 µl of BSA (10 mg/ml), 8 µl of *Sfi* I (20 U/µl), or *Pfl* MI (8 U/µl).

2 For *Sfi* I digestion, overlay with mineral oil (to prevent evaporation) and incubate at 50 °C for at least 12 h. For *Pfl* MI digestion, incubate at 37 °C for at least 12 h.

3 Run 10 µl of the reaction mixture on a 2% (w/v) agarose gel containing ethidium bromide and use undigested PCR product as control. The digested product should be clearly smaller. If the digestion is not complete, add more restriction enzymes and incubate for a longer time.

4 Purify the restricted PCR product with a Qiagen PCR purification kit and elute the DNA in 50 µl buffer EB.

5 Estimate the concentration of the purified *VH* or *VL* fragment by comparison on an ethidium bromide-stained agarose gel with linearized DNA concentration standards. Store fragments at −20 °C until ready for ligation into pSD3 phagemid vector.

Protocol 3

Preparation of phagemid vector pSD3 and construction of an scFv phage display library

Equipment and reagents

- T4 ligase and 10× One-Phor-All PLUS buffer (Pharmacia)
- TG1 electroporation competent *E. coli* cells (Stratagene)
- YTE: 15 g Bacto-agar, 8 g NaCl, 10 g Tryptone, 5 g yeast extract in 1 l distilled water
- DYT: 16 g Bacto-tryptone, 10 g Bacto-yeast extract, 5 g NaCl in 1 l distilled water
- YTEG: YTE containing 2% (w/v) glucose
- YTEGa: YTEG containing 100 µg/ml of ampicillin

- DYTG: DYT containing 2% (w/v) glucose
- DYTGa: DYTG containing 100 µg/ml ampicillin
- DYTa: DYT containing 100 µg/ml ampicillin
- QIAquick gel extraction kit (Qiagen)
- QIAquick PCR purification kit (Qiagen)
- Mineral oil (Sigma)
- EB buffer, 10× restriction buffers for *Sfi* I and *Pfl* MI, restriction enzymes *Sfi* I and *Pfl* MI and BSA (see *Protocol 2*)

Method

1 Digest pSD3 vector with *Pfl* MI as described in *Protocol 2*.

2 Run about 0.5 µg of digested product on a 0.6% (w/v) agarose gel to check for complete digestion.

3 Purify the digested product after running on a 0.6% (w/v) agarose gel with Qiagen gel extraction kit.

4 Estimate the concentration of the purified products as described in *Protocol 2*.

5 Set up the following ligation reaction: 8 µl of water, 20 µl of vector (200 ng/µl), 60 µl of *VH* insert (20–40 ng/µl), 10 µl of 10× One-Phor-All buffer PLUS, 2 µl of T4 DNA ligase (6 U/µl).[a]

6 Incubate the reaction at 12 °C for 1 h.

7 Purify the ligation product using a Qiagen PCR purification kit and elute the DNA in 30 µl of buffer EB.

8 Digest the ligated product with *Sfi* I as described in *Protocol 2*.

9 Run about 0.1 µg digested product on a 0.6% (w/v) agarose gel to check digestion.

10 Purify the digested product using a Qiagen gel extraction kit and elute the DNA in 50 µl of buffer EB.

11 Estimate the concentration as described in *Protocol 2*.

12 Set up the following ligation reaction: 39 µl of water, 20 µl of vector (100 ng/µl), 30 µl of *VL* insert (20–40 ng/µl), 10 µl of 10× One-Phor-All buffer PLUS, and 1 µl of T4 DNA ligase (6 U/µl), and incubate at 12 °C for 1 h.

13 Purify the ligated product using a Qiagen PCR purification kit and elute the DNA in 30 µl of buffer EB. Check the final DNA concentration of the purified products as described in *Protocol 2*.

14 Transform 40 µl of *E. coli* TG1 electroporation-competent cells with 1–2 µl of the Qiagen-purified ligation products by electroporation according to the manufacturer's instructions.

15 In order to estimate the size of the library, plate 0.01, 0.1, 1, and 10 µl of the cells onto YTEGa. Plate the rest of the cells onto large plates (244 × 244 mm). Incubate overnight at 30 °C. The library size should be at least 10^7 cfu/µg ligated DNA. This immune library can be used to select binders with very good affinities. Add 6 ml of DYTGa to each plate and loosen the cells from the agar surface, mix with glycerol to give a final glycerol concentration of 15–20% (v/v), divide into 100 µl or 200 µl aliquots and store at −80 °C.

[a] In order to minimize potential loss of repertoire due to the presence of either *Sfi* I of *Pfl* MI sites in *VH* or *VL*, respectively, the ligations should be performed as *VH* then *VL* and as *VL* then *VH*.

3 Affinity selection (panning) of scFvs from a phage display library

Numerous methods for the isolation of scFvs from phage display libraries have been developed (see also Chapter 10). The method of choice may be dictated by the property of the target or by the requirements of the desired antibody. For targets that are available in a purified form, such as single proteins or conjugated haptens, the isolation of scFvs can be carried out by either of two basic methods. In the first, the target is immobilized on a solid phase, such as the well of an ELISA plate, an immunotube or magnetic beads. In the second, the target is maintained in solution during primary binding. This can be achieved by using, for example, a biotinylated target and then precipitating the phage antibody-target complex using streptavidin-coated magnetic beads (37). Recently, a number of selection systems based on complex solid phase antigens, such as viable bacterial cells (38), human erythrocytes (39), or whole fungal germlings (40) have been described. These systems highlight the use of antibody phage display as a tool for the selection of antibodies to specific, individual cell surface antigens.

As an alternative to these conventional screening/selection strategies, a method for isolating specific antibodies based upon biological selection, similar to the two-hybrid principle, has been described (41). In the selectively infective phage (SIP) method the antibody library is displayed on truncated protein III, lacking the N-terminal domain, that is non-functional in F pilus binding and so these phages cannot infect host cells. The target antigen is then expressed in bacterial cells as a fusion with the N-terminal domain of phage protein III. Antibody-antigen interaction reconstitutes a functional protein III and so those phages displaying functional antibody are able to infect host cells. *Protocols 4* and *5* describe preparation of the library and panning of phase antibodies (see also *Chapter 10*).

Protocol 4

Preparation of the phage display library for selection

Equipment and reagents

- DYTak: DYT (see *Protocol 3*) containing 100 μg/ml ampicillin and 50 μg/ml kanamycin
- DYTGa (see *Protocol 3*)
- PEGS: 20% (w/v) polyethylene glycol 8000, 2.5 M NaCl in distilled water
- TE: 10 mM Tris–HCl pH 7.4, 1 mM EDTA
- VCS M13 helper phage (Stratagene)
- YTEGat: YTEGa (see *Protocol 3*) containing 12.5 μg/ml tetracycline
- XL-1-Blue *E. coli* strain (Stratagene)

Method

1. Dilute the library glycerol stock in 50 ml of DYTGa in a 250-ml flask to give a final OD_{600} of 0.025–0.08, and then grow at 37 °C until OD_{600} is in the range of 0.5–0.8 (this takes about 1.5–2.5 h).

2. Infect the cells with 4×10^{11} pfu VCS M13 (helper phage:cell = 20:1), and incubate at 37 °C for 30 min without shaking followed by gentle shaking (200 rpm) for 30 min at 37 °C.

3. Pellet the cells by centrifuging at 3000 g for 10 min at room temperature. Discard as much of the supernatant as possible.

4. Resuspend the cells in 100 ml DYTak in a 500 ml flask and then grow the cells at 30 °C for 12–16 h with vigorous shaking.

Protocol 4 continued

5 Remove the cells by centrifuging at 8000 g for 15 min at 4 °C. To the supernatant add 25 ml of PEGS, mix well, and incubate on ice for 30 min.

6 Precipitate phage by centrifuging at 8000 g for 10 min at 4 °C.

7 Discard the supernatant and resuspend the pellet in 10 ml TE. Add 2 ml of PEGS and incubate at 4 °C for 10 min.

8 Precipitate phage by centrifuging at 4000 g for 10 min at 4 °C and then resuspend the phage in 1 ml TE.

9 Remove bacterial debris by centrifuging at 12 000 g in a microcentrifuge for 5 min.

10 Titre the phage library by infecting fresh mid-log phase XL-1 Blue cells[a] and plating on YTEGat. This protocol will give $1-5 \times 10^{13}$ cfu phage particles. Store the phage at 4 °C for up to 1 month. The best results are obtained after storage at 4 °C for up to 1 week.

[a] *E. coli* XL-1 Blue should be used because the F' pili can be selected and maintained by growing in medium containing tetracycline.

Protocol 5

Affinity selection (panning) of phage antibodies[a]

Equipment and reagents

- 75 × 12 mm maxisorb immunotubes (Nunc)
- PBS: 10 mM phosphate buffer, pH 7.4, containing 2.7 mM KCl and 138 mM NaCl
- PBSEM: PBS containing 3% (w/v) skimmed milk powder and 3 mM EDTA
- PBST: PBS containing 0.1% (v/v) Tween 20
- DYTG and YTEGat (see *Protocols 3* and *4*, respectively)
- DYTGat: DYTGa (see *Protocol 3*) containing 12.5 μg/ml tetracycline
- 100 mM HCl
- 1 M Tris–HCl, pH 8.3
- 50% (v/v) sterile glycerol
- XL-1-Blue *E. coli* strain (Stratagene)

Method

1 Coat a maxisorb immunotube with 0.5–1.0 ml of target antigen at a concentration 10–100 μg/ml[b] in PBS or 50 mM sodium hydrogen carbonate pH 9.6 at 4 °C or room temperature overnight.

2 Rinse the tube three times with PBS and block the remaining protein binding sites by filling the tube with PBSEM and incubating at 37 °C for 2 h.

3 Dilute the antibody phage stock 1:1 in 2× PBSEM to give a total of $10^{12}-10^{13}$ cfu, and incubate at room temperature for 0.5–1 h.

4 Rinse the immunotube three times with PBS.

5 Add 0.5–1 ml of the diluted antibody phage stock to the immunotube and incubate for 15 min with shaking at room temperature followed by a 1–3 h incubation at 37 °C or room temperature without agitation.

6 Rinse the immunotube 10–40 times with PBST and then 10 times with PBS.

7 To elute the bound antibody phage, add 0.5–1 ml of 100 mM HCl and incubate at room temperature for 5 min with gentle tapping. Remove the eluate to a tube containing 0.2 ml of 1 M Tris–HCl pH 8.3 and mix immediately.

Protocol 5 continued

8 Infect 5 ml of freshly prepared mid-log phase XL-1 Blue cells in DYTG containing 12.5 μg/ml tetracycline with 0.3–0.6 ml of eluate. Incubate at 37 °C for 30 min without agitation and then for 30 min, shaking at 200 rpm.

9 Plate 0.1, 1, and 10 μl of the culture on YTEGat to determine the phage output. Concentrate the rest of the cells to 1 ml by centrifuging at 3000 g for 10 min at room temperature, plate onto large YTEGat plates and incubate at 37 °C for 12–16 h.

10 Scrape the colonies into 5 ml DYTGat. After adding glycerol to a final concentration of 15% (v/v), the cells can be stored at −80 °C or used for preparing antibody phage for the next round of selection as described in *Protocol 4*.

11 The rescue/panning procedures described in *Protocols 4* and *5* are repeated as required (usually two to five rounds).

a This basic protocol is for selection of scFv using target antigen immobilized on an immunotube. All other methods can be modified from this.

b In order to isolate very high affinity antibodies (for example, with affinities in the picomolar range), two modifications of the above protocol may be made. The concentration of the immobilized target could be reduced 10–100-fold for each round of selection. Also, 'off-rate selection' may be applied. This could involve the inclusion of antigen in the wash steps (42), the inclusion of a relatively low affinity binder directed against the target in the wash steps (43) or the use of very extended washes (up to 3 h).

4 Isolation and analysis of monoclonal phage antibodies

4.1 Phage ELISA

Monoclonal phage antibodies can be isolated following the final round of affinity selection by taking single colonies from the YTEGat plates and screening for antigen-binding activity by phage antibody ELISA on 96-well plates. Here (*Protocols 6* and *7*), we describe general methods for the preparation and screening of phage antibodies for binders against antigen coated directly onto ELISA plates (see also *Chapter 10*).

Protocol 6

Preparation of phage antibodies for ELISA

Equipment and reagents

- Recipes for YTEGa, DYTGa and DYTak are given in *Protocols 3* and *4*
- VCSM13 helper phage (Stratagene)
- Sterile 96-well microtitre plate (Nunc).

Method

1 Plate out *E. coli* host cells containing phagemid coding for the selected antibody on YTEGa plates[a] and grow overnight at 30 °C.

2 Pick individual colonies and transfer into a 96-well microtitre plate containing 100 μl DYTGa per well and grow overnight at 30 °C.

Protocol 6 continued

3 Inoculate a second microtitre plate containing 96×100 µl DYTGa with the overnight cultures from the original plate. To this original plate add 50 µl of 50% (v/v) glycerol into each well and store at $-80\,°C$. Grow the second plate to mid-log phase at $37\,°C$ with shaking.

4 Add 10 µl of VCSM13 helper phage (at 5×10^{10} pfu/ml) to each well and incubate at $37\,°C$ without shaking for 30 min followed by an incubation at $37\,°C$ with shaking for 30 min. Centrifuge at 8000 g for 10 min. Resuspend the cells in DYTak and grow overnight with shaking at $30\,°C$.

5 Centrifuge at 8000 g for 10 min and transfer 100 µl of each of the supernatants to a fresh microtitre plate.

[a] If XL1-Blue $E.\ coli$ cells are used, tetracycline (12.5 µg/ml) can be included in the media.

Protocol 7

Detection of antibody phage binding by ELISA

Equipment and reagents

- PBS and PBST (see *Protocol 5*)
- PBSM: PBS containing 3% (w/v) skimmed milk powder
- Rabbit anti-fd antibody (Sigma)
- Alkaline phosphatase-conjugated goat anti-rabbit immunoglobulin (Sigma)
- *p*-Nitrophenyl phosphate substrate (Sigma)
- 96-well maxisorb ELISA plates (Nunc)

Method

1 Coat a 96-well ELISA plate with 10 µg/ml of target antigen diluted in PBS overnight at room temperature.[a]

2 Rinse the wells three times with PBS and block the remaining protein-binding sites by adding 400 µl per well of PBSM for 2 h at $37\,°C$.

3 After 1 h, pre-block the phage, prepared as in *Protocol 6*, by adding 20 µl per well of 6x PBSM and incubating for 1 h at $37\,°C$.

4 Wash the antigen-coated ELISA plate three times in PBS and transfer 100 µl of pre-blocked phage into each well. Incubate at $37\,°C$ for 1 h.

5 Wash the wells three times for 2 min with PBST and then three times with PBS.

6 Add 100 µl of a rabbit anti-fd antibody (1/1000 dilution in PBSM) to each well and incubate at $37\,°C$ for 1 h.

7 Wash as in step 5. Add 100 µl of an alkaline phosphatase-conjugated goat anti-rabbit immunoglobulin (1/5000 dilution in PBSM) to each well and incubate at $37\,°C$ for 1 h.

8 Wash as in step 5. Add 100 µl of *p*-nitrophenyl phosphate substrate solution to each well and incubate at room temperature until sufficient colour develops. Read absorbance at 405 nm (this should be in the range 0.2–1.5).[b]

[a] Other antigen coating conditions can be used including variations in antigen concentration, coating in 50 mM $NaHCO_3$ (pH 9.6), and variations in temperature (see also *Chapter 10*).

[b] Other anti-fd detection systems are available. It is recommended to include 'no phage' and 'no antigen' controls to assess background binding of the detection system and that of individual phages.

4.2 Assessment of antibody phage diversity

Following affinity selection some individual antibodies will be represented many times. The diversity of the final selected phage antibodies can be determined by DNA sequencing (33). For scFv fragments, a much quicker and more convenient method is the analysis of *Bst* NI restriction enzyme digest patterns (*Protocol 8*). However, it should be noted that it is possible for different scFvs to generate identical band patterns and a more thorough analysis would require DNA sequencing.

Protocol 8

Assessment of clone diversity by *Bst* NI digestion

Equipment and reagents

- Taq DNA polymerase and 10× Taq DNA polymerase buffer (Bioline)
- dNTP mixture containing dATP, dGTP, dCTP, and dTTP, each at a concentration of 10 mM in sterile water.
- Mineral oil (Sigma)
- *Bst* NI restriction enzyme, BSA (10 μg/μl) and 10x restriction enzyme buffer (New England Biolabs)
- 0.2 ml microcentrifuge tubes
- Thermal cycler

Method

1. Transfer a small amount of the colonies of interest into 0.2-ml microcentrifuge or PCR tubes containing 20 μl aliquots of: 1× Taq DNA polymerase buffer, 1× Primer mix[a] (500 nM each), dNTP mix (200 μM each), 10 U of Taq DNA polymerase, and sterile ddH$_2$O to 200 μl total volume.

2. Overlay each of the mixtures with mineral oil (if thermal cycler is not equipped with a heated lid), and amplify using 30 cycles of 94 °C for 1 min, 55 °C for 1 min, and 72 °C for 2 min, followed by 1 cycle of 72 °C for 10 min.

3. Analyse 5 μl of the PCR mix on a 1.5% (w/v) agarose gel to check that a band of the correct size has been amplified (for scFvs this should be approximately 1 kb).

4. To the aqueous PCR mix (under oil) add: 1× *Bst* NI buffer, BSA (to 100 μg/ml) and 5 U of *Bst* NI. Incubate at 60 °C for 2 h.

5. Analyse digestion products on a 3% (w/v) agarose gel.

[a] The primers should be homologous to the phagemid DNA flanking the antibody gene insert.

5 Expression and characterization of soluble scFvs

Although phage antibodies can be used directly as immunoreagents for a range of applications, soluble scFv proteins have broader applications and may be more easily incorporated into conventional immunochemistry studies. The production of soluble scFv protein following the selection of a suitable phage antibody is fairly straightforward because phagemids routinely include an amber stop codon between gene III and the scFv gene, allowing the production of soluble scFv protein in non-amber suppresser strains of *E. coli* (e.g. HB2151). The phagemid constructs also usually include tags such as c-myc, for detection of soluble scFv, and poly-histidine, allowing affinity purification of the scFv on nickel columns (see *Protocol 9*).

The method of producing soluble scFvs, once the phagemids have been transferred into a non-amber suppresser host strain, is a variation on *Protocol 6*. The following alterations to the method should be made: (1) At step 3, grow cells to mid-log phase at 30 °C; (2) at step 4 do not superinfect with helper phage, but centrifuge the samples at 8000 g for 10 min, resuspend the cells in DYTa (see *Protocol 3*) containing 1 mM isopropyl β-D-thiogalactoside and grow overnight with shaking at 30 °C. Functional analysis of soluble scFv proteins can be carried out by ELISA as in *Protocol 7* using supernatant containing soluble scFv instead of phage and using mouse anti-c-myc antibody (9E10 from Pierce) in step 6 and alkaline phosphatase-conjugated goat anti-mouse immunoglobulin (Sigma) in step 7.

Protocol 9

Preparation of *E. coli* periplasmic fraction and purification of soluble scFv by immobilized metal affinity chromatography (IMAC)

Equipment and reagents

- PPEB: 50 mM Tris–HCl pH8.0, 20% (w/v) sucrose, 1× Complete™ EDTA-free proteinase inhibitor (Boehringer) and 15 mM imidazole
- Ni^{2+} agarose (Qiagen), prepared according to manufacturer's instructions
- IMAC washing buffer: 50 mM NaH$_2$PO$_4$ pH 8.0, 300 mM NaCl, 20 mM imidazole
- Complete™ proteinase inhibitor (Boehringer)

- IMAC elution buffer: 50 mM NaH$_2$PO$_4$ pH 8.0, 300 mM NaCl, 250 mM imidazole
- Centriplus-10 (Millipore/Amicon)
- DYTGa: see *Protocol 3*
- DYTai: DYTa (see *Protocol 3*) containing 1 mM isopropyl β-D-thiogalactoside
- HB2151 *E. coli* strain

Method

1 Transform *E. coli* HB2151 by infection with monoclonal phage. Use a single colony to inoculate 5 ml DYTGa and incubate at 37 °C with shaking overnight.

2 Sub-inoculate 500 ml of DYTGa with 5 ml of the overnight culture, incubate at 37 °C with shaking until OD$_{600}$ = 0.5–1.0.

3 Pellet the cells at 3000 g for 10 min at room temperature. Resuspend the cells in pre-warmed DYTai, incubate at 25 °C with shaking overnight.

4 Harvest the cells by centrifuging at 3000 g for 15 min at 4 °C. Resuspend the cell pellet in 100 ml of periplasmic protein extraction buffer (PPEB), incubate on ice for 15 min with shaking.

5 Precipitate the cells at 8000 g at 4 °C for 30 min, carefully remove the supernatant (periplasmic fraction) and filter it through a 0.2-μm filter.

6 Add the periplasmic fraction at 1 ml/min to a column containing 2 ml of Ni^{2+} agarose.

7 Wash the column with 100 ml of IMAC washing buffer.

8 Elute the poly-histidine tagged protein with 25 ml of elution buffer. Collect 0.9 ml fractions into microcentrifuge tubes containing 0.1 ml of 10× Complete™ proteinase inhibitor.

9 Concentrate the eluted protein to about 1 ml with a Centriplus-10 filter.

10 Store the purified scFv at 4 °C for up to 2 weeks.

5.1 Affinity ranking of soluble scFv proteins

For some applications scFvs with the highest affinities will be required. An indication of monoclonal phage antibody affinity can be obtained by competition ELISA (44). However, a more reliable assessment of affinity can be achieved by k_{off} ranking as determined by surface plasmon resonance (SPR) using soluble scFv (*Protocol 10*).

Protocol 10

Preparation of soluble scFv for k_{off} ranking

Equipment and reagents

- DYTGa: see *Protocol 3*
- DYTai: see *Protocol 9*
- SPR chips (Biacore)
- SB: 10 mM HEPES pH 7.4, 150 mM NaCl, 3.4 mM EDTA, 1 mg/ml chicken egg lysozyme
- HB2151 *E. coli* strain

Method

1 Transform *E. coli* HB2151 by infection with monoclonal phage. Use a single colony to inoculate 5 ml of DYTGa and incubate at 37 °C overnight.

2 Inoculate 10 ml DYTGa with 100–200 µl of the overnight culture and incubate at 37 °C with shaking until OD_{600} = 0.5–1.0.

3 Pellet the cells at 3000 g for 10 min at room temperature. Resuspend the cells in DYTai and incubate at 30 °C with shaking overnight.

4 Pellet the cells at 3000 g for 15 min at 4 °C. Resuspend the cells in 0.5 ml of sonication buffer (SB) and incubate at 37 °C for 30 min. Lyse cells by sonicating for pulses of 15 s with 15 s pauses between each sonication.[a]

5 Centrifuge at 12 000 g at 4 °C for 10 min to precipitate the cell debris.

6 Apply the supernatant to an SPR chip coated with about 1000 RU (resonance U) of target. The binders with the lowest k_{off} are more likely to be those with a higher affinity.

[a] A time course of protein release during sonication should be carried out to optimise cell lysis.

5.2 Affinity maturation of scFvs

The affinities of scFvs can be accurately determined using SPR methods (45). For scFvs isolated from immune phage display libraries, affinities are normally expected to be in the nanomolar range. With a rabbit immune phage display library of around 10^7 individual colonies and close to 100% diversity, as determined by *Bst* NI digestion, one should be able to isolate scFvs with affinities in the sub-nanomolar range using the optimized isolation protocol (see *Protocol 5*, note b). For naïve or semi-synthetic libraries, affinities of isolated scFv proteins will be very dependent upon the size and diversity of the library. If an scFv is isolated that has the desired specificity, but an affinity that is not adequate for the intended use, it is possible to improve affinity by *in vitro* maturation. Several methods, including CDR walking mutagenesis (43), error prone PCR, and the use of mutator strains of bacteria (24, 42), have been described for this purpose.

6 The use of isolated phage antibodies as immunoreagents

Antibodies isolated from phage display libraries can be used as immunological reagents, either directly, displayed on phage, or as soluble scFv proteins or fusion proteins. Phage antibodies have been used for ELISA (see *Protocol 7*), western blotting (33) and for immunocytochemistry, in both fluorescence (40) and electron microscopy (46). *Figure 2* shows the use of phage scFv to visualize the distribution of cognate epitopes on the surface of the plant pathogenic fungus *Phytophthora infestans*. The phages displaying the scFvs were produced

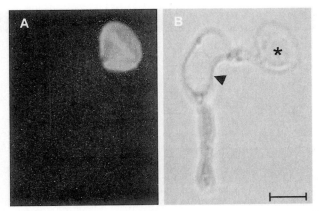

Figure 2 Mapping of a surface epitope of *Phytophthora infestans* germlings with a monoclonal phage displayed antibody. Germlings were immobilized on an ELISA plate and detected using 10 μl of monoclonal phage displayed scFv purified by polyethylene glycol precipitation. Binding was detected using a rabbit anti-fd antibody and a goat anti-rabbit immunoglobulin fluorescein conjugate. Panel A shows the fluorescent image and panel B the corresponding image taken under normal light conditions. The zoospore body (asterisk) and appressorium (arrow head) are labelled and a scale bar of 10 μm shown (see plate 2).

as in *Protocol 4* and used as primary antibody to stain *P. infestans* germlings. The bound phages were visualized by incubation with rabbit anti-fd immunoglobulins followed by fluorescein conjugated goat anti-rabbit immunoglobulins.

To enhance their utility as immunological reagents, scFvs can be produced as fusion proteins. Numerous fusion partners, including those that aid scFv protein isolation, scFv detection or scFv immobilization, have been employed. Alkaline phosphatase-scFv fusions make convenient dimeric immunoreagents that are readily detectable (47, 48). In addition, the coupling of scFvs to a fusion partner may facilitate the correct folding of the scFv in the *E. coli* periplasm and therefore lead to an increased yield of functional antibody (48).

7 Conclusion

The surface display of proteins on bacteriophage, which links proteins presented for binding with the DNA encoding them, is revolutionizing methods for the production of antibodies. In many application areas, including the plant sciences, the demand for efficient generation of functional antibody fragments will be expected to increase. To meet this demand, it is anticipated that there will be an increase in the number, quality, and availability of naïve and synthetic phage libraries. In the near future it may become routine to remove an aliquot of such a library from the freezer, screen it with the antigen of interest, and have antibodies available for use within a couple of weeks. In the meantime, although large prefabricated antibody libraries are increasingly being used as a source of new antibodies, for many applications it is still necessary to exploit the diversity of the immune system to generate extensive panels of high affinity antibodies against a given target. Here, immune antibody phage display libraries offer an economical and rapid alternative to hybridoma technology.

References

1. Smith, G. P. (1985). *Science*, **228**, 1315.
2. de la Cruz, V., Laa, A. and McCutchan, T. (1988). *J. Biol. Chem.*, **263**, 4318.
3. Cortese, R., Monaci, P., Nicosia, A., Luzzago, A., Felici, F., Galfre, G., Pessi, A., Tramontano, A. and Sollazzo, M. (1995). *Curr. Opin. Biotechnol.*, **6**, 73.

4. Kay, B. K. (1995). *Perspect. Drug Discovery Des.*, **2**, 251.
5. Bottger, V., Bottger, A., Howard, S. F., Picksley, S. M., Chene, P., Garcia-Echeverria, C., Hochkeppel, H. K. and Lane, D. P. (1996). *Oncogene,* **13**, 2141.
6. Liang, O. D., Preissner, K. T. and Gursharan, S. C. (1997). *Biochem. Biophys. Res. Commun.*, **234**, 445.
7. Dyson M. R. and Murray K. (1995). *Proc. Natl Acad. Sci. USA*, **92**, 2194.
8. Gough, K. C., Cockburn, W. and Whitelam, G. C. (1999). *J. Virol. Meth.*, **79**, 169.
9. Kishchenko, G., Baltiwala, H. and Makowski, L. (1994). *J. Mol. Biol.*, **241**, 208.
10. Scott, J. K. and Smith, G. P. (1990). *Science*, **249**, 386.
11. Kay, B. K., Adey, N. B., He, Y-S., Manfredi, J. P., Mataragnon, A. H. and Fowlkes, D. M. (1993). *Gene*, **128**, 59.
12. Winter, J. (1996). In *Protein engineering. Principles and practice* (ed. J. L. Cleland and C. S. Craik), p. 349. J. Wiley and Sons, New York.
13. Felici, F., Castagnoli, L., Musacchio, A., Jappelli, R. and Cesareni, G. (1991). *J. Mol. Biol.*, **222**, 301.
14. Luzzago, A., Felici, F., Tramontano, A., Pessi, A. and Cortese, R. (1993). *Gene*, **129**, 51.
15. Ma, J. K-C. and Hiatt, A. (1996). In *Transgenic plants: A production system for industrial and pharmaceutical proteins* (ed. M. R. L. Owen and J. Pen), p. 229. J. Wiley and Sons, Chichester.
16. Whitelam, G. C. and Cockburn, W. (1998). In *Transgenic plant research* (ed. K. Lindsey), p. 219. Harwood Academic Publishers, Netherlands.
17. Owen, M., Gandecha, A., Cockburn, W. and Whitelam, G. C. (1992) *Bio/Technol.*, **10**, 790.
18. Firek, S., Draper, J., Owen, M. R. L., Gandecha, A., Cockburn, B. and Whitelam, G. C. (1993). *Plant Mol. Biol.*, **23**, 861.
19. Artsaenko, O., Peisker, M., zur Neiden, U., Fiedler, U., Weiler, E. W., Muntz, K. and Conrad, U. (1995). *Plant J.*, **8**, 745.
20. Schouten, A., Roosien, J., deBoer, J. M., Wilmink, A., Rosso, M. N. and Bosch, D. (1997). *FEBS Lett.*, **415**, 235.
21. Nagata, T., Okada, K., Kawazu, T. and Tabeke, I. (1987). *Mol. Gen. Genet.*, **207**, 242.
22. Fiedler, U. and Conrad, U. (1995). *Bio/Technol.*, **13**, 1090.
23. Phillips, J., Artsaenko, O., Fiedler, U., Horstmann, C., Mock, H-P., Muntz, K. and Conrad, U. (1997). *EMBO J.*, **16**, 4489.
24. Hoogenboom, H. R., de Bruine, A. P., Hufton, S. E., Hoet, R. M., Arends, J-W. and Roovers, R. C. (1998). *Immunotechnol.*, **4**, 1.
25. Winter, G., Griffiths, A. D., Hawkins, R. E. and Hoogenboom, H. R. (1994). *Ann. Rev. Immunol.*, **12**, 433.
26. Nissim, A., Hoogenboom, H. R., Tomlinson, I. M., Flynn, G., Midgley, C., Lane, D. and Winter, G. (1994). *EMBO J.,* **13**, 692.
27. www.mrc-cpe.cam.ac.uk/phage/index.html.
28. Vaughan, T. J., Williams, A. J., Pritchard, K., Osbourn, J. K., Pope, A. R., Earnshaw, J. C., McCafferty, J., Hodits, R. A., Wilton, J. and Johnson, K. S. (1996). *Nature Biotechnol.*, **14**, 309.
29. Ghahroudi, M. A., Desmyter, A., Wyns, L., Hamers, R. and Muyldermans, S. (1997). *FEBS Lett.,* **414**, 521.
30. Li, Y., Cockburn, W., Kilpatrick, J. B. and Whitelam, G. C. (2000). *Biochem. Biophys. Res. Commun.*, **268**, 398–404.
31. Barbas III, C. F., Bain, J. D., Hoekstra, D. M. and Lerner, R. A. (1992). *Proc. Natl Acad. Sci. USA*, **89**, 4457.
32. Gram, H., Marconi, L-A., Barbas III, C. F., Collet, T. A., Lerner, R. A. and Kang, A. S. (1992). *Proc. Natl Acad. Sci. USA*, **89**, 3576.
33. Marks, J. D., Hoogenboom, H. R., Bonnert, T. P., McCafferty, J., Griffiths, A. D. and Winter, G. (1991). *J. Mol. Biol.*, **222**, 581.
34. Jackson, R. H., McCafferty, J., Johnson, K. S., Pope, A. R., Roberts, A. J., Chiswell, D. J., Clackson, T. R., Griffiths, A. D., Hoogenboom, H. R. and Winter, G. (1992). In *Protein engineering. A practical approach* (ed. A. R. Rees, M. J.E. Sternberg and R. Wetzel), p. 277. Oxford University Press, Oxford.
35. Ridder, R., Schmitz, R., Legay, F. and Gram, H. (1995). *Bio/Technol.*, **13**, 255.
36. Li, Y., Cockburn, W., Kilpatrick, J. B. and Whitelam, G. C. (1999). *Food Agric. Immunol.*, **11**, 5.
37. Hawkins, R. E., Russell, S. J. and Winter, G. (1992). *J. Mol. Biol.*, **226**, 889.
38. Bradbury, A. and Cattaneo, A. (1993). *Bio/Technol.*, **11**, 1565.

39. Marks, J. D. and Winter, G. (1993). *Bio/Technol.*, **11**, 1145.
40. Gough, K. C., Li, Y., Vaughan, T. J., Williams, A. J., Cockburn, W. and Whitelam, G. C. (1999). *J. Immunol. Meth.*, **228**, 99.
41. Spada, S., Krebber, C. and Pluckthun, A. (1997). *J. Biol. Chem.*, **378**, 445.
42. Schier, R., McCall, A., Adams, G. P., Marshall, K. W., Merritt, H., Yim, M., Crawford, R. S., Weiner, L. M., Marks, C. and Marks, J. D. (1996). *J. Mol. Biol.*, **263**, 551.
43. Yang, W. P., Green, K., Pinz-Sweeney, S., Briones, A. T., Burton, D. R. and Barbas , C. F., III (1995). *J. Mol. Biol.*, **254**, 392.
44. Li, Y., Owen, M. R. L., Cockburn, W., Kumagai, I. and Whitelam, G. (1996). *Protein Eng.*, **9**, 1211.
45. Nieba, L., Krebber, A. and Pluckthun, A. (1996). *Anal. Biochem.* **234**, 155.
46. Zebedee, S. L., Barbas, C. F. III, Hom, Y. L., Caothien, R. H., Graff, R., DeGraw, J., Pyati, J., LaPolla, R., Burton, D. R., Lerner, R. A. and Thornton, G. B. (1992). *Proc. Natl Acad. Sci. USA*, **89**, 3175.
47. Gandecha, A., Owen, M. R. L., Cockburn, W. and Whitelam, G. C. (1994). *Protein Express. Purif.*, **5**, 385.
48. Griep, R. A., van Twisk, C., Kerschbaumer, R. J., Harper, K., Torrance, L., Himmler, G., van der Wolf, J. M. and Schots, A. (1999). *Protein Express. Purif.*, **16**, 63.

Section 3 Functional analysis *in vivo*

Chapter 12

Spatial and temporal measurements of calcium ions in living plant cells

Rui Malhó, Luísa Camacho, and Ana Moutinho

Department Biologica Vegetal, FCL Bloco C2, University of Lisbon, Campo Grande, 1749-016 Lisbon, Portugal

1 Introduction

Signalling pathways operating in plant cells are complex and recruit many molecular building blocks to modulate physiological responses. Cytosolic free calcium ($[Ca^{2+}]_c$) is now recognized to be the most common second messenger in eukaryotic cells [1, 2]. To assign such a role to one ion implies that its concentration must be measured in living cells with an accurate spatial and temporal resolution. This implies that any imaging method must not interfere with cell physiology and it must allow precise quantification. The development of fluorescent Ca^{2+}-sensitive dyes [3] made this possible and triggered a technological revolution in imaging systems. Later, transgenic approaches have allowed bioluminescent measurements of $[Ca^{2+}]_c$ in cells expressing the calcium-sensitive photoprotein aequorin [4]. These two methodologies have undergone significant improvements over the years and measurements have been extended to several sub-cellular compartments. Technologies are still not without problems and throughout this chapter we will discuss some of the limitations, as well as relative advantages and enormous potential for signal transduction studies.

2 Fluorescent probes for intracellular Ca^{2+}

2.1 Choosing the correct dye

Ca^{2+} dyes are one of the most powerful tools for quantifying the spatial and temporal dynamics of $[Ca^{2+}]_c$. These compounds have been used in a variety of forms (free acids, AM esters, or conjugated with high molecular weight dextrans) and loaded by different methods (Section 3). The list of Ca^{2+} dyes is extensive, and the reader can find a detailed description of their characteristics and spectral properties in the *Molecular Probes* handbook [5]. When choosing a dye for a particular experiment, the following parameters should be considered:

1 *Excitation and emission wavelengths*: these must be compatible with the imaging system and must avoid problems of sample autofluorescence at the wavelengths to be used.

2 *Quantum efficiency*: the dye should absorb and emit photons efficiently. In other words, it should be as bright as possible.

3 *Dynamic range*: this is the range over which concentrations can be measured. Generally, indicators have a detectable response in the concentration range from approximately $0.1 \times K_d$ to $10 \times K_d$ [5].

4 *Photobleaching*: the rate of loss of fluorescence by photo-irradiation determines the temporal resolution of the experiment and may distort the data collected.

5   *Kinetics of ion-dye interaction*: this is usually not a problem because the time scale of ion-dye
    binding and dissociation is faster than image collection with spatial detectors (Section 4.3).

2.2 Single and dual wavelength dyes

Ion-sensitive dyes can be divided into two main classes: single wavelength (SW) or dual
wavelength (DW) dyes. SW dyes (e.g. Fluo-3 and Calcium Green-1) have one major excitation/
emission peak at which the fluorescence emitted is proportional to the ion concentration,
thus making them suitable to measure relative changes in $[Ca^{2+}]_c$, but not absolute concentrations (3); SW dyes are excited by visible light, which accounts for their popularity
among laser confocal microscope users. However, due to intracellular asymmetries in
viscosity, dye concentration, and photobleaching (all affecting the emission intensity of the
dye), SW dyes cannot be used for absolute quantification of $[Ca^{2+}]_c$. Consequently, DW dyes
(such as Fura-2 and Indo-1), which exhibit spectral shifts upon ion binding and can be used for
more precise ratiometric quantification of $[Ca^{2+}]_c$, are preferable in most systems (3). An
alternative procedure to perform ratio measurements is the simultaneous use of two dyes (6,
7), e.g. a Ca^{2+}-insensitive dye that acts as a volume indicator (such as Texas Red or Rhodamine
B), which is used in ratio against a Ca^{2+}-sensitive dye (such as Fluo-3 or Calcium Green-1). This
technique is prone to many artefacts (Section 5.3) and its efficiency varies greatly with the
biological system (8), so results should be interpreted with caution. However, it does allow
ratiometric measurements without UV excitation, which is required for most DW dyes and
that standard confocal microscopes do not have.

Ratio methods should be used whenever a quantification of $[Ca^{2+}]_c$ is essential. However,
this is not always the case and SW dyes do have some advantages over DW dyes, which can
make them useful reporters of changes in $[Ca^{2+}]_c$. The advantages are:

1   *Higher quantum yield* and, hence, reduced cytotoxicity (less dye is required). The buffering
    effect of the dyes may not be enough to perturb large vital processes, but enough to mask
    small $[Ca^{2+}]_c$ changes. *In situ*, free dyes usually bind to proteins, which may increase their
    K_d 4-fold or more, making it more difficult to detect minor changes (9). This may vary with
    loading time, concentration, and also choice of indicator. The use of dextran-coupled dyes
    may partially overcome this problem, since they have a much lower mobility.

2   *Higher stability and reduced photobleaching,* allowing more time for the experiment. The use
    of dextran-conjugates is particularly meaningful because they have long retention times
    within the cytosol compared with the free dyes that tend to become sequestered in the
    vacuole after a certain time.

3   *Higher K_d*, meaning sensitivity to a wider range of ion concentration.

4   *Excitation with visible light*, which allows imaging on a standard confocal microscope, avoids
    UV damage, and permits the simultaneous use of UV-photo activated caged probes if
    desired (10).

Furthermore, difficulties in calibration of SW dyes (Sections 5.3 and 5.4) can be ameliorated
by ratioing the signal emitted by the cell at two different (but close) time intervals (2, 10). The
obtained ratio is an accurate indication of changes in ion concentration experienced by the
cell in that period of time, even though it gives no information on $[Ca^{2+}]_c$ concentration.
However, this is probably the most meaningful information, because even with ratio dyes a
perfect calibration is impossible to achieve (Section 5.4).

3 Loading of Ca^{2+}-sensitive dyes into cells

Introducing Ca^{2+}-sensitive dyes into plant cells is, at best, a complex task involving several
types of controls. Dyes are either electrically charged or coupled to large molecules (such as

dextrans or lipophilic chains), which renders them impermeable. Thus, special loading techniques had to be developed. Here, we will refer only to the most significant ones.

Important points to consider when selecting a loading method are:

1 Appropriateness of the method to the type of experiment, e.g. the number of cells that must be loaded in a given time.

2 Type of equipment and reagents available and their compatibility with live cell experimentation.

3 Loading and retention characteristics of the probe within the subcellular compartment of interest (e.g. cytosol, mitochondria, plasma membrane).

4 Dye concentration required for optimal imaging and which does not compromise cell viability, e.g. high concentration of free acids may increase the ion diffusion rate, while high dextrans may cause the opposite effect.

The three major forms in which dyes are available are as free acids, AM esters, and dextran conjugates. *Table 1* summarizes our experiences with each of them. Nevertheless, it should be

Table 1 Results obtained with different dye forms and loading methods in *Agapanthus umbellatus* pollen tubes.

Dye form	Loading method	Comments
Free acids	Ionophoresis: good % of success but dependent on dye viscosity; induces changes in membrane potential	Permits loading of only a few cells
		Short time window for experiments because dye is sequestered
	Pressure microinjection: % of success reduced and dependent on dye viscosity; difficulty increased by co-injection of dyes.	Perturbs cell growth (arrest after loading), requires control of physiological parameters
		Ratio procedure required to observe $[Ca^{2+}]_c$ gradient[a]
		Ratio procedure not required to observe $[Ca^{2+}]_c$ changes
AM esters	Ester loading (incubation in the medium): % of success extremely variable, depends on dye and loading conditions. Dye must be freshly prepared from frozen stock solutions.	Permits loading of large number of cells
		Short time window for experiments because dye is sequestered
		High concentration in cell possible—good SNR ratio; potential buffer effect
		Ratio procedure required to observe $[Ca^{2+}]_c$ gradient[a]
		Ratio procedure not required to observe $[Ca^{2+}]_c$ changes
Dextrans	Ionophoresis: introduces primarily low-molecular weight contaminants; not recommended	Permits loading of only a few cells
		Long imaging times possible—no sequestration
	Pressure micro-injection: % of success very reduced and dependent on dye; difficulty increased by co-injection of dyes.	Concentration of dye in cell dependent on volume injected (variable)
		Perturbs cell growth
		Ratio methods provide good spatial resolution, but is not essential to observe $[Ca^{2+}]_c$ gradient[a] or $[Ca^{2+}]_c$ changes.

SNR—signal-to-noise ratio.

[a] Growing pollen tubes possess a tip-focused $[Ca^{2+}]_c$ gradient restricted to the first 5–10 μm of the cell tip with apical levels ~1 μM and basal levels ~200–250 nM (49, 50).

noted that, for each dye and each type of cell, unique situations will probably arise. In these circumstances, a careful reading of the extensive literature available, and inquiries to other research groups is probably the best solution.

3.1 AM-ester loading

Ester loading (7, 11–14) consists of exposing the cells to an acetoxymethyl esterified form of the dye, which is ion-insensitive and non-fluorescent. This 'lipophilic' form readily crosses the plasma membrane and once inside the cytosol is cleaved by endogenous esterases, releasing the ion-sensitive free acid form of the dye. This method allows loading of a large number of cells in a rapid and non-expensive manner and has become very popular in animal cell research. In plant cells, there are several problems associated with the use of AM-esters (15):

1 Inability to achieve sufficient loading for imaging. Due to the presence of apoplastic and/or extracellular esterases in some cell types loading is not possible. Attempts to circumvent this problem include: (i) the use of permeabilizing agents such as dimethyl-sulfoxide (DMSO) and Pluronic F-127 (Molecular Probes) to help dye dispersion (16); and (ii) loading at low temperatures (17).

2 False loading by accumulation of dye in the apoplast. This artefact can be tested by confocal sectioning or by the addition of a small concentration of the membrane-impermeant EGTA. This molecule binds to Ca^{2+} and thus decreases the dye fluorescence.

3 Intense sequestration of dye into any organelle that has esterases. Because organelles have high Ca^{2+} concentrations, sequestration can be seen as bright spots giving the image a 'patchy' appearance; loading in sub-resolution organelles may not be visible. As a consequence, sequestration makes cytosolic measurements impracticable (Section 7.1). Anion channel blockers such as probenecid have been used to reduce sequestration, but without significant success (12). The high levels of dye sequestration in plants may, however, be used to load organelles with low affinity Ca^{2+} dyes (e.g. mag-fura-2) and to measure Ca^{2+}-fluxes across membranes (18).

4 Cytotoxicity: the AM-esters and/or end-products of ester hydrolysis are toxic, particularly after photo-irradiation.

5 Altered spectral properties and background fluorescence induced by incomplete cleavage of the esters.

Some of these problems (3–5) also affect animal cells, albeit to a lesser extent (19). In conclusion, the use of AM dyes in plant cells is possible though tricky and may be useful for very short time experiments, where a population of cells needs to be analysed or to obtain preliminary data. In other circumstances, AM dyes should be avoided.

3.2 Acid loading

A different strategy to introduce small free acid dyes into cells consists of lowering the pH of the incubation medium (to pH 3–5) during the loading period (7, 12–14). At this extracellular pH the dye becomes protonated and is able to cross the plasma membrane. Once in the cytosol (where pH is higher) the dye dissociates and gets trapped. This method has some problems in common with AM-ester loading (difficulty of loading cells with enough dye, binding to the cell wall, sequestration), as well as some other drawbacks:

1 A lowering of extracellular pH may induce physiological changes in the cells.

2 The free dye is fluorescent, so its complete removal from the extracellular medium is imperative before initiating an experiment.

For these reasons this method has been progressively abandoned by most plant research groups.

3.3 Electroporation

Electroporation methods have been adapted to introduce fluorescent probes inside cells (20), but so far its application to plant cells has been restricted, possibly because of difficulties in developing an optimized protocol. One limitation of the technique is the relatively small size of the molecules (<40 kDa) that can be introduced without compromising cell viability.

3.4 Micro-injection

Micro-injection is a difficult methodology that allows only a reduced number of cells to be loaded, and which requires specialized expensive equipment. However, in most cases it is the only reliable loading technique.

Two micro-injection methodologies are usually employed: ionophoretic and pressure injection (*Table 1*; 7, 12). Ionophoresis is used to load small charged molecules, such as free acids, and consists of the application of a small electrical current between the injection pipette and the cell to be injected. This can cause cell damage (e.g. changes in membrane potential) and furthermore, the size of the molecules that can be delivered is restricted (<10 kDa). Ionophoresis can also induce sequestration (15; *Figure 4A*), which limits the experimentation time to around 30 min (dependent on cell type). However, if carefully performed, success rates of 80–90% can be achieved. This method requires a power supply to deliver the current (e.g. Clark Electromedical Instruments), which can also be easily manufactured in-house.

Pressure micro-injection is the most difficult loading method and with the lowest percentage of success. It is, however, the only method that allows reliable micro-injection of large dextran dyes. It consists of pressure injecting (e.g. using water or compressed air) a volume of solution into the cell and requires injection needles wider than those used for ionophoresis. The maximum size of the apical aperture is determined by the cell tolerance to impalement, which limits the size of the molecules that can be introduced (in our experience up to 170 kDa in 10-μm diameter pollen tubes). Pressure micro-injection devices vary from very sophisticated (and expensive, e.g. from Eppendorf) to very simple (12), but ultimately the most important factor is the delivery of the dye in a careful and controlled manner by the operator. Particular attention should be paid to:

1   *The injected volume*: should be as low as possible, which means that the dye concentration in the needle should be very high.
2   *A potential buffering/toxic effect* may occur when the dye enters the cell: the injected volume should not be delivered all at once.
3   *Osmolarity*: the dye should be diluted in an injection buffer with an osmolarity and ionic composition similar to the cytosol.
4   *Cell turgor pressure*: the needle should be pressurized before injection to avoid rapid loss of turgor pressure upon impalement.

The main disadvantages of micro-injection are:

1   Equipment involved: good micromanipulators (e.g. from Narishige) and needle pullers (e.g. from Sutter Instrument Co.) are essential. Often the biggest limitation in micro-injection is in obtaining good injection needles. We currently pull needles from capillary glass with inner filaments and with an outer diameter of 1.2 mm (Clark capillaries). Microscope stability is also an important requirement and a vibration free table can be necessary. However, a good solid table with thick rubber feet and a marble top is often sufficient.

2 Damage induced by impalement and (more critical) following removal of the needle from the cell. We find needle removal to be the most difficult step and the time required vary from 5 s to 2 min. Patience, as well as skill, are essential.

3 Only cells larger than 5 μm in diameter can be injected.

4 Microscope and imaging setting

4.1 Fluorescence microscopes

With the exception of bioluminescence methods (see later), any spatial imaging of ion dynamics requires the use of a conventional fluorescence microscope or of a laser confocal microscope. Fluorescence microscopes use incident illumination (epi-fluorescence) to deliver light of a defined wavelength to the specimen; the excitation source is usually a xenon/mercury lamp and the choice of wavelengths is done by attached filters (*Figure 1*, left panel; 5, 7, 13, 15). In laser confocal microscopes, the light source is a laser emitting at defined wavelengths; through variable pinhole apertures, the light detected is confined to the focal plane (*Figure 1*, right panel; 21). When buying a new microscope or modifying a pre-existing one the researcher should consider (*Table 2*):

1 Compatibility with required excitation source (lamp/laser) and other peripherals (motorized stage, focus motor). Most manufacturers provide compatible and/or custom made solutions to adapt confocal or wide-field systems. If a laser scanning confocal microscope

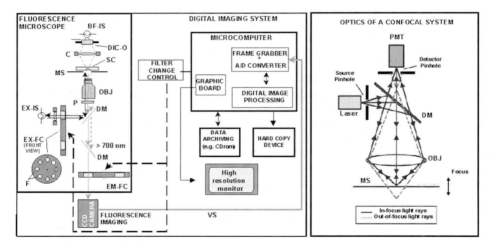

Figure 1 (Left panel) Conventional fluorescence imaging system allowing bright-field and epifluorescence imaging (adapted from 7). The set-up in our laboratory consists of an Olympus IX-50 inverted microscope equipped with long working distance objectives, Ludl BioPoint filter wheels, and a V-scan cool CCD (Photonic Science). The V-scan camera has a frame transfer architecture and requires no shutter. Integrated control of filter wheel and image acquisition is achieved by Image-Pro Plus 4.0 and Scope-Pro 3.1 (Media Cybernetics). Settings for image acquisition (camera exposure time, filters, time interval, and storing modes) are determined by custom-made macros. Hard copies are produced with an Epson Stylus Photo on Epson photo paper. (Right panel) Simplified optics of a laser confocal microscope; pinhole apertures at the excitation light source and at the light detector reduce out-of-focus light thus increasing z-axis resolution. The set-up in our laboratory consists of a Bio-Rad MRC600 equipped with enhanced PMTs and a Kr–Ar laser, coupled to a Nikon Microphot FX-A upright microscope. BF–IS = bright-field illumination source; C = condenser; DIC–O = DIC optics; EM–FC = emission filter wheel L EX–IS = excitation light source; EX–FC = excitation filter wheel; F = filters; MS = microscope stage; OBJ = objective; P = prism; PMT = photomultiplier tube; SC = specimen chamber; VS = video signal.

Table 2 Comparison of measurement techniques used to monitor spatial ion imaging in plant cells

Technique	Comments
Confocal microscopy	Monitors single or population of cells in intact tissue. The (x,y) and (z) resolution are relatively well defined (possible misrepresentation due to spherical and chromatic aberrations when imaging planes buried deep in the sample). The fastest temporal resolution is dependent on the instrument and the volume sampled; typically 0.1–1 s for a 2D section, and seconds to minutes for 3D data stack.
	Simultaneous dual excitation and dual emission experiments easy to implement, but lower optical transfer efficiencies require stronger irradiation or more probe inside the cells.
	Excitation wavelengths are limited by available lasers and most probes suitable for ratio imaging of Ca^{2+} require a UV laser system.
	Very expensive.
Video imaging	Two main types: tube cameras and CCD cameras.
	Monitors single or a population of cells in 'thin' tissues. Subcellular regions typically down to 0.3–0.4 μm in (x,y), but (z) is poorly defined. Simultaneous dual excitation and dual emission experiments is possible with a single camera, but it requires split-view optics.
	Sampling interval possible every 0.1–0.2 s (at maximum resolution) but depends greatly on excitation source and emitted quantum yield. Sampling may need to be intermittent to reduce photobleaching.
	High sensitivity (particularly in cooled CCDs); recent architectures allow bioluminescence imaging at the cellular level. Extended integration to increase SNR possible. Autofluorescence subtraction is relatively easy.
	Should be combined with deconvolution software and computer-controlled filter wheels, shutters and z-focus.
	Expensive, but price can be substantially reduced if components are bought separately.
Multi-photon imaging	Monitors single or population of cells in intact tissue. The (x,y) and (z) resolution are relatively well defined, but the depth penetration into thick tissues is much better than a conventional confocal or epifluorescent system. UV irradiation is not required, thus minimizing tissue damage. However, the energy used can 'microwave' the specimen. Excitation is restricted to the focal point, thus minimizing photobleaching. Temporal resolution equivalent to a conventional confocal system. Simultaneous dual excitation is very difficult to implement.
	Excitation spectra are much broader and not yet well defined.
	Extremely expensive.

(LSCM) is the choice of imaging system it is necessary to check its wavelength excitation and emission characteristics, as well as its suitability for sub-second real time imaging.

2 Suitable emission and excitation ports for attachment of peripherals (filter wheels, cameras). Most microscopes come with universal C-mount adapters.

3 IC optics: all major manufacturers have now microscopes with infinity-corrected (IC) optics. Although this is not essential, it greatly improves image quality and allows easily correlative microscopy with DIC (differential interference contrast) optics. In IC optics the rays of the light beam leaving the objective are parallel. The advantage of these objectives is that the necessary devices can be inserted into the microscope tube, under the tube lens, without altering the optical characteristics.

The choice between an upright or inverted system depends on the type of experiments to be performed. Inverted systems are usually preferred for ion imaging because it is easier to perform micro-injection and to insert perfusion chambers (often required in live cell experiments). The main disadvantage is when the objectives to be used require immersion (Section 4.2).

4.2 Objectives

Often neglected, the choice of a proper objective can be one of the most important factors limiting image quality. Some parameters for choosing an objective are outlined below:

1   A high numerical aperture (NA) is important for optical resolution (axial and lateral) and for greater efficiency in light collection. However, if IC optics are not being used, high NA usually means a short working distance, problematic for micro-injection.

2   Whenever possible, and for visible wavelengths use 'Plan Apo' (Planar, Apochromatic) objectives. 'DIC' or 'Ph' (phase contrast) objectives are simply not suitable for fluorescence imaging. 'Fluo' (fluorescence) objectives are made of quartz and should be used when UV wavelengths are being used. For UV confocal microscopy specialized lenses corrected for chromatic aberration are required.

3   Correction for chromatic aberration and coverslip thickness. Most objectives have corrections for these two parameters so the best approach is to contact the manufacturer and ask for the right type of objective for your experiment (e.g. objectives are available that do not require the use of coverslips).

4   Maintenance of the objectives is vital, particularly with immersion objectives mounted in inverted microscopes. Avoid excessive oil/water and clean any spillages immediately. Always clean your objectives after finishing your experiment. For general cleaning, use only appropriate lens tissue and avoid corrosive liquids. Pure ethanol is usually the best agent for cleaning but only when the objective is very dirty.

4.3 Detectors for spatial imaging

This chapter deals only with the spatial imaging of ions. Therefore, we will not refer to non-spatial fluorescence detectors such as fluorimetry, flow cytometry, and micro-photometry. Details of these can be found elsewhere (13).

Plant cells pose particular problems when it comes to spatial imaging of ions and molecules. These cells often have asymmetric organelle distribution, large vacuoles, and geometrical forms that result in uneven probe distribution. This stresses the importance of using ratio methods combined with LSCM optical sectioning or deconvolution-based sectioning for ion imaging. The simultaneous use of ratio imaging with LSCM has not been easy. LSCM systems with UV lasers (needed to excite the most common DW dyes) have only recently become available and are extremely expensive. Furthermore, many biological systems do not tolerate well the high doses of UV irradiation necessary for imaging. The recently introduced multi-photon imaging systems may help to overcome this problem, but the potential of the technique for ion imaging is still to be determined. Multi-photon systems are also extremely expensive, which may prevent the rapid employment of such systems. Therefore, DW dyes have been used mainly with epifluorescence microscopes equipped with video detectors (e.g. CCDs—charged coupled device cameras), while LSCM has been restricted to experiments with SW dyes (alone or together with a volume indicator). The main disadvantage of non-confocal systems—collection of out-of-focus light—can be partially circumvented by the use of deconvolution software (e.g. from Autoquant). However, application of deconvolution procedures in quantitative ion imaging must be done with caution because the algorithm used may alter pixel intensity.

When choosing a spatial detector one should look carefully at the following characteristics:

246

1 *Quantum efficiency*: the ratio between the number of photons arriving at the detector and those actually detected.

2 *Signal-to-noise ratio*: sensitivity and noise cannot be evaluated separately. Generally better in cooled CCDs. These cameras work at minus 50–20 °C, thus reducing intrinsic random noise to a minimum (see Section 5.1).

3 *Linearity of the response at different wavelengths*: most detectors (CCD cameras or photo-multiplier tubes used in LSCM) work less well at shorter wavelengths; special coatings may be required in some cases.

4 *Spatial and temporal resolution*: these are inversely proportional so one must establish priorities. Image-intensified CCDs can work at video rate, but sampling rate in currently available cooled CCDs is fast enough for most plant Ca^{2+} responses. LSCM systems have good spatial resolution, but a high temporal resolution may result in photodamage due to high doses of irradiation.

If the information supplied by the manufacturers is not straightforward the best solution is to contact research groups who have experience and who can give you a 'hands-on' opinion. Independently of the detector used, a series of rules should be followed for optimal imaging:

1 Careful alignment of optical parts, excitation and emission light (improves the amount of photons detected).

2 Correct gain and black level settings to avoid detector saturation (increases the life-time of the detector).

3 Make preliminary tests on tolerance of cells to irradiation, temperature, and general growth conditions under the microscope.

4 Routine checks on detector sensitivity (some manufacturers provide a protocol to test this parameter), age of excitation source (lamp hours, laser voltage output).

5 When performing 3D measurements using a LSCM determine the axial resolution of your imaging system, i.e. the recommended minimum thickness for an optical section (*Protocol 1*). Otherwise artefacts on the *z*-axis measurements will occur.

6 Keep all parts of the system clean and protected from dust, and get a maintenance contract from the manufacturers.

Protocol 1

Determining the optical resolution of an LSCM using sub-resolution fluorescent beads[a]

Equipment and reagents

- LSCM equipped with an electronic stepper motor to control Z-focus position.
- 0.1% poly-L-lysine solution
- Glass coverslips
- 175 nm Subresolution fluorescent beads (Molecular Probes)

Method

1 Place coverslips in 0.1% poly-L-lysine solution for 10–15 min; rinse briefly with water.

2 Add 10–20 μl of a 1:2000 aqueous dilution of the bead stock solution. Allow to settle for a few minutes, pour off excess solution, and add 20–40 μl of water. Cover with a second coverslip (if using an inverted microscope; otherwise use a glass slide) and place the sample on the microscope stage.

Protocol 1 continued

3 After focusing, close down the confocal aperture to the desired level, set the zoom to the maximum value, and focus on the medium focal plane of the bead ($z = 0$). Set the gain and black level to provide a non-saturated signal.

4 Using the 'line scanning mode' perform line scanning across the centre of the bead in 0.1 μm steps from below the focal plane to above the focal plane of the bead (e.g. -10 μm $< z < +10$ μm). The image formed will look like a long vertical ellipse; a spherical bead strongly deformed in the (z) axis (axial resolution), but not in the (x,y) axis (lateral resolution). This image represents the 'point spread function' (PSF) (21).

5 Using the 'length' command, measure the bead distortion in both axes (zero distortion is equivalent to the bead diameter, 175 nm in this protocol).

6 Repeat steps 1–5 for 5–10 beads using different objectives and different (z) steps. The values obtained provide an estimation of the lateral and axial resolution of your system.

[a] When LCSM is used in fluorescence microscopy, the theoretical limits of the spatial resolution can be obtained in practice. In conventional microscopy, this is very difficult to obtain.

4.4 Choosing your own filters

When setting up a new system, or making an upgrade to an imaging system, the researcher will find an immense array of light filters available for microscopy. It is therefore important to know exactly what the wavelength requirements of the experiment(s) are. First and foremost is to use appropriate excitation, dichroic, and emission filters for the fluorophore(s) to be used. The main criteria for choosing and installing filters are outlined below:

1 Autofluorescence is usually a problem when studying plant cells. If one knows the autofluorescence wavelengths, a barrier filter can be acquired, and installed in the filter block (e.g. to cut off chloroplast autofluorescence when looking at Green Fluorescent Protein fluorescence, one can use a 600 nm cut-off filter, which removes all fluorescence emissions above 600 nm).

2 Narrow bandwidth filters allow measurements of fluorescence at very precise wavelengths and may be very useful to reduce detector cross-talk (Section 5.2) and other types of signal contamination. However, they also cut significantly the amount of light passing through. In some cases, a 10 nm increase in half bandwidth (HBW) may increase the number of photons passing through by 70–80%. Thus, the HBW of a filter should be carefully chosen; shorter if signal specificity is required, longer if signal contamination is not a problem.

3 Dichroic mirrors (used to separate excitation from emission light) must be selected carefully because most systems do not allow a rapid change of dichroic mirror. So when imaging in dual channel mode (e.g. for fluo-3 and Texas Red), one must choose a dichroic mirror that matches the excitation/emission spectra of the probe(s) that is being used. Double and triple dichroics can be custom made from various manufacturers (e.g. Omega Optical and Glen Spectra).

4 Filters for excitation have different coatings from emission filters, even if from the same wavelength. It is advisable to consult the manufacturer before moving a filter from the emission to the excitation port and vice-versa.

5 The filters have to be installed in one specific direction. This is usually indicated in the filter holder, but if not consult the manufacturer for further instructions.

6   Before proceeding with imaging, make sure that only the required filters are in the light path. Most epifluorescence microscopes also have DIC optics incorporated and the light polarizer can reduce significantly the number of photons that reach your detector.

The filter wheels (e.g. from Ludl and Sutter) must allow a rapid (100 ms or less) selection of the different excitation/emission wavelengths and be synchronized with image capture. Software packages for image acquisition, which include drivers that support filter wheel and detector control are available, and the compatibility of all the items must be carefully examined when designing a customized system.

4.5 Hardware and software

Computer hardware and software for image capture and processing are essential tools. Often these are included in a package together with your fluorescence detector (Section 4.3), but frequently the software provided does not perform all the tasks desired. Dedicated software packages for Ca^{2+} ratio imaging are available, but these are expensive. Alternatively, more general software can be used to write macros for specific tasks. In these cases, the researcher should check for this possibility and also the user-friendliness of the macro language used by the software. We currently use Image-Pro Plus (Media Cybernetics), and the modules Scope-Pro and Fluo-Pro, which allow full automation of image acquisition, filter exchange, camera exposure times, and $x/y\,z$ focus.

5 Optimizing imaging performance

5.1 Signal-to-noise ratio

The signal-to-noise ratio (SNR) is a valuable measure of image quality. Although several theoretical definitions exist for SNR, in practice it is an indicator of the smallest signal, which can be detected in a reliable way, above the background noise generated by the imaging system. It is also a useful measure of the smallest differences that can be distinguished, i.e. the level of precision.

Presently, cooled CCDs are the detectors that allow the best SNR. Experimentally, an estimate of the noise in an image can be obtained by capturing consecutive images with the light path blocked. In a perfect situation, no photons should be recorded in these conditions, even with the detector at maximum sensitivity.

The degree of the detector's intrinsic noise (21) cannot be controlled. Intrinsic noise, leading to random variation in individual pixel values, arises from several sources during the process of digital imaging (22). One of the principal sources of such random variation between pixel values is the Poisson distributed variation in photons detected from the specimen, known as 'shot' noise (21), which imposes a fundamental limit upon the SNR of images with respect to the number of photons detected at each pixel. Here, we are not considering any type of signal coming from the sample (autofluorescence, stray background fluorescence, out-of-focus light) as noise. That depends on each particular example and cannot be objectively determined as a parameter to compare performance of the imaging system. However, it should be noted that during data processing (Section 7.1) this noise can be quantified and subtracted from the overall signal. The easiest way to do this is to capture an image of the cells before 'loading' and use the value obtained as the 'total' noise recorded in the experiment. For bioluminescence this procedure is adequate but for Ca^{2+}-sensitive dyes it depends on the type of cell analysed. If the experiment consists of imaging a thin monolayer of cells it may be sufficient, but if a multi-layer and/or thick cells are being studied, then the PSF of the imaging system should be taken into account. Nonetheless, the out-of-focus light originating from the probe will always limit the final resolution of the image, even in LSCM.

As a general guide, the magnitude of the minimum detectable signal needs to be at least three or four times that of the background noise level.

5.2 Filter block cross-talk

Ca^{2+}-sensitive probes have broad emission spectra, which may lead to 'bleed-through' or cross-talk between the two imaging channels when performing dual labelling or ratio imaging. Thus, care must be taken to ensure that no signal 'bleed-through' occurs. Otherwise, cross-talk between the two detected wavelengths could result in artefactual data. We tested this problem for the Bio-Rad MRC600 LSCM using Calcium Green-1/Rhodamine B, a combination commonly used for $[Ca^{2+}]_c$ ratio imaging. Independently of the laser used (argon or krypton–argon) for excitation, it was found that fluorescence from Rhodamine B was only detected in the 'red' channel (>600 nm), even at maximum sensitivity for green emission (maximum gain level and confocal aperture fully open). On the contrary, fluorescence of Calcium Green-1 could be detected in both channels. The signal contamination in the 'red' channel was worse when using an argon laser for excitation, a consequence of the characteristics of the filter sets. This signal contamination can, nevertheless, be eliminated (either with the Ar or the Kr–Ar laser) simply by decreasing the gain level in the >600 nm channel. Under the experimental conditions tested, we found that an approximately linear correlation could be established between the signal from the Ca^{2+}-sensitive dye detected in the 'green' channel and the maximum gain allowed in the 'red' channel for minimum cross-talk (*Figure 2*; *Protocol 2*).

Protocol 2

Correcting cross-talk in dual emission imaging in a LSCM

Equipment and reagents

- LSCM equipped with a suitable filter set for dual channel imaging (excitation: 488/568 nm; emission: 530/>600 nm)
- Calcium Green-1 free acid and Rhodamine B (Molecular Probes Inc.) stored at −20 °C.
- Degassed mineral oil
- Ultrapure HPLC water

Method

1 Prepare 50 μM solutions of Calcium Green-1 and Rhodamine B in ultrapure water.

2 Add ~1 μl of Calcium Green-1 to 30 μl of mineral oil and agitate vigorously for 1 min.

3 Pour the mineral oil in a coverslip and observe in the microscope. Numerous aqueous droplets of dye with different diameters will form. Choose droplets with a diameter similar to the cells under study.

4 Image the dye spheres in suspension in dual channel mode and under different settings (laser intensity, gain, pinhole aperture).

5 Measure fluorescence intensity for the different settings using the histogram function and plot the data (e.g. with Microsoft Excel 7.0); determine the settings at which cross-talk is zero (*Figure 2*).

6 Repeat steps 1–5 for Rhodamine B. A similar protocol can be applied to other dyes and also to wide-field systems.

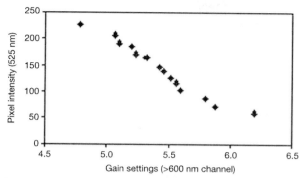

Figure 2 Cross-talk between filter blocks for Calcium Green-1 excited by a krypton/argon laser. Y axis—average pixel intensity detected in the 525 nm channel for a constant gain setting. X axis—gain in the detector channel (>600 nm) at which the cross-talk is zero. As pixel intensity in the 525 nm channel goes up, the sensitivity of the >600 nm channel must be reduced accordingly.

5.3 Calibration and behaviour of intracellular fluorescent probes

The concentration of probe introduced in the cell should be kept low to minimize buffering of the ion to be measured and to reduce potential side effects. At the same time, it must be high enough to provide a good SNR. Plant cells have high autofluorescence so there is an inherent difficulty in using very low dye concentrations because the SNR may be low.

Estimation of intracellular probe concentration is usually based on indirect quantification; comparison of intracellular levels with fluorescence emitted by cell-sized *in vitro* solutions with known concentrations of probe (Section 5.2). Although a useful measure, this type of quantification is very difficult because of photobleaching, and different dye behaviour *in vivo* and *in vitro*. More elaborate quantifications (23) involving computer modelling and analysis of PSF are difficult to incorporate in daily research and, for the most part, are not required. Usually, the best procedure is to estimate the upper limit of dye loading consistent with minimal disruption of cell physiology (growth, division, elongation rates, etc.).

Intracellular dye response is even more difficult to determine. It is well established that interaction of the probe with its target is influenced by the concentration of other ions, its own concentration, viscosity, temperature and hydrophobicity (19, 24, 25). For example, Blumenfeld and co-workers (26) reported that $[Ca^{2+}]_c$ just under the membrane surface cannot be accurately recorded with Fura-2. This is particularly relevant in cells with asymmetric cytoplasmic distributions (e.g. apical growing cells) and in organelle targeting experiments (Section 6.1). The relationship between fluorescence emission and ion concentration may be mathematically described (3, 27), but in most cases, application of this formula in daily research is not practical. For instance, the calibration of ratiometric dye pairs (consisting of two ion-sensitive probes) is complicated because two different K_d values for Ca^{2+} binding are involved (8). This means that the standard equations for calibration are no longer valid, although experimentally derived calibrations can be used (Section 5.4). Presently, the only way to circumvent this problem is to select equivalent Areas of Interest (AOI; Section 7.1).

5.4 Calibration of $[Ca^{2+}]_c$

Calibration of $[Ca^{2+}]_c$ can be achieved in many ways: *in vitro*, *in situ*, and semi-*in situ* (artificial cytosol). *In vitro* is the simplest way but also the one that is likely to produce a higher degree of error (*Figure 3*). *In situ* calibration is more difficult to perform and often impossible to do accurately because it relies on the use of ionophores, and on the assumption that these molecules will determine extra- and intra-cellular concentrations identically. Furthermore, the use of ionophores results in cell death, which in plants usually leads to cytoplasmic rearrangements. A more complete description of the problems associated with this method (as well as detailed protocols) can be found in Thomas and Delaville (27) and Fricker *et al.* (13). Therefore, semi-*in situ* calibration (*Protocol 3*) is often preferred because it is easy to perform and yields similar data to the *in situ* measurements. In all circumstances, accurate curve fitting to the data depends on a multiple point calibration. If a more accurate calibration is essential, other measurement techniques (e.g. ion selective microelectrodes) should be employed (28). However, it must be stressed that, regardless of the type of calibration, the numbers obtained should be regarded only as estimates. Accuracy is dictated primarily by how the dye behaves intracellularly and how well the dye response is calibrated.

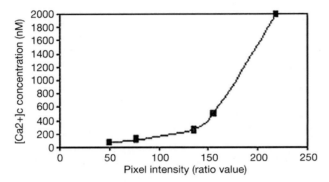

Figure 3 *In vitro* Ca^{2+} calibration of Fluo-3/cSNARF-1 free acid imaged by LSCM. Measuring fluorescence intensity in an aqueous solution in response to different concentrations of Ca^{2+} is a simple and fast way to perform a calibration. However, this type of calibration ignores the physical and chemical nature of the intracellular environment.

Protocol 3

Semi-*in situ* calibration

Equipment and reagents

- Fluorescence imaging system
- Incubation chamber/coverslips
- Fluorescent dye(s): Calcium Green-1/Rhodamine B free acids (Molecular Probes, Inc.); other dyes can be used.
- Artificial cytosol: 25% (v/v) ethanol, 60% (w/v) sucrose, 10 mM MES, 10 mM HEPES, 100 mM KCl, 20 mM NaCl, and 1 mM $MgSO_4$ (Sigma).
- Calcium Calibration Buffer Kit (Molecular Probes, Inc.)

Method

1 Prepare artificial cytosol (29) with the different Ca^{2+} concentrations provided in the calibration kit (other intermediate values are also recommended).

2 Prepare 10 μl of a mixture of 1 mM Calcium Green-1 and 1 mM Rhodamine B.

   On the coverslip or in the incubation chamber add 1 μl of dye solution to 99 μl of artificial cytosol (the dye concentration inside the cells is usually estimated to be 1–20 μM).

Protocol 3 continued

3 Image in dual channel mode, extract numerical information (Section 7.1) and perform ratio calculation. Repeat steps 1–4 for different Ca^{2+} concentrations.

4 Plot data in calibration curve (e.g. *Figure 3*).

5.5 Image acquisition

The successful monitoring of ions and molecules inside living cells relies on the technological tools available, but also on the good use we make of them. Although the acquisition of an image varies with the type of imaging system (*Table 3*), there is a series of general points that the researcher should consider *a priori*:

1 The actual pixel size (with respect to the specimen) should be set at a third of the optical resolution of the imaging system (30; Section 7.1).

2 Adjust gain detector to avoid saturation. Any signal above the detectors maximum sensitivity will not be recorded, thus decreasing the amplitude of your measurements and leading to artefacts. The reader can find a simple but very comprehensive review on this subject in Silver *et al.* (31).

3 Within a defined range, increasing average fluorescence brightness values (e.g. from 25 to 150 on a 0–255 scale) may result in an improvement in precision of >150%. This can be achieved either by increasing detector gain or intensity of excitation light, although the latter may cause photo-irradiation.

4 Kalman filtering (or equivalent) also results in an improvement in precision over direct image collection (increases SNR). However, Kalman filtering implies higher doses of irradiation and a significant decrease in temporal resolution. It is difficult to apply to fast growing cells or when photo-bleaching is a problem.

5 Accumulative filtering to increase signal strength also improves precision, but is inferior to the direct capture of images at a signal strength equivalent to the accumulated image

Table 3 Optimized imaging parameters for confocal ratio imaging

Variable	Settings[a]	Significance
Laser power	1–3% (25 mW argon ion laser)	Ensures minimum photo-irradiation
Scan speed	1–3 s per frame (512 lines)	
Low signal enhance[b]	Signal integrated over full pixel dwell-time	Only available in slow scan mode
Detector sensitivity	Over the range 55–75% to give fluorescence intensity between 80 and 180 (in a 0–255 scale) in cell resting levels	Ensures the detector is not saturated
Black-level setting	Dark signal pixel intensity of ~ 10	Ensures no signal is eliminated
Accumulate filter	Up to 3 (addition of subsequent scans)	Improvement in SNR
Kalman filter	Up to 3 (cumulative averaging of subsequent scans)	
Objectives	40× dry (NA 0.95) or 60× oil (NA 1.4)	Determines optical resolution and pixel size
Confocal apertures	Pinhole ~ 30% open for both channels	
Electronic zoom	3–4×	

[a] Settings refer to the Bio-Rad MRC-600 confocal microscope (see 21 for a more detailed explanation).

[b] Slow scan speed and low signal enhance are routinely used for imaging.

(plus all the disadvantages associated with Kalman filtering). Should therefore be used only if the signal is very weak.

6   Slower scans allow better spatial resolution, but decrease temporal resolution and increase photo-irradiation. Highest temporal resolution can be achieved by fast line scanning.

7   Most imaging systems allow electronic zooming that does not improve image quality. Optical zooming (using a higher magnification objective) is preferable unless it implies a significant loss of photons (Section 4.2).

8   Standardized image capture makes application of systematic statistical analysis easier and increases data reproducibility. However, cell-to-cell variation is an inherent problem (2, 32).

6   Bioluminescence indicators

Aequorin is a Ca^{2+}-sensitive luminescent protein, cloned from the coelenterate *Aequorea victoria*. Recombinant apoaequorin is a 22 kDa polypeptide that can be reconstituted into the active photoprotein by adding the luminophore coelenterazine to the incubation medium. Upon Ca^{2+} binding (there are three EF-hand domains in its structure), aequorin undergoes an irreversible conformational change accompanied by the emission of blue light (emission maximum ~ 466 nm). In the physiological pCa range, the emitted light is proportional to the $[Ca^{2+}]_c$ and measurements of aequorin-derived bioluminescence enables direct quantitation of Ca^{2+} transients inside the transformed cell.

The advantages of aequorin over fluorescent indicators include (i) high specificity for Ca^{2+} ions, (ii) non-toxicity, (iii) pH stability, (iv) a much larger dynamic range and signal-to-noise ratio, and (v) detection without illuminating the sample (thereby eliminating interference from autofluorescence and allowing the simultaneous use of caged probes). Moreover, the aequorin cDNA can be engineered to include defined signal sequences that enable the targeting of the probe to specific locations inside the cell, namely the endoplasmic reticulum, the nucleus, mitochondria, or plastids (13, 32). Organelle targeting is a clear advantage of this technology over the presently available fluorescent dyes. The major drawback of this technique is the extremely low light levels emitted by the photoprotein, which seriously compromise spatial imaging. This problem is augmented by the fact that emission of a photon from aequorin by reaction with Ca^{2+}, irreversibly discharges the molecule. High-sensitivity cooled CCD camera can be used for this purpose, but still require long integration times thus precluding imaging of fast Ca^{2+} responses. Sub-cellular imaging is still not possible to attain and measurements on individual cells can be performed only in very specific conditions (e.g. guard cells, pollen tubes). A comprehensive analysis with useful cross-references about luminescent techniques can be found in Fricker *et al.* (13).

6.1   Aequorin inside plant cells

For transient assays, the purified photoprotein can either be microinjected in living cells or the cDNA transiently transfected in plant protoplasts. However, the possibility of stably expressing the recombinant protein in transgenic systems eliminates the traumatic loading procedures, as well as improving the levels of expression. Recombinant aequorin was first introduced in plant cells by Knight and co-workers (4).

A plasmid containing the cloned gene in a binary vector is commercially available from Molecular Probes. It encodes a cytosolic form of apoaequorin under the control of the cauliflower mosaic virus (CaMV) 35S promoter, which is generally used for expression in the majority of plant tissues (although not all, like pollen, where it has a very poor expression). Engineered forms of aequorin have been reported (33) and a chloroplast-targetting expression vector can also be acquired from Molecular Probes. Targeting is normally accomplished by using peptide leader sequences or the whole polypeptidic sequence of proteins localized to a chosen location. In order to maintain aequorin fully active, fusion proteins should, if possible, be made at the N-terminus of apoaequorin since the C-terminal proline residue is essential for long-term stability (34).

Agobacterium-mediated transformation is the most currently used procedure to introduce foreign genes in plant cells (35). It is always advisable to test the constructs prior to stable transformation, especially if we are dealing with fluorescent indicator proteins, whose cellular behaviour is ultimately dependent on protein structure and conformation. Transient expression of aequorin can be accomplished by a variety of methods, namely biolistics, electroporation, and polyethyleneglycol (PEG)-mediated transformation of protoplasts.

The model plants *Arabidopsis thaliana* and *Nicotiana plumbaginifolia* have been successfully used to report Ca^{2+} transients by luminescence methods (4, 33). However, it should be noted that the level of light emission from recombinant plants is low (only pg of protein are expressed per mg of fresh weight), limiting the spatial resolution, and frequently necessitating multi-cellular analysis. A luminescence detector must be able to resolve light signals over multiple orders of magnitude with fast time resolution. Luminometers present the most affordable and simple technology for measuring Ca^{2+} transients in whole tissues. Another option is a low-light-level camera, which provides additional spatial information (Section 6.3), but sensitivity is a problem.

6.2 Luminescence measurements

Luminometry involves detecting blue light emitted from a sample with a photomultiplier placed adjacently to the housing device, while a photon counting system provides temporal information. The type of photomultiplier detector, the chamber characteristics, and the reconstitution protocol are all of great importance, and should be optimized for each particular need. The basic equipment is available on the market, and a cooled photomultiplier tube, with low dark current should perform well in the blue, providing the sample housing is tightly protected from background contamination. Any luminescent materials (like certain types of plastic and glass) must be avoided from the sample holder. Several calibrations and background measurements should be performed before making meaningful experiments.

The plants to be used in the experiments must be carefully chosen among all the recombinant lines obtained from the initial transformation process. The resistance to an antibiotic is a simple and effective method to screen for transformants. Afterwards, the different plant lines can be screened by western blotting for the presence of aequorin protein, but the most important screening is for those which actually yield detectable amounts of light. (*Protocols* 4 and 5).

Once the best plant lines have been identified, it is recommended to optimize the reconstitution protocol (namely, the type and concentration of coelenterazine to be used), the incubation time and the time course stability for reconstituted aequorin. The degree of reconstitution increases with the concentration of coelenterazine used until around 10 μM (4). Various types of synthetic coelenterazine analogues are available from Molecular Probes (36), which have different Ca^{2+} affinities and spectral properties.

Protocol 4

Reconstitution of aequorin *in vitro* and *in vivo*

Equipment and reagents

- Micropestle or glass homogeneizer
- Coelenterazine (Molecular Probes, Inc.); the coelenterazine should be kept at −80 °C, aliquoted in appropriate working volumes (e.g. 50 μl of a 200 μM stock solution in methanol, vacuum dried after being aliquoted, and resuspended just before use).[a]

- Liquid nitrogen.
- Extraction buffer (10 mM Tris, pH 7.4, 0.5 M NaCl, 5 mM EDTA, 5 mM mercaptoethanol, and 0.1% gelatin)
- Assay buffer (200 mM Tris, pH 7.0, 0.5 mM EDTA)

Methods

(A) In vitro

1 If working with primary transformants collect a leaf from a mature plant; for F1 or F2 generations use green 7-day-old seedlings selected on nutrient media plates supplemented with the appropriate antibiotic. Place in a microcentrifuge tube and snap freeze in liquid nitrogen. Controls should include a wild type equivalent.

2 Grind the tissue to powder.

3 Add 0.1 ml of extraction buffer and continue grinding. Add an extra 0.4 ml buffer, homogenize, and leave on ice until all samples have been processed.

4 Centrifuge at 13 000 g for 10 min, at 4 °C. Transfer the supernatant to a clean microcentrifuge tube.

5 Transfer 0.1 ml to a new microcentrifuge tube and add resuspended coelenterazine to a final concentration of 1 μM (e.g., add 1 μl of 100 μM solution). Pipette up and down to mix and aerate. Incubate in the dark for 3 h at room temperature. To controls add an equivalent volume of methanol.

6 Before measuring, dilute from 5- to 50-fold with assay buffer, in a total volume of approximately 0.5 ml.

(B) In vivo

1 Float the plant material[b] in a minimal volume of water (only sufficient not to dry the specimen).

2 Add coelenterazine to a final concentration of 2.5 μM (general recommendation) and incubate in the dark, at room temperature, for the appropriate period of time.

[a] Being a luminophore, it should be protected from the light at all times (e.g. wrap in foil and use in dim light). Special care must be taken when preparing the reconstitution incubations.

[b] Remember that the amount of tissue used is directly correlated with the volume of reconstitution and thus with the amount of coelenterazine needed.

6.3 Luminescence imaging

Transgenic plants containing aequorin can be imaged by placing the biological sample inside a black chamber, attached to a low-light-level camera. As well as good spatial resolution, the controlling software should permit the capture of Ca^{2+} transients within fractions of a second, as well as integrate signals over 1 or more sec. The reconstitution protocol used is similar for both luminometry and imaging (*Protocol 5*). It is strongly recommended to select the most highly luminescent plant lines for this type of experiment. Before setting the camera to the photon-counting mode of luminescent detection it is advisable to collect bright-field images of the specimen under analysis, as the digital image is expected to be dark and spotted. The image collection should only be started once the sample chamber is closed. As the levels of emitted light are extremely low, luminescence imaging requires extremely sensitive cameras operating at maximum levels of sensitivity. However, even in optimum conditions, sub-cellular resolution is still not possible.

6.4 Calibration of aequorin measurements

The kinetic profile of intracellular Ca^{2+} concentration changes can be obtained by converting data from each luminescence curve into Ca^{2+} values, according to calibration algorithms. There are several ways to calibrate the aequorin light signal, including *in vitro* and *in vivo* methodologies (13). There are significant differences, however, between *in vivo* and *in vitro* calibrations. The basic parameters to be measured are:

1   The light intensity at each time interval L is obtained by integrating the luminometer output over these time intervals.

2   L_{total}, the sum of all L values (total area under the curve), including the light values recorded at the final discharge.

3   $L_{cumulative}$, the sum of all L values from time zero to any given point.

4   L_{max}, corresponding to $L_{total} - L_{cumulative}$ at each time point, representing the remaining aequorin to be discharged; this value decreases as aequorin is consumed during the experiment.

The relation between all these parameters can be found in calibration formulas, which are specific for the type of aequorin used (37, 38). Note that average background values should be deducted from all data points before any further calculation. The final discharge of remaining aequorin by high Ca^{2+} should be recorded until a three-magnitude drop in the counts has been achieved. If the levels of aequorin are relatively low, the emitted light intensity does not respond linearly to Ca^{2+} because not enough protein is present to compensate for the aequorin, which is irreversibly discharged. A further problem with the calibration of aequorin originates from asymmetries in the expression level of apo-aequorin in different cells and different decay rates of bioluminescence over time. The solution for these problems would be to engineer a ratio form of aequorin.

Protocol 5

Aequorin luminescence measurements

Equipment and reagents

- Luminescence detector (luminometer/CCD camera)
- Discharging solution (e.g. 2 M $CaCl_2$ with 20% ethanol)
- Sample holders (dependent on the type of sample chamber and plant material, e.g. luminometer cuvettes or glass coverslips)

Method

1   Determine the background counts using (a) a wild type equivalent and (b) water only. The illumination in the room should be dimmed to the minimum possible before switching on the equipment, in order to reduce background light contamination.

Protocol 5 continued

2   Place a freshly reconstituted specimen inside a luminometer cuvette (or alternative sample holder) containing water and allow to settle. Stirring is not recommended and controls must be performed for any perturbation in the system, including temperature variations, mechanical perturbation, and light entry when injecting any substances through the port.

3   Start to record the luminescence before administering the treatment. The frequency of recording depends on the type of experiment and should be adjusted accordingly (e.g. every 0.1 s for rapid responses; 1–10 s for slower responses).

4   At the end of the assay discharge the unconsumed aequorin by injecting 1 vol of 2 M $CaCl_2$, 20% ethanol, and determine the total light output.

7 Data processing

Extracting useful data from images is not always a linear process and there is a wide range of different analytical techniques that can be applied. Basically, they can be divided into three main categories:

1   Pixel-by-pixel ratio imaging to produce an image that compensates (in principle) for varying cell thickness, probe levels, leakage, and bleaching. This method has found wide applications in ion imaging, but it is not available in all software packages. It provides a good visual indicator of the magnitude of the response and the level of spatial heterogeneity within or between cells. However, the use of a ratiometric procedure is not error-proof. If the collected signal falls below a certain limit, the ratio no longer exhibits a linear response and artefactual gradients may appear (39).

2   Data extraction in equivalent AOIs.

3   Statistical analysis.

7.1 Numerical data extraction and statistical analysis

Ideally, these two parameters should be considered together. However, most imaging software packages do not contain statistical functions and only a few allow repetitive data extraction in a straightforward manner. As a result, the researcher often has to perform these analyses on individual images and using his/her own criteria.

When extracting numerical data from an image, the researcher should consider the following points:

1   The value of each pixel in an image relates to the number of photons detected from that region of the sample. However, because of the random variation between pixels, each pixel in an image does not accurately represent ion concentration. Thus, examination of numerical data at the individual pixel is misleading. The variation between individual pixel values within an image becomes increasingly important when extracting quantitative data from small image areas (on the order of 100's of pixels) and when making comparisons between the values from different regions within the same image. Therefore, a means of quantifying the magnitude of the variation between pixel values is required to provide estimates of (i) the precision of quantitative data for particular sample sizes (and thus the spatial resolution) and (ii) confidence limits to allow meaningful comparison between different regions of the same image. A minimum AOI must be defined and the value obtained for that AOI averaged (*Protocol 6*).

2   When performing a ratio calculation (7), numerical data should be extracted from equivalent AOIs in the original raw images and not from the final ratio image. We are not

aware of any software package that includes this option, so most researchers use just the ratio image. Values in the ratio image include a noise-dependent difference, which becomes more significant in measurements closer to the resolution limit. This procedure may require advanced statistical analysis (39).

3 The variation in the fluorescence signal is proportional to the mean fluorescence intensity. Therefore, the precision of measurement itself varies with ion and probe concentration, quantum yield of the probe, photobleaching, and autofluorescence of the cell.

Protocol 6

Determination of minimum AOI for extraction of numerical data and precision of Ca^{2+} measurement

Equipment and reagents

Identical to *Protocol 3*

Method

1 Repeat steps 1–3 of *Protocol 3* for different Ca^{2+} concentrations (the success of this protocol depends on the accuracy of the ion concentration). In theory, the samples prepared are uniform and fluorescence is even.

2 Image in dual channel mode and save the different images (2–4 for each Ca^{2+} concentration).

3 For each Ca^{2+} concentration, extract numerical information in AOI of decreasing size (>5000–20 pixels) and perform ratio calculation.

4 Compare the values obtained for each AOI with your calibration curve (*Protocol 3*) and determine the minimum size of AOI that still fits your calibration. *In situ*, the size of the minimum AOI is likely to be higher because the sample is no longer uniform. However, if your probe is uniformly distributed (Section 3) this protocol produces a close match.

7.2 Image processing and enhancement

Data presentation can be regarded as the final and most important part of any work. Currently, digitized images may be visually enhanced in numerous ways with the sole purpose of better demonstrating a hypothesis. Image processing software is becoming so powerful that soon the limits of what we can do will be just ethical. Therefore, it is important that researchers (and journal editors) pay special attention to this issue. Authors should include in their manuscripts at least one unprocessed image ('raw'; *Figure 4*) so that readers can make a clear evaluation of the work. A list of the most common enhancement 'tools' and their characteristics can be found in *Table 4*.

Once processed, images can be saved as TIFF, JPG, or BMP files (just to mention the three most common formats) and exported to drawing/illustrator software packages where lettering, numbering, and the final organization of a plate can be done. In this case, the researcher should choose software accepted by the journal where the work is to be submitted.

Printouts can be done in a variety of forms depending on the type of images. Graphs, line traces, and colour can be produced at low cost by inkjet printers with photo quality and using special glossy/photography paper (use paper from the same manufacturer as the printer).

Black and white images can also be produced by this type of printer, but the quality is poorer. Video or dye sublimation printers may work out better for this type of image, but it is a more costly solution.

Table 4 Most common enhancement 'tools' available in software packages

Tool	Comments
Smoothing/median filtering/averaging	Gives a more even look to images; useful for reducing distracting random speckle. It causes 'rounding' of cell periphery and distorts highly localized asymmetries in pixel values. Reduces image sharpness. Should not be applied before extraction of numerical data.
Contrast enhancement	Gives a more 'crispy' look to images. It increases sharpness particularly at peripheries. Should not be applied before extraction of numerical data.
3D representation/Y_{mod} plots	Consists of representing pixel intensity as height (*Figure 5*). Useful to stress small or localized changes such as ion gradients or hot spots because one can assign minimum and maximum values for pixel intensity (e.g. all pixels values $<20 = 0$). It should be accompanied by an image of the cell.
Pseudo-colouring	Consists of assigning a colour scale to grey levels (usually designated by a look-up table or LUT). Useful to stress small or localized changes such as ion gradients or hot spots. Makes analysis of probe distribution difficult (e.g. sequestration is concealed by this method) and may give rise to over-interpretations of data (LUTs are often stepped).

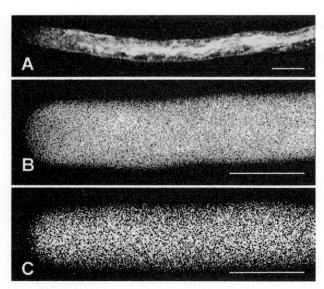

Figure 4 Imaging of different forms of Calcium Green-1 in growing pollen tubes using LSCM. Bar = 10 μm. (A) 10 kDa dextran loaded by ionophoresis, (B) free acid and (C) 10 kDa dextran loaded by pressure micro-injection. The signal emitted by the 'clear cap' significantly differs with the form of dye and loading method; pseudo-colouring would make these differences impossible to visualize. The presentation of 'raw' black and white images is essential for evaluation of dye distribution and behaviour in the cytosol.

Figure 5 Imaging $[Ca^{2+}]_c$ and putative Ca^{2+} channel activity with Indo-1 and the Mn^{2+} quenching technique (49) in growing pollen tubes of *Agapanthus umbellatus* using video microscopy. Each graph represents a Y_{mod} plot of fluorescence intensity from the first 50 μm of one pollen tube. In each case, the tip of the cell is indicated by an arrow. (A) Ratio imaging of Indo-1 in a growing tube reveals the tip-focused $[Ca^{2+}]_c$ gradient. (B) Ratio imaging of Indo-1 in a non-growing tube. The tip-focused $[Ca^{2+}]_c$ gradient was replaced by a uniform distribution. The emergence of channel activity is concomitant with the re-establishment of the $[Ca^{2+}]_c$ gradient.

8 New molecular indicators

Recent developments in fluorescence microscopy include the use of stably expressed indicator proteins to report ion behaviour inside living cells. The Green Fluorescent Protein (GFP), also derived from the jellyfish *Aequorea victoria*, has emerged as an apparently ideal probe to report cellular phenomena. Following the first report of the recombinant expression of GFP (40), it has been widely used as a fusion with many host proteins to produce fluorescent chimeric proteins. Moreover, fluorescence resonance energy transfer (FRET) between new engineered GFP mutants makes possible the dynamic monitoring of protein–protein interactions in living cells. The wild type protein has two main absorption peaks in the blue, at 395 and 475 nm, emitting green light. Nowadays a multitude of GFP mutants are available, with increased fluorescence emissions, and wavelength shifts in both excitation and emission spectra (41, 42) (see *Chapter 13*).

8.1 Ion monitoring using GFP-based indicators

A new family of indicators, based on FRET technology between variants of GFP, has been developed to report Ca^{2+} concentrations *in vivo* (43–45). The rationale behind this type of indicators is to have two GFP mutants connected via a sensitive linker, which acts as a FRET promoter or disrupter. The possibilities are extraordinary, provided correct protein engineering is achieved. At present, research in this field is expanding and various groups are trying to manipulate constructs to produce the perfect indicators. It has proved to be difficult, although new insights are constantly being reported.

Many parameters must be considered prior to attempting the construction or even the use of these type of indicators. First of all, the light levels involved are of vital importance for the success of imaging. Although GFP fluorescence is not species specific, the expression of recombinant forms in different organisms needs to be assessed. Frequently, the light signals are low and are difficult to discern from background contamination. Plant cells pose an additional problem with the autofluorescence of cell walls and chlorophyll (Section 4.3). The different mutants that have been produced enable some of these problems to be overcome by increasing fluorescent levels and manipulating the spectra. Nevertheless, the researcher must be aware of these problems, before embarking in the laborious production of complicated constructs from unreliable GFP clones, and often without possessing the fluorescent equipment required for its observation (Section 4).

The first Ca^{2+} indicator based on this type of technology (43) consisted of two GFP variants joined by the calmodulin (CaM)-binding domain of myosin light chain kinase. Whenever Ca^{2+}-activated CaM is bound to this domain, FRET between the two fluorophores is attenuated. Following this work (44) the initial construct was modified by introducing an intrinsic CaM in the C terminus, resulting in a new indicator that responds directly to Ca^{2+} levels (K_d = 100 nM). The authors reported a six-fold attenuation in fluorescence by disruption of FRET. In parallel, the group of Roger Tsien reported another indicator, named cameleon (45), consisting of a tandem fusion between two GFP variants and CaM. In this case, the binding of Ca^{2+} to the CaM EF hand motifs alters protein conformation, increasing the FRET between the flanking fluorophores. Mutations in the CaM structure make possible Ca^{2+} measurements anywhere from between 10^{-8} to 10^{-1} M, and a 70% increase in the emission ratio of the two GFPs. Although the cameleons are still not commercially available, upon request to the authors several researchers have tried them in their own model systems. In the plant field, some problems have been encountered, namely with respect to light levels, dynamic range and the detection of FRET variations.

8.2 The tools of the future

The future appears dominated by the new cloning techniques and protein engineering strategies that enable the observation of probes to examine function in the living cell. In plant cells, GFP fusion constructs have already been used to study molecules (46) despite some recognized drawbacks. The polypeptide is 27 kDa, which for many applications is too large; it can only be fused to the extremes of the host protein, and is of no assistance for spectroscopic readouts other than fluorescence. Moreover, molecular biology techniques are laborious and frequently limited to model organisms that permit easy transformation. Some pioneer reports (47) incorporating peptide technology coupled to sensitive detection methods are also likely to be extended to plant research.

Molecular engineering of the aequorin structure may also lead to improved bioluminescent indicators. Another significant development for this area would be the use of dual-wavelength coelenterazines, which allow ratio measurements, thus improving Ca^{2+} quantification (48).

In parallel to these advances, Ca^{2+} dyes can and will continue to be used for many purposes. Dye technology improves continually, demanding more and more of monitoring equipment, and pushing resolution limits further. Ca^{2+} dyes have been used to discriminate between Ca^{2+} influx and Ca^{2+} mobilization from internal stores (49) and this strategy will no doubt find further applications. Coupling Ca^{2+} measurements to caged-probe technology has also found many applications (10, 13, 15) and two-photon excitation may circumvent the prior requirement of UV excitation (see 13 for further details on uncaging equipment).

Acknowledgements

Research in R.M. lab is supported by Fundação Ciência e Tecnologia and Centro Biotecnologia Vegetal (Portugal) and by the Human Frontier Science Programme.

References

1. Malhó, R. (1999). *Plant Biol.*, **1**, 487.
2. Trewavas, A. J. and Malhó, R. (1997). *Plant Cell*, **9**, 1181.
3. Grynkiewicz, G., Poenie, M. and Tsien, R. Y. (1985). *J. Biol. Chem.*, **260**, 3440.
4. Knight, M. R., Campbell, A. K., Smith, S. M. and Trewavas, A. J. (1991). *Nature*, **352**, 524.
5. Haugland, R. P. (1996). *Molecular probes: handbook of fluorescent probes and research chemicals*, 6th edn. Molecular Probes, Inc., Eugene.
6. Opas, M. (1997). *Trends Cell Biol.*, **7**, 75.
7. Parton, R. M. and Read, N. D. (1999). In: *Light microscopy in Biology* (ed. A. T. Lacey), 2nd edn, p. 211. Oxford University Press, Oxford.
8. Lipp, P. and Niggli, E. (1993). *Cell Calcium*, **14**, 359.

9. Zhao, M., Hollingworth, S. and Baylor, S. M. (1997). *Biophys. J.*, **72**, 2736.
10. Malhó, R. and Trewavas, A. J. (1996). *Plant Cell*, **8**, 1935.
11. Tsien, R. Y. (1981). *Nature*, **290**, 527.
12. Callaham, D. A. and Hepler, P. K. (1991). In: *Cellular calcium—a practical approach* (ed. J. G. McCormack and R. H. Cobbold), p. 383. IRL Press, Oxford.
13. Fricker, M. D., Plieth, C., Knight, H., Blancaflor, E., Knight, M. R., White, N. S. and Gilroy, S. (1999). In: *Fluorescent and luminescent probes*, 2nd edn, p. 569. Academic Press, London.
14. Gilroy, S. (1997). *Ann. Rev. Plant Physiol. Plant Mol. Biol.*, **48**, 165.
15. Read, N. D., Allan, W., Knight, H., *et al.* (1992). *J. Microsc.*, **165**, 586.
16. Camacho, L., Parton, R., Trewavas, A. J. and Malhó, R. (1999). *Protoplasma*, **212**, 162.
17. Zhang, W-H., Rengel, Z. and Kuo, J. (1998). *Plant J.*, **15**, 147.
18. Hofer, A. M., Landolfi, B., Debellis, L., Pozzan, T. and Curci, S. (1998). *EMBO J.*, **17**, 1986.
19. Moore, E. D. W., Becker, P. L., Fogarty, K. E., Williams, D. A. and Fay, F. S. (1990). *Cell Calcium*, **11**, 157.
20. Obermeyer, G. and Weisenseel, M. H. (1995). *Protoplasma*, **187**, 132.
21. Pawley, J. B. (ed.) (1995) *Handbook of confocal microscopy*, 2nd edn. Plenum Press, New York.
22. Sheppard, C. J. R., Gan, X., Gu, M. and Roy, M. (1995). In: *Handbook of confocal microscopy*, 2nd edn, p. 19. Plenum Press, New York
23. Fink, C., Morgan, F. and Loew, L. M. (1998). *Biophys. J.*, **75**, 1648.
24. Roe, M. W., Lemasters, J. J. and Herman, B. (1990). *Cell Calcium*, **11**, 63.
25. Busa, W. B. (1992). *Cell Calcium*, **13**, 313.
26. Blumenfeld, H., Zablow, L. and Sabatini, B. (1992). *Biophys. J.*, **63**, 1146.
27. Thomas, A. P. and Delaville, F. (1991). In: *Cellular calcium: a practical approach* (ed. J. G. McCormack and R. H. Cobbold), p. 1. IRL Press, Oxford.
28. Felle, H. H. and Hepler, P. K. (1997). *Plant Physiol.*, **114**, 39.
29. Fricker, M. D., Tlalka, M., Ermantruat, J., *et al.* (1994). *Scanning Microsc.*, Supplement **8**, 391.
30. Webb, R. H. and Dorey, C. K. (1995). In: *Handbook of confocal microscopy*, 2nd edn, p. 55. Plenum Press, New York.
31. Silver, R. A., Whitaker, M. and Bolsover, S. R. (1992). *Pflügers Arch.*, **420**, 595.
32. Malhó, R., Moutinho, A., van der Luit, A. and Trewavas, A. J. (1998). *Phil. Trans. R. Soc.*, **B353**, 1463.
33. Knight, H, Trewavas, A. J. and Knight, M. R. (1996). *Plant Cell*, **8**, 489.
34. Watkins, N. J. and Campbell, A. K. (1993). *Biochem. J.*, **293**, 181.
35. Sambrook, J., Fritsch, E. F. and Maniatis, T. (1989) *Molecular cloning: a laboratory manual*, 2nd edn. Cold Spring Harbour Laboratory, Cold Spring Harbour.
36. Shimomura, O., Kishi, Y. and Inuoye, S. (1993) *Biochem. J.*, **296**, 549.
37. Allen, D. G., Blinks, J. R. and Prendergast, F. G. (1976). *Science*, **195**, 996.
38. Cobbold, P. H. and Rink, T. J. (1997). *Biochem. J.*, **248**, 313.
39. Parton, R., Fischer, S., Malhó, R., Papasouliotis, O., Jelitto, T., Leonard, T., and Read, N. D. (1997). *J. Cell Sci.*, **110**, 1187.
40. Chalfie, M., Tu, Y., Euskirchen, G., Ward, W. W. and Prasher, D. C. (1994). *Science*, **263**, 802.
41. Haseloff, J., Dormand, E. L. and Brand, A. H. (1999). *Methods Mol. Biol.*, **122**, 241.
42. Haseloff, J. (1999). *Methods Cell Biol.*, **58**, 139.
43. Romoser, V. A., Hinkle, P. M., Persechini, A. (1997). *J. Biol. Chem.*, **272**, 13270.
44. Persechini, A., Lynch, J. A. and Romoser, V. A. (1997). *Cell Calcium*, **22**, 209.
45. Miyawaki, A., Liopis, J., Heim, R., McCaffery, J. M., Adams, J. A., Ikura, M. and Tsien, R. Y. (1997). *Nature*, **388**, 882.
46. Kost, B., Spielhofer, P. and Chua, N.-H. (1998). *Plant J.*, **16**, 393.
47. Griffin, B. A., Adams, S. R. and Tsien, R. Y. (1998). *Science*, **281**, 269.
48. Knight, M. R., Read, N. D., Campbell, A. K. and Trewavas, A. J. (1993). *J. Cell Biol.*, **121**, 83.
49. Malhó, R., Read, N. D., Trewavas, A. J. and Pais, M. S. (1995). *Plant Cell*, **7**, 1173.
50. Miller, D. B., Callaham, D. A., Gross, D. J. and Hepler, P. K. (1992). *J. Cell Sci.*, **101**, 7.

Chapter 13
Reporter genes and *in vivo* imaging

Angela Falciatore, Fabio Formiggini and Chris Bowler
Laboratory of Molecular Plant Biology, Stazione Zoologica, Villa Comunale, I 80121 Napoli, Italy

1 Introduction

Reporter genes are an important tool in plant research. They have been most commonly used to study the activity of regulatory elements, such as promoters, enhancers, introns, and transcription terminators. Furthermore, when fused to coding sequences of interest they have also been used to study the subcellular localization and trafficking of proteins. The β-glucuronidase (GUS) enzyme reporter was the first to be developed specifically for use in plants, largely as a substitute for the β-galactosidase (LacZ) reporter, which is commonly used in other organisms, but which cannot be used in plants due to high endogenous activities. In addition, firefly luciferase (LUC) has been used with success in several systems and, more recently, the jellyfish Green Fluorescent Protein (GFP) has attracted widespread popularity for subcellular localization studies and potentially for studying protein–protein interactions in real time. In this chapter, the applications of these three reporters in plant cells is described, together with accompanying protocols.

2 β-Glucuronidase (GUS)

2.1 Introduction

β-glucuronidase (GUS) has been the predominant reporter used to study gene expression in plants. GUS is also used as a reporter in other organisms, but it is particularly useful in higher plants due to the absence of endogenous GUS activity in most plant species. The system was developed in 1987 by Richard Jefferson *et al.* (1) and is based on the *Escherichia coli* gene encoding β-glucuronidase (*uidA*). The enzyme β-glucuronidase (GUS) catalyses the hydrolysis of a wide variety of glucuronides. Its substrates consist of D-glucuronic acid conjugated through a β-O-glycosidic linkage to a range of aglycones. The assay was developed initially as a reporter gene fusion marker in *E. coli* and in *Caenorhabditis elegans*, but was then more successfully used as a reporter gene for promoter analysis in transient assays and in stably transformed plants. Higher plants transformed with *GUS* are healthy, develop normally and are fertile. GUS has been used to identify promoter elements involved in many aspects of regulation of gene expression, such as tissue specific and developmental regulation (2), hormonal regulation (3), and photoregulation (4). Additionally, it has been used to develop transformation procedures in some plants (5), as well as in photosynthetic algae (6), and to study mechanisms of *Agrobacterium tumefaciens*-mediated transformation (7).

The advantages of the GUS reporter gene system can be briefly summarized:

1   Absence of endogenous GUS activity in many organisms. Both lower and higher plants, most bacteria, algae, fungi, and many insects are largely, if not completely, devoid of GUS activity. GUS activity can therefore be measured, even in single cells, when *GUS* is used as

a reporter gene in these systems. On the contrary, the β–galactosidase (*lacZ*) gene from *E. coli*, that is one of the most versatile reporter genes, cannot be used in plants because many cell types have high endogenous activities

2   β-glucuronidase catalyses the hydrolysis of a wide variety of β-glucuronides and many substrates are now commercially available (e.g. Clontech: www.clontech.com, and Molecular Probes: www.probes.com). GUS activity can be measured easily and quantitatively by spectrophotometric, fluorometric and histochemical methods (see *Protocols 1–3* below)

3   The β-glucuronidase enzyme is very stable and will tolerate many detergents and variations in ionic conditions. Moreover GUS is active in the presence of thiol reducing agents such as β-mercaptoethanol. GUS is resistant to thermal inactivation (the half life at 55 °C is about 2 h) and its activity can be assayed in simple buffers over a range of pH values (5.0–9.0)

4   The enzyme can tolerate large C-terminal fusions, with normally no significant reduction in activity. GUS can readily traverse membranes when appropriate transit or signal sequences are fused to its amino terminus. For these properties GUS has been used to study targeting of chimeric proteins to chloroplasts (8), mitochondria (9), and endoplasmic reticulum in plant cells (10).

5   A number of vectors designed for the construction of GUS gene fusions are described by Jefferson (11) and can be purchased from Clontech (www.clontech.com). These can be easily used to perform transcriptional and translational fusions with regulatory regions of interest (12). In transcriptional gene fusions the site of the fusion is located within the untranslated DNA sequences. These fusions are often generated for promoter analysis experiments. In this case, transcription of the *GUS* gene is under the control of the upstream promoter. In translational fusions, the site of the fusion is located within the coding regions. In this case it is essential that the correct reading frame is conserved across the junction of the two coding regions. There are three plasmids in the pBI.101 series (Clontech) that provide three different reading frames, relative to the cloning sites, for the construction of N-terminal translational fusions.

Before performing GUS assays in a new species it is important to verify the absence of endogenous GUS activity. Moreover, it is important to note that the GUS activity found in some samples could result from fungal and/or bacterial contamination. It is therefore necessary to perform the assays in sterile conditions and to compare the obtained results with respect to negative controls. The proliferation of bacteria during long incubation times can be prevented by adding sodium azide (0.02%) or chloramphenicol (100 μg/ml) to the X-Gluc solution, as reported by Jefferson and Wilson (13; see *Protocol 1*).

2.2 GUS assay protocols

GUS activity can be measured easily and quantitatively by spectrophotometric and fluorimetric methods. The tissue specific localization of GUS can be visualized in a histochemical assay. The advantages and limitations of these assays are described herein.

Spectrophotometric and fluorimetric assays are performed on cell extracts. The method of lysis will depend upon the nature of the tissue. Usually tissues or cultures can be homogenized in many different buffers and using either the French press, sonication, or grinding with sand or glass beads, or in liquid nitrogen. Phenylmethyl sulfonyl fluoride (PMSF—an inhibitor of serine proteases) can be added in the lysis buffer if proteases are a potential problem. For both assays, total protein contents of extracts must also be determined in order to quantify the GUS content in the sample. The histochemical assay can be performed on whole samples or on sections, as discussed in Section 2.2.3.

2.2.1 Spectrophotometric assay

This assay is very straightforward and is moderately sensitive. The sensitivity can be enhanced by performing very long incubation times (e.g. overnight assays; 14). It is also very cheap and does not require sophisticated instrumentation. Its limitations are the intrinsic lack of sensitivity of methods based on absorption of light (*Protocol 1*) and the problems caused by light absorption of plant pigments present in the extracts. Some of these problems can be overcome by using fluorometric assays (*Protocol 2*).

Protocol 1

Spectrophotometric GUS assay

Equipment and reagents

- Plant tissue
- Equipment for grinding of plant tissue (micropestles, prechilled mortar and pestle, or prechilled homogenizer, liquid nitrogen)
- GUS extraction buffer (50 mM $NaPO_4$ pH 7.0, 10 mM beta-mercaptoethanol, 0.1% Triton X-100), freshly prepared

- Substrate [*p*-nitrophenyl glucuronide (PNPG; Sigma)], dissolved in GUS extraction buffer
- 2-amino, 2-methyl propandiol (Stop buffer; Sigma)
- Spectrophotometer

Method

1 Grind frozen tissue and add 1 ml of GUS extraction buffer.

2 Centrifuge the homogenate in a microcentrifuge at 12 000 g for 5 min and collect the extract supernatant.[a,b]

3 Add 1 mM *p*-nitrophenyl glucuronide (final concentration).

4 Incubate a 37 °C.[c]

5 Stop the reaction by the addition of 0.4 ml 2.5 M 2-amino, 2-methyl propandiol.

6 Measure the absorbance at 415 nm against a substrate blank (if the turbidity of the extract is a problem, measure the absorbance against a stopped blank reaction to which an identical amount of extract has been added). Under these conditions the molar extinction coefficient of *p*-nitrophenol is assumed to be 14,000; thus in 1.4 ml final volume, an absorbance of 0.010 represents 1 nmol of product produced. One unit is defined as the amount of enzyme that produces 1 nmol of product/min at 37 °C. This represents about 5 ng of pure β-glucuronidase.

[a] GUS is remarkably resistant to protease action, with a very long half-life in living cells and in most extracts, but if proteases are a potential problem, PMSF at a final concentration of 25 μg/ml can be included in the lysis buffer with no ill effects on the enzyme. Extracts can be stored indefinitely at −80 °C, with no loss of activity, or at 4 °C, with very little loss. Avoid storage at −20 °C, which can inactivate the enzyme, at least in this lysis buffer. Tissue stored at −20 °C does not seem to lose significant GUS activity.

[b] At this point, it may be convenient to measure total protein content in the sample, e.g. using the kit described in *Protocol 4*.

[c] Because of the remarkable stability of GUS one can enhance the sensitivity by using very long incubation times (e.g. overnight incubation). For long assays, add 0.02% NaN_3 to prevent microbial growth and 100 ng/ml BSA to stabilize the enzyme.

2.2.2 Fluorimetric assay

This assay is the most sensitive and versatile. Fluorescence methods are 100–1000 times more sensitive than colorimetric assays. The substrate 4-methylumbelliferyl β–D-glucuronide (4-MUG) is used to measure GUS activity. GUS catalyses hydrolysis of the substrate 4-MUG to give the products D-glucuronic acid and 4-methylumbelliferone (4-MU). The latter can be assayed fluorometrically with excitation at 365 nm and emission wavelengths at 455 nm. The rate of accumulation of 4-MU is linearly related to the concentration of GUS enzyme and can be used to test its activity. It is important to stop the reaction with Na_2CO_3, which serves also to maximize the fluorescence of 4-MU. It is possible to perform discontinuous (endpoint) assays, in which the fluorescence values of aliquots of a reaction mixture are measured at regular time intervals. This approach is not recommended when a large amount of samples are assayed simultaneously, due to the errors owing to imprecise timing of the end time points. Alternatively, it is possible to perform continuous kinetic assays in microtitre plates with a fluorescent plate reader (e.g. from Bio-Rad: www.bio-rad.com). This permits the analysis of large numbers of samples at the same time, with the obvious advantage that much less time is required; once the components of the reaction have been mixed in the plate and the fluorimeter has been programmed, the reaction can be left unattended. In some cases, the intrinsic fluorescence of the extract can limit the sensitivity of this assay. When this is a problem it is possible to extract the fluorescent compounds prior to analysis or to use one of the other fluorogenic substrates produced by Molecular Probes, Inc. The fluorimetric assay is discussed in detail in the review of Hull and Devic (12).

Protocol 2

GUS Fluorimetric assay

Equipment and reagents

- Plant tissue
- Equipment for grinding of plant tissue (micropestles, prechilled mortar and pestle, or prechilled homogenizer)
- GUS extraction buffer (see *Protocol 1*)
- 1 mM 4-methylumbelliferyl β-D-glucuronide (4-MUG) (Sigma), prepared in GUS extraction buffer

- 0.1 μM and 1 μM standard 4-methylumbelliferone (4-MU) (Sigma) stocks in stop buffer
- 0.2 M Na_2CO_3 (Stop buffer), freshly prepared
- UV/Visible Fluorimeter
- Spectrophotometer

Method (Endpoint Fluorimetric Assay)

1 Grind tissue in 0.4 ml of GUS extraction buffer if grinding in a mortar and pestle or glass homogenizer, and 0.1–0.2 ml of buffer if using a microcentrifuge tube.[a]

2 Centrifuge the homogenate in a microcentrifuge at 12 000 g for 5 min, take the supernatant and place on ice.[b]

3 In a microcentrifuge tube, mix 0.1 ml of sample extract with 0.4 ml of 1 mM 4-MUG.

4 Incubate at 37 °C. After 2 min, remove a 0.1 ml aliquot of the reaction mix and add this to an equal volume of 0.2 M Na_2CO_3. This timepoint is $T = 0$. Repeat the above step four more times at regular intervals.[c]

5 Measure the fluorescence with excitation at 365 nm, emission at 455 nm, plot against time, and calculate the rate of accumulation of the fluorescent product.[d]

Protocol 2 continued

^a The volume of the buffer depends on the amount of tissue and the method of grinding.

^b At this point, it may be convenient to measure total protein content in the sample, e.g. using the kit described in *Protocol 4*.

^c The frequency of the timepoints depends on the level of activity in the sample. As a first assay try 10, 20, 40, 60 min. If the readings in the fluorimeter are off scale, take readings at closer time intervals and/or use less extract in the reaction.

^d Prepare standards of 0.1 μM and 1 μM 4-MU in stop buffer. Calibrate the fluorimeter by using the standards and blanking against the zero time point. The rate of accumulation of the fluorescent product is then converted into pmol 4-MU/min using a calibration curve of fluorescent units (fl.u) against concentration 4-MU. The protein content of each sample is then determined and the rates are standardized and expressed as pmol 4-MU/mg protein/min (12).

2.2.3 Histochemical assay

The chemical reaction that allows the histochemical detection of GUS activity occurs in two steps: β-glucuronidase catalyses the hydrolysis of the substrate X-Gluc and liberates one molecule of 5-bromo-4-chloro-indoxyl, that is not coloured. The indoxyl then undergoes an oxidative dimerization that results in the blue coloration. Dimerization is stimulated by atmospheric oxygen and can be enhanced by using an oxidation catalyst, such as a potassium ferricyanide/potassium ferrocyanide mixture. The ferricyanide/ferrocyanide is also important in the histochemical assay because it slows down diffusion of the indoxyls into neighbouring cells, which is a potential source of artefacts in this assay. The potassium ferricyanide/ potassium ferrocyanide prevents the diffusion of the indoxyls by enhancing dimerization. However, even though false staining is largely avoided, the relative degree of staining in different cells or tissues may not necessarily reflect the concentrations of GUS.

The histochemical assay can be performed either on whole samples or on sections. In the latter the samples must be fixed, embedded, and sectioned (see *Chapter 10*). Good fixation is important to maintain a good structure of the samples, but on the other hand some components of fixative solutions, especially glutaraldehyde, have an inhibitory effect on β-glucuronidase activity. More detailed information about staining procedures are included in the review of Hull and Devic (12) and an example of histochemical GUS staining is shown in *Figure 1*.

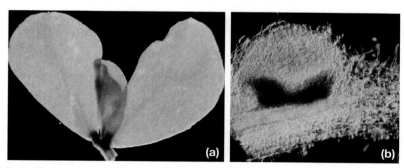

Figure 1 Examples of GUS staining. The images are derived from different lines of *Lotus japonicus* tagged with a promoterless GUS construct. (a) Whole mount staining showing GUS activity at the leaf bases of young leaves. (b) Hand-cut section showing GUS activity within the nodule vascular bundles. Images courtesy of L. Martirani and M. Chiurazzi (IIGB, Napoli, Italy) (see plate 3).

Protocol 3

Histochemical GUS assay

Equipment and reagents

- Plant material
- X-Gluc substrate (5-bromo-4-chloro-3-indolyl-β-D-glucuronic acid (Molecular Probes or Clontech)
- 100 mM Sodium phosphate buffer, pH 7.0

- Oxidative catalyst solutions (50 mM potassium ferricyanide, 50 mM potassium ferrocyanide mixture; Sigma)
- 70% ethanol
- Stereo or transmitted light microscope

Method

1 Dissolve 5 mg X-Gluc (1 mM) in 100 μl dimethylformamide in a chemical hood.

2 Add to 10 ml of 100 mM sodium phosphate buffer, pH 7.0. Wrap the tube in aluminium foil to protect from light and store at 4 °C for up to 1 week. However, it is recommended to prepare fresh X-Gluc solution regularly.

3 Dissolve potassium ferricyanide and potassium ferrocyanide in water at final concentrations of 50 mM. Add 0.1 ml of each stock solution to 10 ml X-Gluc solution for a final concentration of 0.5 mM.[a]

4 Place fresh plant tissues in X-Gluc solution.[b]

5 Incubate the samples in the dark at 37 °C from 1 h to overnight (until the staining is satisfactory).

6 After staining, clear the tissue with 70% ethanol to improve contrast.[c]

[a] This step is optional.

[b] A vacuum infiltration for 5 min can improve the penetration of the substrate.

[c] Optional, only if necessary to remove photosynthetic pigments.

3 Luciferase

3.1 Introduction

Luciferases are enzymes that catalyse light-emitting reactions, a phenomenon known as bioluminescence, and have apparently evolved independently in several phyla. Luciferase genes have been cloned from bacteria, beetles (including the firefly *Photinus pyralis*), coelenterates such as *Renilla* and *Aequorea*, and dinoflagellates (e.g. *Gonyaulax*). Of these, only the luciferases from bacteria, firefly and *Renilla* have found general use as indicators of gene expression.

3.1.1 Bacterial luciferase

The bacterial luciferase from *Vibrio harveyi* (LUX) is a heterodimer of 80 kDa consisting of α and β subunits encoded by two separate genes, *luxA* and *luxB*. These genes are part of an operon and are transcribed from a single promoter (14). The *V. harveyi* luciferase catalyses the oxidation of long-chain aldehydes into carboxylic acids, using $FMNH_2$ (reduced flavin mononucleotide) as a cofactor, and produces photons as one of the reaction products:

$$FMNH_2 + O_2 + \textit{n-}decanal \xrightarrow{\text{luciferase}} FMN + \textit{n-}decanoic\ acid + H_2O + light\ (490\ nm)$$

Utilization of the bacterial luciferase in eukaryotic cells requires the simultaneous expression of the two distinct transcriptional units *luxA* and *luxB*, because eukaryotes are unable to translate prokaryotic polycistronic messages efficiently (15). This can be easily achieved by using a chimeric *luxAluxB* cassette that is translated into a luxAluxB fusion protein. The LUX reporter system has the advantage that the aldehyde substrate (decanal) diffuses readily into tissue when supplied by evaporation from a wick (16). Nevertheless, high decanal levels can cause severe tissue damage; Millar *et al.* (17) found that after titration of substrate concentration, the decanal level required for seedling survival gave less than maximal luminescence.

3.1.2 Firefly luciferase

Firefly luciferase (LUC) is by far the most extensively used of the bioluminescent reporters. The enzyme is composed of a single polypeptide of 61 kDa, which is active in the monomeric form. The cloning of the *LUC* gene in 1987 from the firefly *Photinus pyralis* and its expression in cells from a range of different organisms has generated great interest in the possible applications of the gene as a tool in biological studies (18). It was first used as a reporter for monitoring promoter activity in mammalian as well as in plant cells (19).

Firefly luciferase (E) catalyses the ATP-dependent oxidative decarboxylation of beetle luciferin (LH_2), a heterocyclic carboxylic acid, to generate oxyluciferin (P), and light, as shown below in the following three steps:

1. $LH_2 + ATP + E \xrightarrow{Mg^{2+}} E\bullet LH_2 - AMP + Ppi$

2. $E\bullet LH_2 - AMP + O_2 \rightarrow E\bullet P^* + CO_2 + AMP$

3. $E\bullet P^* \rightarrow E\bullet P + \text{light (560 nm)}$

The quantum yield of the reaction (0.88) is the highest known of any bioluminescent reaction (19). Upon mixing with its substrates, firefly luciferase produces an initial burst of light (flash) that decays over about 15 s to a low level of sustained luminescence. Coenzyme A (CoA) may also participate in the reaction: the inclusion of CoA in the assay mixture transforms the flash into a plateau lasting 30 s or more, followed by a gradual decrease in light emission (17). This obviously makes the assay more convenient for routine use.

LUC is proving to be a highly versatile reporter (20). Detection of luciferase can be accomplished either *in vitro* (after extraction of protein from expressing cells) (see *Protocol 4*) or *in vivo* (see *Protocol 5*). The luciferase gene has been used to study promoter activity in many organisms including bacteria, yeast (21), mammalian cells (18), and plants (17). One notable example is that of Millar *et al.* (22), who used a fragment of the *Arabidopsis CAB2* promoter that mediates circadian-regulated transcription of the *LUC* reporter to screen for circadian clock mutants in *Arabidopsis* seedlings.

LUC is a preferred reporter system in plants for the following reasons:

1. Background luciferase activity does not exist in plant tissue and plants do not usually emit light.

2. In contrast to the bacterial enzyme, the firefly luciferase is active as a monomer and requires the expression of just one gene. It also has the highest quantum yield of known luciferases and, unlike LUX, it is not dependent upon reduced flavin mononucleotide ($FMNH_2$), a substrate not readily available in plant cells.

3. Luciferase assays can be non-invasive, unlike GUS. Cells transformed with luciferase will luminescence in the presence of the substrate and will remain viable. Because magnesium, ATP and oxygen appear to be present in plant cells at concentrations sufficient to saturate the enzyme (23), only luciferin needs to be applied to cells assayed *in vivo*.

4   LUC is relatively unstable because it is sensitive to degradation by proteases, and therefore has a half-life of about 3 h (24). The rapid turnover of LUC protein makes this reporter an excellent candidate for studying dynamic changes in gene expression *in vivo*, in contrast to GUS and LacZ reporters that are stable for many hours.

5   The LUC assay is very sensitive. Light signals are recorded using a luminometer or by liquid scintillation counting. Alternatively, the light signals can be detected by low-light video imaging, in which case spatial information can also be obtained. The spatial resolution of the activity is dependent upon the sensitivity and resolution of the camera, and the optics employed.

The major disadvantages of the LUC reporter system are the investments required to purchase the equipment; luminometers and imaging equipment are not cheap. However, the 'stable' assay performed in the presence of CoA (17), retains the benefits of the original assay, but the luminescence can be recorded with a scintillation counter. Another limitation of the *in vivo* assay is delivery of the substrates into the cells. The inefficient entry of luciferin into cells is not surprising, since it is a carboxylic acid, which is ionized at physiological pH (19). To optimize import of luciferin into cells, the assay utilizes a low pH buffer (5.0–5.5) to protonate the substrate, as well as 10% DMSO or Triton X-100 to permeabilize the cell membrane. This pH-sensitivity could be a problem for the utilization of the *in vivo* LUC assay in organisms that live in conditions where the pH is neutral or basic, such as for marine organisms (sea water pH is around 8.0).

The native enzyme is targeted to the peroxisome in plants, animals and fungi (25). Usually, with high levels of LUC activity the peroxisomes become saturated resulting in enzyme remaining in the cytoplasm. Sometimes the LUC compartmentalization can be problematic because the relative enzyme stability can differ in these two compartments, leading to concentration-dependent biases in the correlation between protein synthesis and quantifiable luminescence. There is also some evidence that impairment of normal peroxisomal activity can lead to destabilization of transgenic cell lines expressing the luciferase gene. To overcome these problems, a modified form of the *LUC* gene, called *LUC⁺*, has been developed which is optimized for reporter gene applications. The pSP-luc⁺ plasmid from Promega is a luciferase cassette vector containing the engineered firefly LUC⁺ with the following modifications:

1   A C-terminal tripeptide has been removed to eliminate peroxisome targeting of the expressed protein.

2   Codon usage has been improved for expression in plant and animal cells.

3   Two potential sites of N-glycosylation have been removed.

4   Several DNA sequence changes have been made to disrupt extended palindromes, remove internal restriction sites and eliminate sequences that may cause spurious transcriptional activation.

3.1.3 *Renilla* luciferase

Renilla luciferase is a 31 kDa monomeric enzyme that catalyses the oxidation of coelenterazine to yield coelenteramine and blue light of 480 nm (26), with a quantum yield of 0.055%. The host organism, *Renilla reniformis*, is a coelenterate that creates bright green flashes upon tactile stimulation, apparently to ward off potential predators. Although *Renilla* and *Aequorea* are both coelenterates that luminescence based on coelenterazine oxidation, and although both luciferases are associated with a green fluorescent protein, their luciferases are structurally unrelated. In particular, *Renilla* luciferase does not require calcium for the luminescence reaction (see also *Chapter 12*).

Renilla luciferase is generally not recommended as an alternative to firefly luciferase, although it shares many of the same benefits. Its primary limitation is the presence of a low level of non-enzymatic luminescence, termed autoluminescence, which reduces the utility of the assay. Nevertheless, *Renilla* luciferase has recently become popular as a companion reporter for experiments where two different reporters are needed. Promega have developed an integrated assay format, called the Dual-Luciferase Reporter Assay to allow rapid sequential quantification of both the firefly and *Renilla* luciferases in the same sample (27).

3.2 Assays

In this section we include protocols for *in vitro* and *in vivo* LUC assays, commonly used in plant cells. In case of problems a visit to the Promega web site (www.promega.com) is highly recommended. The site contains a lot of useful information about luciferase assays, detection methodologies and available vectors.

3.2.1 *In vitro* LUC assay

Luciferase *in vitro* assays are performed after protein extraction from expressing cells. A disadvantage of the *in vitro* assay for luciferase activity is that the sample is destroyed in the process (as in the case with most other reporter gene systems).

Promega have developed a Luciferase Assay System designed specifically for reporter quantification in mammalian cells, and which is broadly applicable for a range of experimental systems, including plants (28) and eukaryotic algae (6). Briefly, for the *in vitro* assay, frozen tissue powder or cells are extracted in a specific buffer, an aliquot of the cleared extract is diluted in assay buffer containing ATP and the peak of luminescence following luciferin addition is measured in a luminometer or scintillation counter. Using the Promega kit the light intensity is constant for more than 60 s because CoA is incorporated into the reaction buffer.

Protocol 4

In vitro LUC assay

Equipment and reagents

- Plant tissue
- Equipment for grinding of plant tissue (micropestles, prechilled mortar and pestles, or prechilled homogenizer, liquid nitrogen)
- Luciferase assay kit (Promega), containing luciferase assay substrate, luciferase cell culture lysis reagent (CCLR), and luciferase assay buffer (LAR)
- Purified firefly luciferase (Boehringer Mannheim)
- DC protein assay kit (Bio-Rad)
- 1 mg/ml bovine serum albumin (BSA)
- Luminometer or scintillation counter
- Visible wavelength spectrophotometer

Method

1 Quick freeze the tissue in liquid nitrogen, grind the frozen tissue to a powder and resuspend at room temperature in a minimal volume of 1× CCLR, enough to cover the cells.

2 Remove the debris after cell lysis by centrifugation at 4°C for 1 min at full speed in a microcentrifuge (12 000 g). Transfer the supernatant to a fresh microcentrifuge tube and repeat centrifugation. Transfer supernatant to a fresh microcentrifuge tube on ice.[a]

3 Measure the protein content of the samples using the DC (Detergent Compatible) protein assay kit (Bio-Rad), according to the manufacturer's instructions (www.bio-rad.com) (see also *Chapter 8*).

Protocol 4 continued

4 Dispense 50 μl aliquots of LAR into luminometer cuvettes.[b]

5 Add 10 μl extracts to the cuvettes. Mix by swirling and measure the luminescence.[c]

6 Normalize the luminescence intensity to specific luciferase activity using a standard curve prepared with serial dilutions of purified firefly luciferase (Boehringer Mannheim) in 1× CCLR, containing 1 mg/ml BSA. In this system, 100 fg of LUC is routinely detected and luminescence is linearly related to LUC concentration over at least eight orders of magnitude.

[a] The extract can be stored indefinitely at −80 °C; however, luciferase activity will decrease as a result of repeated freeze/thaw cycles.

[b] It is important that the LAR is equilibrated at room temperature before beginning the measurements because the temperature optimum for the LUC assay is approximately 20–25 °C.

[c] Use the highest of ten successive 2 s counts, beginning about 10 s after swirling the reaction mixture.

3.2.2 *In vivo* LUC assay

An important application of firefly luciferase is as an *in vivo* reporter of gene expression. It is possible to treat cells with a specific stimulus and to assess the effects on gene expression by monitoring real-time changes in *LUC* reporter gene expression. The major limitation of the *in vivo* assay is the delivery of substrate into the cells, as described previously. However, several reports have described measurements of *in vivo* luminescence in animal, plant, bacterial cells, and tissues. Sometimes acidic buffers or agents, such as DMSO, ionic detergents, or combinations have been used to facilitate the intracellular accessibility of luciferin.

Most typically, plants used for the *in vivo* assay are pre-sprayed with luciferin before beginning the experimental time course, in order to better detect the variations in gene expression (17).

LUC luminescence in the *in vivo* assay can be recorded with a photomultiplier tube in a luminometer or in microtitre plates.

To image luciferase activity with a low light imaging system researchers can choose between a slow-scan cooled CCD or an intensified CCD such as are available from Hamamatsu (29, 30). We suggest assessing several different camera systems, as well as all the necessary components to collect and analyse low-light images because these systems are expensive and because the choice of a particular system will be governed by many user-specific parameters.

Protocol 5

In vivo LUC assay

Equipment and reagents

- Plant tissue
- D-luciferin (Sigma)
- Triton X-100
- Luminometer
- Video imaging system and components for image acquisition and processing

Method

1 Pre-spray plant material with 5 mM D-luciferin in 0.01% Triton X-100 three times at 6-h intervals before starting the experiment, and approximately 20 min prior to each harvesting or imaging time point (if a time course is to be performed).

2 Collect luminescence for approximately 30 min[b] in the low-light imaging system or in the luminometer, 20 min after luciferin application.

[a] Luciferin dissolved in sterile water can be stored in the dark at $-20\,°C$.

[b] The time required depends on expression level.

4 Green Fluorescent Protein (GFP)

4.1 Introduction

Green Fluorescent Protein (GFP) was discovered by Morin and Hastings as a component of the bioluminescence of some marine coelenterates in 1971 (31). Subsequently, the two proteins responsible for the green bioluminescence of the jellyfish *Aequorea victoria*, aequorin and GFP, were characterized (32, 33) (see also *Chapter 12*). *In vivo*, these two small proteins interact in a spectacular way: the aequorin, a calcium activated photoprotein, emits blue light which is converted by energy transfer into green light by GFP, a fluorescent protein that is excited by blue light and which emits green light. Some other organisms, e.g. *Renilla*, *Mitrocomia*, and *Ptilosarcus*, also have green fluorescent proteins, although very little information is available with regard to the structure of these proteins. In contrast, there has been an enormous effort to characterize the *Aequorea* GFP, and this has led to precise information about structure and function. For this reason, usually, the abbreviation GFP refers to the *Aequorea* green fluorescent protein.

Another unrelated autofluorescent protein denoted drFP583 or dsRed, isolated from a marine *Anthozoa* species, has also been described (34). Due to its red fluorescence emission it is likely to replace GFP-based assays in some experimental systems, particularly in animal cells, which do not have red chlorophyll autofluorescence. In plant cells, its use is still limited to specific applications (35), although a new variant, denoted E5, may prove useful because its colour changes with time (36).

GFP is an 'autofluorescent' protein of 238 amino acids (MW 26 kDa) whose native structure is essentially an 11-stranded β-barrel ('β-lantern'), which wraps the imidazolinone chromophore, guarding it very efficiently (37). The chromophore is formed in the presence of oxygen by a slow (4 h) reaction of autocatalytic cyclization (38) of the three amino acids Ser 65, Tyr 66, and Gly 67 (39) lying on a central helical segment. These features explain some fundamental characteristics of GFP:

1 No external cofactors, except for oxygen, are required for fluorescence (40).

2 Very high stability with respect to the effects of denaturants, proteases, temperature (T_m = 78 °C) and pH (41).

Immediately following isolation of the GFP cDNA from *Aequorea victoria* the potential uses of the protein were clear (42). Indeed, GFP has now been used as a fluorescent marker of many proteins *in vivo* to follow their cellular fate. Furthermore, it has been possible to study the interaction between different proteins and other structures (e.g. membranes) by exploiting the same energy transfer process that occurs naturally in the jellyfish between aequorin and GFP (see Section 4.5.2). In all these applications, the GFP cDNA is fused to the cDNA of the protein of interest to obtain a labelled fusion protein, which often retains all the features of both the constituting proteins. These hybrid cDNAs can be expressed potentially in any cells, if carried by a suitable expression vector. Moreover, GFP has been used as an *in vivo* transcriptional reporter, by placing the cDNA under the control of regulatory elements (e.g. promoters and enhancers), to determine the activity of these elements with time and space resolution.

4.2 Advantages and disadvantages of GFP

Several features of GFP and derivatives make them very useful as *in vivo* reporter genes with often quite superior characteristics with respect to the more classical reporters such as GUS and LacZ. For example:

1 GFP, being a fluorescent molecule, provides very high sensitivity.
2 GFP is non-toxic and its visualization does not normally perturb cell metabolism.
3 GFP is cost and time effective.
4 The existence of many different GFP spectral variants allows multiple labelling of different structures in the same cell.

Because a fluorescent molecule that is continuosly excited by light will emit many photons during its lifetime (before photobleaching occurs), it resembles an enzyme that, in the presence of its substrate, generates large amounts of product. This amplification is quite relevant because it produces such high levels of detectable product (light) that it can be detected even in a single cell. Moreover, GFP is a very good fluorescent molecule thanks to its high stability and quantum yield, and low photobleaching rate. These features allow one to easily study GFP or, more interestingly, GFP-tagged protein localization at the subcellular level, in a normal epifluorescence microscope.

GFP-based applications offer some advantages with respect to enzymatic- or antibody-based assays. In the latter cases the administration of specific substrates or antibodies to the cells is required, and it is often accompanied by some step of cell fixation and/or permeabilization which can lead to some artefacts. Moreover, these compounds can be quite expensive. On the contrary, GFP fluorescence requires only an inexpensive excitation light that is relatively non-toxic and which is usually capable of 'permeating' the cells easily without any perturbation. As a consequence, GFP allows reliable *in vivo* real-time experiments to be performed in single cells or whole organisms. Furthermore, GFP use is easy because it involves only the application of standard recombinant DNA techniques to place the GFP cDNA in the desired expression vector. Other methodologies, e.g. those involving labelling of proteins with detectable groups, require extensive use of biochemical techniques and consequently are both money- and time consuming.

Finally, many GFP spectral variants are now available (see Section 4.3). It is therefore possible to label different proteins with different GFPs having different light excitation/ emission characteristics to follow their behaviour simultaneously in the same cell, and to study their interactions (see Section 4.5.2).

One disadvantage lies in the nature of the detectable signal, light. Sometimes the organism under study is opaque or its own colour or fluorescence of endogenous compounds can prevent or complicate the detection of the light signal coming from GFP. Moreover, the inherent stability of GFP does not allow its universal use as a reporter gene (see Section 4.5.3). Another problem is to demonstrate that the physiological activity of the fusion protein with GFP is conserved; however, this is a problem with many techniques involving protein labelling. Finally, it should be mentioned that, in spite of the high sensitivity of GFP-based markers, enzymatic assays nonetheless allow higher sensitivity.

4.3 GFP spectral variants

Wild-type GFP (wtGFP) has two peaks of excitation (corresponding to two wavelengths of blue light); the major one has a maximum at 395 nm and the minor at 475 nm. On the contrary, there is only one emission peak with a maximum at 507 nm (corresponding to green light; *Table 1, Figure 2*). As mentioned before, many GFP spectral variants are now available (supplied from various companies such as Clontech). Indeed, numerous mutations in and around the

TABLE 1 Spectroscopic and biochemical properties of some green fluorescent protein mutants

Common name	Class	Mutations	λEx max/Ext. coeff. (mM⁻¹ cm⁻¹)	λEm max/ Quant. yield	Rel. Fluor. (37°C)[a]
Wt, BioGreen	1		395/21.0 475/7.1	508/0.77	1
GFP-S65T	2	S65T	489/52.0	511/0.64	21
EGFP	2	F64L, S65T	488/55.9	507/0.60	35
ECFP, W1B	5	F64L, S65T, Y66W, N146I, M153T, V163A	434/32.5 452	476/0.4 505	
EYFP, 10c	4	S65G, V68L, S72A, T203Y	514/83.4	527/0.61	35
EBFP, BFP2	6	F64L, S65T, Y66H, Y145F	383/26.3	447/0.26	1
mGFP4	1	Plant codon usage	395/21.0 475/7.1	508/0.77	
mGFP5	1	Plant codon usage, V163A, I167T, S175G	395[b] 473[b]	508	39 111

[a] GFP mutants were expressed in E.coli at 37°C from the same vector in similar conditions. These values represent not only the intrinsic fluorescence brightness, but also the folding efficiency at 37°C.

[b] The two extinction coefficients have not been reported.

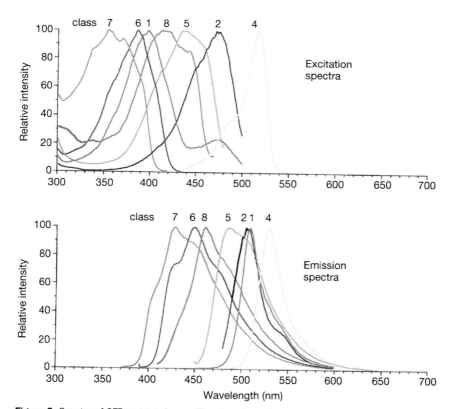

Figure 2 Spectra of GFP mutant classes. The class 3 spectra are very similar to those of class 1, except for the missing peak B (475 nm) in the excitation spectrum.

chromophore alter the spectral features of wtGFP by shifting its excitation or emission peaks. Before going into detail, it is useful to mention that other mutations have been introduced into the wtGFP sequence to obtain GFP variants with an enhanced overall fluorescence in different conditions or living hosts. These mutations change codon usage, eliminate intron splicing sites or enhance the folding of GFP, but in all cases the spectra do not change.

Roger Tsien (43) has grouped all the GFP mutants into seven classes with similar excitation and emission spectra, denoted EGFP variants. Their spectra are represented in *Figure 2*, and other useful characteristics of some mutants are shown in *Table 1*. From the spectra it is easy to see the excitation and emission wavelength shifts of the GFP variants. Essentially, both blue- and red- shifted mutants with respect to the spectra of wtGFP are available and, in addition, the excitation spectra are simplified because only one peak is present. Given the wide availability (e.g. from Clontech) of so many different mutants that are selectively detectable by using suitable monochromators or optical filters (Omega Optical, Inc. www.omegafilters.com, or Chroma Technology corp. www.chroma.com), it is now possible to choose GFP variants, that allow multiple labelling applications, such as:

1   Monitoring gene expression from multiple promoters in the same cell, tissue or organism (see Section 4.5.3).
2   Intracellular localization of multiple proteins (see Section 4.5.1).
3   Real time analysis of protein–protein interactions using FRET (see Section 4.5.2).
4   FACS analysis of mixed cell populations (see Section 4.5.4).

4.4 Hints for the construction of GFP-tagged proteins

Although there are many articles in which GFP fusion proteins have been used, it is not easy to prescribe a recipe for always obtaining functional fusion proteins in which both proteins retain their activity. This is because it is not easy to predict the effect of the fusion of two proteins on the two individual structures. Nevertheless, in many cases, the fusion proteins are functional.

Generally, it is worthwhile to produce both N- and C-terminal fusion proteins. Sometimes, both will be active, sometimes only one will be. If a terminal portion of a protein is known to be essential for its activity, it is a good idea to use the other terminus for the fusion.

Another aspect is the introduction of a short peptide linker sequence between the two proteins. This can facilitate the correct folding of both proteins, especially if it is flexible. Flexibility can be obtained by introducing glycine and proline residues. Another possibility is to introduce a stretch of the same amino acid residue. For example, a 10-alanine-residue linker between an actin ACT1 and GFP resulted in a better fusion protein with respect to no linker or a four-residue-linker construct (44). Moreover, this result evidenced the importance of linker length. Based on many publications, a good linker length would range from 6 to 25 amino acid residues. However, there are many reports of functional GFP fusion proteins in which no linker was introduced or no care was given to the composition of the linker, other than the necessity of placing convenient restriction enzyme sites between the two fused proteins.

After having produced the fusion protein a very important step is to test the activity of both the fusion partners. While testing GFP fluorescence is trivial, it can be more complicated to test the other partner. It is sometimes possible to determine some *in vitro* activity based on an enzymatic or binding assay, but the most important test is to demonstrate that the fusion protein can complement a knock-out mutation.

4.5 Applications

4.5.1 Protein localization

The unique features of GFP variants (see Section 4.2) make them ideal to study *in vivo* localization and trafficking of fluorescent GFP-tagged proteins in time and space (e.g. 45, 46) at the cellular and subcellular levels using normal fluorescence microscopes, which can be coupled with video-imaging systems to obtain a full range of information from the sample under study. Moreover, very high quality images are obtainable by confocal microscopy (47). A confocal microscope allows highly resolved thin optical sections of thick fluorescent specimens to be obtained thanks to a point source of excitation light obtained by a laser and a suitable arrangement of the optical parts of the microscope, which overcome one of the limitations of epifluorescence microscopy, namely the lower quality of the images due to the out of focus blur that arises from thick specimens such as whole cells (see Chapter 12). Although this can be partially overcome by video-imaging systems and deconvolution algorhythms, such technologies are expensive. However, confocal microscopes are also expensive. As an example of images obtainable by confocal microscopy, *Figure 3* shows a stack of 'optical slices' from a cultured *Nicotiana benthamiana* cell transfected with mGFP4 (48; a GFP variant that is very popular in plants).

The major limitation of confocal microscopy is that the standard laser beam (488 nm) used for excitation is not compatible with many of the GFP variants (see *Table 1*, *Figure 2*). If funds allow, this problem can be resolved by using a tunable laser.

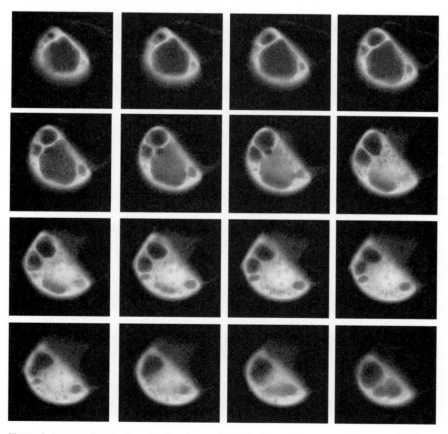

Figure 3 Images of a *Nicotiana benthamiana* cell transiently transformed with mGFP4, captured by a confocal microscope with a water immersion 63× objective and 2 μm of z distance between optical sections.

A further possibility for visualizing GFP is to use a stereo microscope fitted with a fluorescent excitation light source. Such microscopes are useful for macroscopic analysis and are now available from all the major microscope manufacturers.

4.5.2 Fluorescence Resonance Energy Transfer (FRET)

Fluorescence Resonance Energy Transfer (FRET) is a very promising application for GFP. FRET is the process that can occur between two different chromophores, denoted donor and acceptor, if they are in *close proximity* and in a suitable orientation. The donor is a fluorescent molecule. After donor excitation, the acquired excitation energy can be transferred in a non-radiative way to the acceptor molecule if its excitation spectrum is at least partially *overlapped* to the emission spectrum of the donor. The efficiency of FRET (E), i.e. the fraction of excitation energy absorbed by the donor, which is transferred to the acceptor, is defined in the Förster equation (49):

$$E = R_o^6/(R^6 + R_o^6)$$

where R is the distance between donor and acceptor and R_o is the distance at which E is 50% (usually between 30 and 60 Å). A higher R_o value means higher efficiency of energy transfer at any R value. Because from the equation, E depends on the distance R between FRET partners, it is a measurement of their proximity. Actually, because simple group co-localization is not able to lead to FRET owing to too short a life of donor-acceptor proximity, the FRET estimate will be related to a real group interaction. When the FRET partners are tags of two interesting proteins then it is possible to study the protein interaction, thus making the phenomenon very useful for some applications.

The fluorescence energy transfer is manifested by two classes of events, one related to the donor and the other to the acceptor:

1   Donor-related events—reduction in the fluorescence intensity and in the excited state lifetime and increase in the photobleaching time constant.

2   Acceptor-related events—if the acceptor is a fluorophore, it will re-emit its characteristic longer wavelength light which therefore will increase. If the acceptor is only an absorbing group, this event does not occur and the species is called a quencher.

The efficiency of energy transfer, 'FRET efficiency', is calculated starting from the measurement of the above events. This is not a trivial matter for the following reasons:

(1)   limitations in the available monochromators, especially if optical filters are used;

(2)   donor and acceptor excitation and emission spectra are not completely separated even for the best FRET pairs;

(3)   advanced instrumentation is sometimes required.

Because of the different excitation wavelengths of the GFP variants, confocal microscopy cannot normally be used for FRET, unless the microscope is equipped with a tunable laser. Most systems therefore employ epifluorescence microscopes fitted with filter wheels and a sensitive CCD camera for fluorescent light detection. It is not possible to discuss the problems of quantitation of FRET in this chapter, but a list of excellent articles is provided below for anyone interested in the subject:

(1)   a detailed explanation on how to perform the quantitation of FRET using epi-fluorescence microscopy and optical filters (50);

(2)   the measurement of the donor photobleaching rate by epi-fluorescence microscopy and optical filters to determine FRET efficiency (51, 52);

(3)   a description of the theory and instrumentation of FLIM (Fluorescence Lifetime IMaging) microscopy in which the lifetime of the donor is used to detect FRET (53).

To theoretically decide whether two groups are a good or a bad FRET pair, it is necessary to consider the parameters from which the R_o value (in Å) depends:

$$R_o = 9.79 \times 10^3 (\kappa^2 Q J n^{-4})^{1/6}$$

where κ is the orientation factor (the value used is usually 2/3, corresponding to freely mobile components); Q is the quantum yield (number photons emitted/number photons absorbed) of the donor in the absence of acceptor; J is the overlap integral, essentially expressing the degree of spectral overlap between donor emission and acceptor absorption $J = \int_\infty{}^0 F_D(\lambda)\varepsilon_A(\lambda)\lambda^4 d\lambda$ (ε_A extinction coefficient of the acceptor and F_D is the normalized fluorescence of the donor); and n is the refractive index of the intervening medium.

Among them, the most important parameters associated to the nature of the donor and acceptor are Q and J. For practical purposes, the Q values for selected GFP spectral variants are shown in *Table 1*, while J is estimable from the comparison of spectra such as those in *Figure 2*. For all these parameters, the higher the value the better the FRET pair, i.e. giving a higher sensitivity in the detection of the interaction between the paired species.

At this moment, there are still very few reports of FRET-based assays using GFP in the literature. Among them, ECFP-EYFP and EBFP-EGFP (54) are FRET partners constituted by two GFP spectral variants, which have been successfully used as FRET pairs. These results evidence the great potential of the technique (55).

Another interesting application exploits intra-molecular FRET. In one example, the two FRET partners (CFP and YFP) have been linked (fused) to the opposite sides of a calmodulin fused to the calmodulin-binding domain of myosin light chain kinase (M13) to obtain new *in vivo* calcium sensors, designated cameleons (54; see Chapter 12). In the absence of calcium the molecule is in a relaxed conformation and the E value is very low (because R is higher than R_o). When there is an increase in calcium concentration, calcium can bind calmodulin and induce its linkage to the M13 domain, which in turn produces a contraction of the sensor which brings CFP and YFP closer together, thereby increasing the E value.

4.5.3 GFP as a reporter of gene transcription

The accurate measurement of changes in gene expression requires that a reporter protein turns over rapidly, otherwise it will be impossible to detect any down-regulation and subsequent up-regulation, because of a too high background reporter protein level.

wtGFP and the other mutants discussed up until now are bad reporter proteins, because of their high *in vivo* stability (their half-life is at least 24h; 56). To partially overcome this problem, a destabilized EGFP (derived from Enhanced GFP; see spectral variants) has been generated, the so-called dEGFP, by fusing the degradation domain of mouse ornithine decarboxylase (PEST domain) to EGFP (57). The half-life of dEGFP is about 2 h and is therefore comparable to the firefly luciferase (about 4 h). Zhao *et al.* (58) compared EGFP and dEGFP as reporters of up- and down-regulated genes using firefly luciferase as control. The reliability of the destabilized EGFP was remarkable whereas poor results were obtained with EGFP. Indeed, EGFP did not allow the detection of down-regulation and revealed up-regulation only partially. In spite of the fact that the utility of dEGFP appears quite comparable to firefly luciferase, it is evident that luciferase is still a more reliable reporter because of the long folding time required for GFP to become a fluorescent protein, which can lead to incorrect results. Indeed, in a time course study using dEGFP as reporter, after induction of gene expression a *delay* in induction was observed compared with the luciferase assay (58).

In conclusion, dEGFP can be used as a fluorescent reporter for gene transcription for *in vivo* real time analysis of single cells by exploiting all the advantages derived from the use of GFP, although it is necessary to bear in mind that the delay between expression of GFP and the establishment of its fluorescence can sometimes influence the observed results.

4.5.4 Flow cytometry

Flow cytometry involves the optical analysis of biological particles and cells forced to flow within a fluid stream through the focus of an intense source of light, thereby absorbing and scattering light. The emitted light (fluorescence and scattered light) is detected using photomultipliers and photodiodes as intensity versus time curves, and the data are stored on a cell-by-cell basis or in the form of population frequency distributions. The optical characteristics of cells are determined on the basis of the light scattered at the angle close to the illumination axis and orthogonally to detect preferentially the emission light (e.g. fluorescence).

Flow cytometry allows the sensitive measurement of the fluorescence arising from cells in which a certain fluorochrome has been introduced. Because the light source is usually constituted by an air-cooled argon laser emitting only light of 488 nm, there is some limitation in the choice of GFP variants. However, fully equipped flow cytometers provided with tunable lasers emitting a full range of selectable wavelengths from 400 to 800 nm can allow the analysis of all the GFP variants.

Flow cytometry can be used for quantitative measurements of cell fluorescence on a single cell basis for obtaining accurate single cell information in a whole cell population in a short time and in a statistically meaningful fashion. Moreover, by the sorter function, it is possible to isolate cells with precise optical characteristics. This makes it possible to screen whole cell populations for a particular fluorescent marker to quantify and isolate the labelled cells. Although the use of flow cytometry in plants has historically been limited, some recent reports demonstrate the power of such systems when combined with GFP (59).

Acknowledgements

We would like to thank the European Union (ERBF-CT98-0243 and QLRT-1999-30357), the CNR target project in biotechnology, and the Italian Ministry of Agriculture (MiPAF) for financial support.

References

1. Jefferson, R. A., Kavanagh, T. A. and Bevan, M. W. (1987). *EMBO J.*, **6**, 3901.
2. Twell, D., Yamaguchi, J. and McCormick, S. (1990). *Development*, **109**, 705.
3. Rogers, J. C. and Rogers, S. W. (1992). *Plant Cell*, **4**, 1443.
4. Luan, S. and Bogorad, L. (1992). *Plant Cell*, **4**, 971.
5. Gordon-Kamm, W. J., Spencer, M. T., Mangano, M. L., *et al.* (1990). *Plant Cell*, **2**, 603.
6. Falciatore, A., Casotti, R., Leblanc C., Abrescia, C. and Bowler, C. (1999). *Mar. Biotechnol.*, **1**, 239.
7. Mozo, T. and Hooykaas, P. J. J. (1992). *Plant Mol. Biol.*, **19**, 1019.
8. Kavanagh, T. A., Jefferson, R. A. and Bevan, M. W. (1988). *Mol. Gen. Genet.*, **215,** 38.
9. Schmitz, U. K. and Lonsdale, D. M. (1989). *Plant Cell*, **1,** 783.
10. Iturriaga, G., Jefferson, R. A. and Bevan, M. W. (1989). *Plant Cell*, **1**, 381.
11. Jefferson, R. A. (1988). In *Genetic Engineering* (ed. J. K. Setlow), Vol. 10, p. 247, Plenum Press, New York.
12. Hull, G. A. and Devic, M. (1995). *Methods Mol. Biol.*, **49**, 125.
13. Jefferson, R. A. and Wilson, K. J. (1991). *Plant Mol. Biol. Man.*, **B14**, 1.
14. Belas, R., Mileham, A., Cohn, D., Hilmen, M., Simon, M. and Silverman, M. (1982). *Science*, **218**, 791.
15. Kirchner, G., Roberts, J. L., Gustafson, G. D. and Ingolia, T. D. (1989). *Gene*, **81**, 349.
16. Koncz, C., Olsson, O., Langridge, W. H. R., Schell, J. and Szalay, A. A. (1987). *Proc. Natl Acad. Sci. USA*, **84,** 131.
17. Millar, A. J., Short, S. R., Hiratsuka, K., Chua, N-H. and Kay, S. A. (1992). *Plant Mol. Biol. Rep.*, **10**, 324.
18. De Wet, J. R., Wood, K.V., DeLuca, M., Helinski, D. R. and Subramani, S. (1987). *Mol. Cell. Biol.*, **7**, 725.
19. Gould, S. J. and Subramani, S. (1988). *Anal. Biochem.*, **175**, 5.

20. Alam, J. and Cook, J. L. (1990). *Anal. Biochem.*, **188**, 245.
21. Wood, K. V. and DeLuca, M. (1987). *Anal. Biochem.*, **161**, 501.
22. Millar, A. J., Carré, I. A., Strayer, C. A., Chua, N-H. and Kay, S. A. (1995). *Science*, **267**, 1161.
23. Afalo, C. (1991). *Int. Rev. Cytol.,* **130**, 269.
24. Thompson, J. F., Hayes, L. S. and Lloyd, D. B. (1991). *Gene*, **103**, 171.
25. Keller, G-A., Gould, S., DeLuca, M. and Subramani, S. (1987). *Proc. Natl Acad. Sci. USA*, **84**, 3264.
26. Lorenz W. W., McCann, R. O., Longiaru, M. and Cormier, M. J. (1991). *Proc. Natl Acad. Sci. USA*, **88**, 4438.
27. Sherf, B. A. and Wood, K. V. (1993). *Promega Notes Magazine*, **44**, 18.
28. Millar, A. J., Short, S. R., Chua, N-H. and Kay, S. A. (1992). *Plant Cell*, **4**, 1075.
29. Millar, A. J., Smith, K. W., Anderson, S. L., Brandes, C., Hall, J. C. and Kay, S. A. (1994). *Promega Notes Magazine* **49**, 22.
30. Kost, B., Schnorf, M., Potrykus, I. and Neuhaus, G. (1995). *Plant J.*, **8**, 155.
31. Morin, J. G. and Hastings, J. W. (1971). *J. Cell. Physiol.*, **77**, 305.
32. Shimomura, O. and Johnson, F. H. (1975). *Proc. Natl Acad. Sci. USA*, **72**, 1546.
33. Blinks, J. R. (1976). *Pharmacol. Rev.*, **28**, 1.
34. Matz, M. V., *et al.* (1999). *Nature Biotechnol.* **17**, 969.
35. Màs, P., Devlin, P. F., Panda, S. and Kay, S. A. (2000). *Nature*, **408**, 207.
36. Terskikh, A., Fradkov, A., Ermakova, G., *et al.* (2000). *Science*, **290**, 1585.
37. Ormo, M., Cubitt, A. B., Kallio, K., Gross, L. A., Tsien, R. Y. and Remington, S. J. (1996). *Science*, **273**, 1392.
38. Heim, R., Prasher, D. C. and Tsien, R. Y. (1994), *Proc. Natl Acad. Sci. USA*, **91**, 12501.
39. Cody, C. W., Prasher, D. C., Westler, W. M., Prendergast, F. G. and Ward, W. W. (1993). *Biochemistry*, **32**, 1212.
40. Cubitt, A. B., Heim, R., Adams, S. R., Boyd, A. E., Gross, L. A. and Tsien, R. Y. (1995). *Trends Biochem. Sci.*, **20**, 448.
41. Ward, W. W. (1982). *Photochem. Photobiol.*, **35**, 803.
42. Prasher, D. C. (1995). *Trends Genet.*, **11**, 320.
43. Tsien, R. Y. (1998). *Ann. Rev. Biochem.*, **67**, 509.
44. Doyle, T. and Botstein, D. (1996). *Proc. Natl Acad. Sci. USA* **93**, 3886.
45. Flach, J., Bossie, M., Vogel, J., Corbett, A., Jinks, T., Willins, D. A. and Silver, P. A. (1994). *Mol. Cell. Biol.*, **14**, 8399.
46. Wang, S. and Hazelrigg, T. (1994). *Nature*, **369**, 400.
47. *Methods in enzymology* (1999). Vol. 307, *Confocal Microscopy* (ed. J. N. Abelson & M. I. Simon). Academic Press, San Diego.
48. Haseloff, J., Siemering, K. R., Prasher, D. C. and Hodge, S. (1997). *Proc. Natl Acad. Sci. USA*, **94**, 2122.
49. Förster, T. (1948). *Ann. Phys. (Leipzig)*, **2**, 55.
50. Gordon G. W., Berry G., Liang X. H., Levine B. and Herman B. (1998). *Biophys. J.*, **74**, 2702.
51. Gadella, T. W. J and Jovin, T. M. (1995). *J. Cell Biol.*, **129**, 1543.
52. Rocheville, M., Lange, D. C., Kumar, U., Sasi, R., Patel, R. C. and Patel, Y. C. (2000). *J. Biol. Chem.*, **275** (11), 7862.
53. Gadella T. W. J., Jr (1999). In *Fluorescent and luminescent probes for biological activity* (ed. W. T. Mason) 2nd edn. Academic Press, San Diego, p. 467.
54. Miyawaki, A., Llopis, J., Heim, R., McCaffery, J. M., Adams, J. A., Ikura, M. and Tsien, R. Y. (1997). *Nature* **388**, 882.
55. Mahajan, N. P., Linder, K., Berry, G., Gordon, G. W., Heim, R. and Herman, B. (1998). *Nature Biotechnol.*, **16**(6), 547.
56. Wood, K. V. (1998). *Curr. Opin. Biotech.*, **6**, 50.
57. Li, X., Zhao, X., Fang, Y., Jiang, X., Duong, T., Fan, C., Huang, C.-C. and Kain, S. R. (1998). *J. Biol. Chem.*, **273**, 34970.
58. Zhao, X., Duong, T., Huang, C-C., Kain, S. R. and Li, X. (1999). In *Methods in enzymology* (ed. P. M. Conn), Vol. 302, p. 32. Academic Press, San Diego.
59. Kovtun, Y., Chiu, W. L., Zeng, W. and Sheen, J. (1998). *Nature*, **395**, 716.

Chapter 14
Moss gene technology

Celia D. Knight, David J. Cove and Andrew C. Cuming
Centre for Plant Sciences, University of Leeds, Leeds, UK

Ralph S. Quatrano
Department of Biology, Washington University at St Louis, USA

1 Introduction

The moss *Physcomitrella patens* has been developed as a model for studying plant development (1, 2). As a bryophyte, it is a member of the earliest group of land plants, thus occupying a basal position in the plant phylogeny. The dominant vegetative form of the moss is haploid, affording easy isolation and recognition of mutant phenotypes, and it is easily cultured under laboratory conditions. Anatomically simple, *Physcomitrella* is ideally suited to the study of plant growth and differentiation at the cellular level, with the developmental transitions between the principal cell types easily observed and manipulated experimentally. In recent years, the molecular genetics of the organism has been developed, culminating in the recent, exciting discovery that *P. patens* has an efficient somatic homologous recombination system. Uniquely among plants, this makes gene targeting by transforming DNA a viable approach to the study of gene function (3–5). When the high degree of DNA sequence similarity between moss and higher plant genes is taken into account (6, 7), it is clear that *Physcomitrella* can be used experimentally to knock-out the homologues of higher plant genes. With the completion of the *Arabidopsis* genome sequencing project (8), and the accumulation of an increasing number of T-DNA and transposon-tagged *Arabidopsis* lines (9), it is likely that it will soon be possible to obtain a null mutant for any given *Arabidopsis* gene. However, being able to undertake a precise allele replacement through homologous recombination offers a much greater level of sophistication to the study of plant gene function, *in vivo*. We term this process gene surgery.

An additional and important attribute of gene targeting in moss is that novel gene functions might be revealed, either by the disruption of a gene sequence from the expressed sequence tag (EST) database, for which there is no known sequence homology, or by the identification of an unexpected phenotype following targeted gene disruption of a higher plant homologue. The latter approach is of particular importance for the understanding of the evolution of plant gene function. For example, there are moss homologues of the genes that regulate floral initiation and organ specification, and yet mosses do not flower. We anticipate that the targeting of moss homologues of other genes, whose role in higher plants would suggest that no parallel process exists in moss, will reveal the ancestral functions of these genes and yield insights into the ways that they have been recruited to their present functions in angiosperms.

In this chapter, we provide the protocols necessary for the use of *Physcomitrella* as a model organism. These include methods for culture, nucleic acid isolation, vector construction, protoplast transformation, and some examples of phenotype analysis. We also describe the molecular resources being made available through the Biotechnology and Biological Sciences Research Council (BBSRC)-supported *Physcomitrella* EST Programme (PEP) and provide details of how these may be obtained and utilized.

2 Moss development and tissue culture

The life-cycle and cell types of *Physcomitrella* have been described in detail in Cove *et al.* (10) and it is not the purpose of this chapter to reiterate these. However, example figures of the appearance of cultures will be used to demonstrate some of the protocols. For example, *Figure 1* shows all of the gametophytic cell types, i.e. the filamentous protonema (chloronema and caulonema), and the leafy shoots or gametophores. All of these cells are haploid, the diploid sporophyte is not represented.

Protocol 1

Media preparation

Use analytical grade inorganic chemicals where possible and sterilize all solutions by autoclaving or filtration, unless it is otherwise stated.

Equipment and reagents

- Minimal 'BCD' medium
- Stock solution B: 25g $MgSO_4.7H_2O$ (or 12 g anhydrous $MgSO_4$) per 1 l distilled water
- Stock solution C: 25 g KH_2PO_4 per 1 l distilled water (make up in 500 ml water, adjust pH to 6.5 with minimal volume of 4 M KOH, and make up to 1 l with additional distilled H_2O)
- Stock solution D: 101 g KNO_3, 1.25 g $FeSO_4.7H_2O$ per 1 l distilled water

- Trace element solution (Hoagland's A–Z)[a]: 614 mg H_3BO_3, 389 mg $MnCl_2.4H_2O$, 55 mg $Al_2(SO_4)_3.K_2SO_4.24H_2O$, 55 mg $CoCl_2.6H_2O$, 55 mg $CuSO_4.5H_2O$, 55 mg $ZnSO_4.7H_2O$, 28 mg KBr, 28 mg KI, 28 mg LiCl, 28 mg $SnCl_2.2H_2O$
- Agar: 8 g per litre
- $CaCl_2$ stock solution: 500 mM

Method

1 Store stock solutions at 4 °C. Only the $CaCl_2$ stock solution need be autoclaved.

2 Add 10 ml of each of stock solutions B, C, and D, and 1 ml of the trace element stock solution to 1 l of distilled water and mix. Avoid mixing the concentrates before adding to water.

3 Add 8 g Sigma Agar #A9799[b] and dissolve by heating.

4 Distribute as required into glass bottles or flasks, and autoclave at 121 °C for 15 min.

5 Before use, add $CaCl_2$, mix and pour into Petri dishes in a laminar flow cabinet or clean room. For appropriate $CaCl_2$ concentrations, see protocols that follow.

[a] Other trace element solutions may also be used.

[b] Other high grade agars may be used, but the concentration needed and affect on pH should be tested.

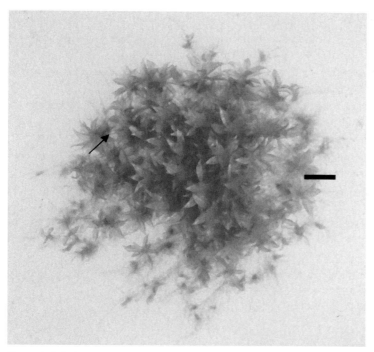

Figure 1 A 3–4-week culture of wild-type *Physcomitrella*, growing on a 50 mm agar plate, showing leafy shoots and protonema (arrow). Scale bar = 2 mm (see plate 4).

Protocol 2

Routine sub-culture of moss tissue

Sub-culture moss routinely on Petri dishes containing BCD + CaCl$_2$. This medium (using nitrate as the nitrogen source) and method allows development of all gametophytic cell types to be observed. Growth on medium containing ammonium as the nitrogen source for the generation of larger quantities of tissue is described in *Protocol 3*. All moss subculturing and manipulation is best carried out in a laminar flow cabinet.

Equipment and reagents

- 50 mm (deep) Petri dishes (Scientific Laboratory Supplies—SLS) containing approximately 15 ml, or 90 mm Petri dishes containing approximately 30 ml of BCD + 1 mM CaCl$_2$ medium (see *Protocol 1*)
- Fine forceps (SLS)
- Moss tissue or spores

- Plant growth facility:
- Constant temperature room or cabinet set to give a temperature under the lights of 24–26 °C.
- Light from fluorescent tubes at an intensity of between 5 and 20 W/m^2.

Methods

(A) Inoculation from a plate containing moss somatic tissue

1 Use flame-sterilized (and cooled) fine forceps to pick up a clump of cells (approximately 1–2 mm in diameter and containing about 50–200 cells) from the growing edge of the colony. For analysis

287

Protocol 2 continued

of growth and phenotype, this should be taken from a young (7–10-day) colony, but for initiating growth, old cultures are suitable, provided that green cells are present.

2   Inoculate a new plate with five well spaced inocula per 50 mm Petri dish, making sure to implant the inoculum into the agar, but without complete submergence.

3   Incubate the Petri dish at 25 °C under continuous white light (a 16 h light/8 h dark cycle may be used, but results in slower growth).

4   Observe growth and development from approximately 3 days.

5   Mature colonies (3–4 weeks old, see *Figure 1*) can be preserved for 6–9 months by taping the plate with paraffin film and storing at 15 °C. Longer-term storage methods have been tested (e.g. cryopreservation (11) and in sterile distilled water at 4 °C).

(B) Inoculation from spores

1   Pipette 50–100 μl of a spore suspension (obtained by crushing a mature spore capsule in 1 ml sterile distilled water; spores may be stored at 4 °C in a dry or wet state) onto a plate of BCD + 1 mM $CaCl_2$ medium.

2   Spread the spore suspension evenly under sterile conditions using a glass spreader.

3   Incubate, as described in *Protocol 2A*.

4   Spore germination is visible after about 5 days.

Protocol 3
Moss culture using ammonium as a nitrogen source

Some protocols require larger quantities of tissue, e.g. protoplast isolation (*Protocol 4*) and nucleic acid extraction (*Protocols 7–9*). For this purpose, it is better to culture moss with added ammonium, which favours chloronemal development; however, this is not a good medium to observe all developmental stages. We normally have cultures growing on both kinds of media, for different purposes.

Equipment and reagents

- A 2–3 week moss culture grown on BCD + 1 mM $CaCl_2$ + 5 mM ammonium tartrate
- 90 mm Petri dishes containing BCD + 1 mM $CaCl_2$ + 5 mM ammonium tartrate
- Cellophanes (see http://www.moss.leeds.ac.uk for suppliers) autoclaved dry (interleave with filter paper).
- Sterile distilled water.

- Blender and vessels
- We use an old Atomix MSE blender, which has glass vessels with individual blenders. This has worked very well, but is no longer on the market. Other types of tissue blender can be used, e.g. IKA Ultra-Turrax homogenizers (Jencons-PLS), providing the head is sterilized between cultures.

Method

1   Using sterile forceps, transfer a sterile cellophane disc to each agar plate. Try to do this while the discs and/or the agar plates are slightly wet, so that the discs will flatten on to the agar. If discs are left to hydrate for at least an hour, it is easier to straighten any creases in them.

2   With fine forceps, pick off all of the moss tissue, and blend in about 10 ml sterile distilled water.

3 Blend until the tissue clumps have been fragmented, but not macerated. The tissue should be easily to pipette, but contain clumps of about 100 intact cells.

4 Pipette 1–2 ml of tissue suspension onto each plate.

5 Incubate for 7 days as in *Protocol 1*.

6 The moss should grow as a lawn of cells (see *Figure 2*); if the density is low from the first sub-culture, this tissue may be scraped from the cellophane overlay, blended, and plated out as before.

3 PEP: the *Physcomitrella* EST Programme

The utility of *Physcomitrella* as a vehicle for the systematic study of gene function ('functional genomics') has recently been recognized by both industrial and public-sector bodies. In the UK, the Biotechnology and Biological Sciences Research Council (BBSRC) has initiated the *Physcomitrella* EST Programme (PEP), a public-access collaboration between Leeds University and Washington University at St Louis, USA (http://www.moss.leeds.ac.uk) and in the private sector a major commercial initiative has been initiated in Germany, at the University of Freiburg, under the direction of Professor Ralf Reski (http://www.plant-biotech.net/).

The PEP collaboration is developing a collection of *c.* 30,000 ESTs from *Physcomitrella*. This will facilitate the identification of moss gene homologous with those of higher plants, whose function can subsequently be investigated using gene targeting technology. In addition to the EST sequences, which are being deposited directly into GenBank, the programme will distribute *Physcomitrella* clones, cDNA, and genomic libraries to researchers worldwide. Other resources available through PEP are vectors for moss transformation, and training in the techniques of moss culture, transformation, and phenotypic and molecular analysis through a series of short residential workshops. Alternatively, the programme offers a transformation service: a researcher who has isolated a *Physcomitrella* sequence and modified it for gene targeting can submit this for delivery into moss tissue, and will be provided with the ensuing, stable transformants for subsequent molecular and phenotypic analysis.

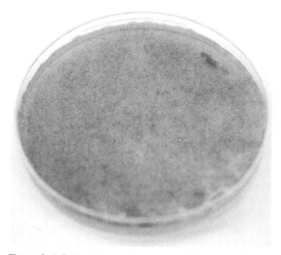

Figure 2 A 7-day homogenate culture of *Physcomitrella*, growing on a 90 mm agar plate, overlaid with cellophane, and showing protonemata only (see plate 5).

A brief list of the resources available through PEP is given below. Full details of the programme, the resources available and how to access these are available on the PEP web site (http://www.moss.leeds.ac.uk). The site includes a BLAST search facility with which the EST collection can be screened for sequence homologues.

3.1 PEP Resources

3.1.1 cDNA libraries

- The 'PPU' library: Physcomitrella Protonemal—Untreated
- The 'PPN' library: Physcomitrella Protonemal NAA (auxin)—treated
- The 'PPA' library: Physcomitrella Protonemal Abscisic acid (ABA)—treated
- The 'PPG' library: Physcomitrella Patens Gametophore

3.1.2 Genomic libraries

- Bacteriophage library: genomic DNA from *Physcomitrella* wild-type protonemata, cloned in the bacteriophage vector λ Fix II (Stratagene) and containing an average insert size of about 17 kb.
- BAC library: a BAC library (20,000 clones) constructed at the Southern Illinois University, Carbondale, USA. The clones average *c.* 100 kb with 10–15% near 200 kb.
- Moss transformation vectors (see Section 4 for details)
- pMBL5: an insertion vector
- pMBL6: a replacement vector

4 Vector construction

Upon identification of a moss homologue of a gene of interest, a genomic or cDNA sequence can be cloned into a moss transformation vector. Once the BAC library has been fully evaluated any EST of interest can be used to probe a grid filter to identify the appropriate BAC clone, from which a region of genomic sequence corresponding to the gene can be sub-cloned into the appropriate transformation vector. Until this refined procedure is available, ESTs are currently used as probes to isolate the genomic clone or full length cDNA from the appropriate libraries. Alternatively, primers from conserved sequences within the EST are synthesized for PCR amplification from genomic DNA.

The moss transformation vectors pMBL5 and pMBL6 were constructed by adaptation of pJIT161 (12). The multiple cloning site from pBluescript was cloned upstream of the 35S-*npt*II-CaMV poly A expression cassette of pJIT161 to generate pMBL5. Plasmid pMBL5 was then further adapted by the insertion of a second multiple cloning site downstream of the *npt*II expression cassette to generate pMBL6. The full sequence of the *npt*II cassette and cloning sites are available at (http://www.moss.leeds.ac.uk).

The insertion of a moss DNA sequence into the multiple cloning site of pMBL5 produces a vector that, when used to transform moss, is most likely to result in insertion of the plasmid by a single cross-over event (thus, pMBL5 is termed an insertion vector, see *Figure 3*). A more sophisticated strategy entails the introduction of a selectable marker gene cassette, excised from the vector pMBL6 into the middle of a moss gene sequence. Such disrupted moss sequences are most likely to result in replacement of the endogenous sequence by a double cross-over event (thus pMBL6 is used to construct a replacement vector, see *Figure 3*). For gene targeting, we recommend the use of replacement, rather than insertion vectors, for ease of confirmation of targeting. However, insertion vectors can be used for allele replacement experiments.

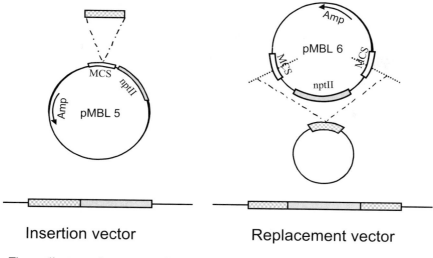

Insertion vector

The *npt*II expression cassette is adjacent to the moss sequence.

Replacement vector

The *npt*II expression cassette is flanked by moss sequences.

Figure 3 Strategies for using pMBL5 and pMBL6 to construct insertion and replacement vectors, respectively

4.1 Construction of a replacement vector using pMBL6: strategic considerations

The frequency with which transforming DNA integrates at the homologous site in the moss genome is high (between 1 and 100% of stable transformants may be targeted in any experiment), but several factors will favour the recovery of targeted transformants. Most important is the length and extent of sequence contiguity between the transforming DNA and its target site. Although cDNA clones have been used successfully for gene targeting (4, 5), the presence of introns within moss genes reduces the overall contiguity with targeting constructs based on cDNAs. For this reason, we strongly recommend the use of genomic fragments, either isolated from a genomic library or amplified by PCR from the moss genome. To facilitate the subsequent molecular analysis, we recommend the use of the 'replacement vector' strategy, if a simple gene disruption is required. Such events can be easily detected by PCR analysis.

We recommend using at least 1.0 kb of cloned moss DNA, into which the 2 kb 35S-*npt*II-CaMV poly A cassette can be inserted. Sometimes, this can be simply excised from pMBL6 by restriction enzyme digestion, and ligated into a convenient, centrally located site in the moss sequence. Alternatively, it may be necessary to introduce such a site into the moss sequence by oligonucleotide-directed mutagenesis, or by the addition of suitable linkers. Another approach, currently under development, is the use of *in vitro* transposition to introduce a selectable marker randomly into a target sequence.

More effective than the simple insertion of a selectable marker into a single site in the targeting construct is to replace an internal fragment of the cloned moss sequence with the selectable marker cassette. This will reduce the chances of the gene being targeted, but not disrupted, due to recombination between multiple copies of the transforming DNA inserted at the targeted locus. This should also be less subject to topological constraints, which might otherwise arise as a consequence of extending the distance between the recombination sites on either side of the selectable marker.

Highly purified DNA is required to achieve high frequencies of transformation. Plasmid DNA purified by caesium chloride centrifugation has proven most efficient in our hands, but the use of a proprietary plasmid purification kit (e.g. Quiagen column or Promega 'Wizard') is an acceptable alternative. Both circular and linear DNA can be used to transform moss by PEG-mediated protoplast transformation (see *Protocol 5*). For several reasons, the use of linearized DNA is preferable. Undigested plasmid DNA will integrate as multiple tandem repeats and furthermore, the presence of plasmid replicon sequences at the integration site will act as targets by which further copies of the transforming plasmid may become integrated. High levels of sequence reiteration at the targeted locus may stimulate sequence rearrangements causing molecular analysis of the integration event to be difficult to interpret.

Ideally, the targeting sequence should be released from its vector backbone by restriction enzyme digestion, or by PCR amplification (although approximately 10 μg DNA are required for each transformation). Although it is not necessary to purify the released targeting fragment from the vector backbone prior to moss transformation, it may be advisable to do so. We have found that when both plasmid backbone and released targeting fragments are delivered, the resulting transformants often contain plasmid sequences integrated at sites other than the (correctly) targeted locus. Clearly, this may have implications for phenotypic analysis of transformants, as it will be necessary to distinguish between a mutant phenotype caused by the knockout of a targeted gene, and one caused by the adventitious disruption of another gene by the independently integrated plasmid backbone.

5 Protoplast transformation

The method that has been used to generate targeted transgenics is that of direct DNA uptake of plasmid DNA by protoplasts in the presence of polyethylene glycol (13). *Protocol 4* describes the isolation and regeneration of protoplasts (14, 15).

Protocol 4

Protoplast isolation and regeneration

Equipment and reagents

- D-mannitol: 8% (w/v) in distilled water and autoclaved
- Driselase solution: 1% (w/v) in 8% D-mannitol solution[a]
- 5–7 day-old blended protonemata grown on BCD + 1 mM CaCl$_2$ + 5 mM ammonium (see *Protocol 3*)
- Stainless steel or nylon mesh filters (TWP; pore size: approx. 100 × 100 μm, for *P. patens*)
- PRMT (Protoplast Regeneration Medium Top layer): 8% (w/v) D-mannitol in BCD medium +

5 mM di-ammonium (+) tartrate + 0.5% (w/v) agar. Sterilize by autoclaving. CaCl$_2$ is added to a final concentration of 10 mM after autoclaving.
- PRMB (Protoplast Regeneration Medium Bottom layer): 6% (w/v) D-mannitol in BCD medium + 5 mM di-ammonium (+) tartrate + 0.8% (w/v) agar. Sterilize by autoclaving. CaCl$_2$ is added to a final concentration of 10 mM after autoclaving.

Method

1 Gently mix (do not shake vigorously) the Driselase[a] and mannitol solution, and leave to stand at room temperature for 15 min, mixing by inversion at approximately 5-min intervals. Centrifuge at 2500 g for 5 min; remove and filter sterilize the clear supernatant.

Protocol 4 continued

2 Scrape the protonemal tissue from the cellophane-overlaid agar plate into the sterile Driselase solution at the rate of about 100 mg (fresh wt.) per ml and incubate for 20–30 min at 25 °C with intermittent gentle shaking. About 10^6 viable protoplasts are obtained per gram of fresh weight tissue.

3 Filter the sterile protoplasts through a sterile stainless steel or nylon mesh and sediment by gentle centrifugation (100–200 g for 3 min).

4 Carefully remove the supernatant and wash twice with sterile 8% (w/v) D-mannitol (use about the same volume for each wash as enzyme solution used for digestion).

5 Resuspend the protoplasts at an appropriate density in sterile, aqueous 8% D-mannitol and add gently to molten PRMT kept at 42 °C in a water-bath (add 1 vol of protoplast suspension per 10 vols of PRMT).

6 Pipette or pour gently, but quickly, the mixture onto PRMB, (overlaid with sterile cellophane, if required; 1 ml of top layer covers a 90 mm Petri dish, 400 μl covers a 50 mm dish).

7 Incubate plates at high light intensities (> 5 W/m^2) for 3–4 days, by which time the protoplasts will have regenerated cell walls and osmoticum is no longer required.

8 Transfer the cellophane and agar overlay to BCD medium containing 1 mM $CaCl_2$ + 5 mM ammonium tartrate, which does not contain mannitol and will allow faster growth.

[a] Batches of Driselase may vary.

Protocol 5

Protoplast transformation

Equipment and reagents

- Protoplasts isolated as in *Protocol 5*, up to stage 4
- 1 M $MgCl_2$
- 1% MES pH 5.6: Use 1% (w/v) 2-[N-morpholino]ethanesulphonic acid) in distilled water. Adjust to pH 5.6 with 0.1 M KOH
- D-mannitol/$MgCl_2$/MES: on day of use, add 150 μl 1 M $MgCl_2$ solution and 1 ml 1% MES pH 5.6 solution to 10 ml 0.91% sterile mannitol solution, and filter sterilize
- 10 μg linear plasmid DNA in 30 μl sterile 10 mM Tris–HCl/l mM EDTA
- 1 M $Ca(NO_3)_2$
- D-mannitol/$Ca(NO_3)_2$ solution: on day of use, add 9 ml 8% (w/v) D-mannitol solution; 1 ml 1 M $Ca(NO_3)_2$ solution; 100 μl 1 M Tris buffer, pH 8.0, and filter sterilize

- PEGT solution: autoclave 2 g PEG 6000 in a glass Universal bottle. On day of use, melt PEG in microwave, add 5 ml D-mannitol/$Ca(NO_3)_2$ solution, and mix well. Leave at room temperature for 2–3 h before use
- 8% (w/v) D-mannitol solution
- 5 ml liquid BCD + 8% mannitol + 10 mM $CaCl_2$ + 0.5% D-glucose + 5 mM ammonium tartrate
- PRMB + 0.5% (w/v) D-glucose medium overlaid with cellophane
- BCD + 5 mM $CaCl_2$ + 5 mM ammonium tartrate ± antibiotic selection (e.g. the *nptII* gene in pMBL6 is selected on G418 at 50 μg/ml; in other plasmids, hygromycin-resistance may be conferred by the *hpt* gene at 30 μg/ml)

Method

1 Use tissue from four 7-day-old cellophane-overlay plates.

2 Follow steps 1–4 of *Protocol 5*.

Protocol 5 continued

3 After the second wash following protoplast isolation, resuspend protoplasts in 10 ml 8% (w/v) D-mannitol solution.

4 Estimate protoplast density using a haemocytometer.

5 Centrifuge protoplasts at 100–200 g for 5 min and resuspend in sufficient D-mannitol/MgCl$_2$/MES solution to give a final protoplast density of 1.6×10^6/ml. This should yield about 2–4 ml of protoplast suspension.

6 Meanwhile, prepare DNA to be used in transformation by dispensing 10 μg of linear DNA (see strategic considerations for vector construction, Section 4; volume no more than 30 μl, but otherwise not critical) into sterile 10 ml tubes. Make sure that the DNA is at the bottom of the tube.

7 Add 300 μl protoplast suspension from 1 to DNA from 2.

8 Add 300 μl PEGT solution.

9 Heat for 5 min at 45 °C.

10 Incubate protoplast suspension at room temperature (20 °C) for 5 min.

11 Add 300 μl 8% D-mannitol solution. Invert gently to mix. Incubate at room temperature (20 °C) for 3 min.

12 Repeat step 11, four more times.

13 Add 1 ml 8% D-mannitol solution. Invert gently to mix. Incubate at room temperature (20 °C) for 3 min.

14 Repeat step 13 four more times.

15 Centrifuge at 100–200 g for 5 min.

16 Remove supernatant and resuspend pellet in 5 ml liquid BCD + 8% mannitol + 10 mM CaCl$_2$ + 0.5% D-glucose.

17 Incubate for 24 h at 25 °C in darkness.

18 Centrifuge at 100–200 g for 5 min, remove supernatant and resuspend in 500 μl 8% mannitol solution.

19 Add 2.5 ml molten PRMT medium.[a]

20 Dispense at rate of 1 ml per 90 mm diameter plate of PRMB + 0.5% (w/v) D-glucose medium overlaid with cellophane.

21 Dilute a sample of the minus-DNA control to 1 in 100 to test for regeneration. The remaining undiluted minus-DNA control can be plated to test for spontaneous mutations, this is not necessary in every experiment as it is should always be negative.

22 Incubate cells in light for 5 days at 25 °C; then transfer protoplasts for all test treatments and the undiluted minus-DNA control on top layer to medium containing antibiotic; the diluted minus-DNA control protoplasts should be transferred to medium without antibiotic.

23 After 14 days, test and control plates should look similar to *Figure* 4. The right-hand plate shows a large number of small colonies and a single large colony, which are antibiotic-resistant. The left hand control plate (no antibiotic) shows how protoplasts regenerate into mature colonies. The majority of colonies on the right hand plate do not have the plasmid stably integrated; however, the strong growing single colony is a good candidate for a stable transformant.

24 To distinguish between stable and unstable transformants and transient expressers of the transgene, sub-culture at 14 days onto medium containing antibiotic. We normally pick off the regenerants onto a 9 cm agar plate (approximately 25 per plate) containing BCD + CaCl$_2$ +

Protocol 5 continued

ammonium tartrate + antibiotic, but without cellophane, using the technique described in *Protocol 2a*.

25 After a further 14 days, those cells showing transient expression of the plasmid die; all those that grow are either stable or unstable transformants.

26 Sub-culture duplicate cultures onto medium with and without antibiotic, as described in step 24.

27 After a further 14 days, sub-culture from the plates without antibiotic, onto medium containing antibiotic.

28 After a further 7–14 days, those cultures that survived a period of growth without selection are the stable transformants.

Therefore, a period of about 8 weeks is required to confirm whether the plasmid is integrated stably. Southern analysis and passage through meiosis would be conclusive proof (13), but this is not done routinely.

^a Kept molten at 42 °C.

6 Molecular analysis of transgenics

As the efficiency of gene targeting is high (up to 100%), only a small number of transgenics need to be confirmed as targeted by molecular techniques. The protocol below describes how DNA can be isolated for PCR analysis. Further useful information is obtained by Southern blot analysis, which requires the isolation of 3 μg DNA per track (see *Protocol 8*) and indicates the number of copies inserted, and whether or not vector sequence is present. Although not essential, this identifies the transgenics that have a high copy number of plasmids inserted, which could present difficulties in the phenotypic analysis.

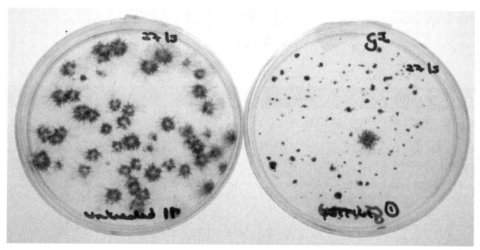

Figure 4 Appearance of transformants on 9 cm agar plates 14 days after transfer of cellophane overlay carrying regenerating colonies to medium containing antibiotic (right hand plate) and to medium without antibiotic (left hand plate; see plate 6).

6.1 Nucleic acid isolation

Protocol 6

Preparation of tissue for nucleic acid extraction

Equipment and reagents

- Protonemal tissue, less than 7 days old, cultured as described in *Protocol 3*. Use of older tissue or tissue stored at low temperature for long periods results in co-extraction of unidentified contaminants,

possibly carbohydrate and phenolics, that are detrimental to subsequent enzyme reactions.
- Filter paper
- Liquid nitrogen

Method

1 It is important to remove as much extraneous liquid as possible from the tissue. Scrape the tissue from the cellophane overlay and place the mass of tissue on two sheets of filter paper. Overlay with another two sheets of paper then press down hard, squeezing liquid out of the tissue.

2 Transfer the squeezed tissue to fresh filter papers and repeat step 1 twice. A lot of the contaminating material is squeezed out of the tissue in this way.

3 Freeze the squeeze-dried mat of tissue in liquid nitrogen. This can then be stored for long periods at $-80\,°C$, or extracted immediately.

6.2 DNA extraction

Commercially available plant DNA extraction kits can be used to isolate DNA from protonemal tissue with little difficulty—we used the Nucleon Phytopure Plant DNA extraction kit (Tepnel Life Sciences) to isolate genomic DNA for the construction of our *Physcomitrella* genomic library.

However, we normally use a modified CTAB extraction method for DNA isolation, which is (i) reliable, (ii) quick, and (iii) cheap.

Protocol 7

Small-scale DNA isolation

This protocol can be used to isolate DNA suitable for PCR.

Equipment

- Extraction buffer: 0.1 M Tris-Cl, pH 8.0; 2% CTAB 1.42 M NaCl; 20 mM Na_2EDTA; 2% PVP-40. Autoclave and store at room temperature. Immediately prior to use, add 7 μl β-mercaptoethanol and 10 mg ascorbic acid to 10 ml buffer stock. Prewarm at 65 °C
- A Kontes-type homogenizer (Sigma: catalogue no. Z35–994–7, Z35–996–3)

- 10 mg/ml RNase A (pre-boiled for 10 min to denature any contaminating DNase)
- Chloroform-*iso*-amyl alcohol (24:1)
- TE buffer: 10 mM Tris-Cl, pH8, 1 mM Na_2EDTA.

Protocol 7 continued

Method

1 Using forceps, pick moss tissue from a colony grown on BCD + $CaCl_2$ + ammonium (*Protocol 2*), minimum size approximately 10 mm in diameter, approximately 5 mg. Squeeze-dry as described in *Protocol 6* and drop it into a 1.5 ml microcentrifuge tube.

2 Freeze the tissue in liquid N_2 and immediately homogenize to a powder using a Kontes-type homogenizer.

3 Add 100 μl extraction buffer and homogenize to produce an even slurry. Add 1 μl RNase and incubate at 65 °C for 5 min.

4 Add 100 μl chloroform-*iso*-amyl alcohol (24:1) and vortex briefly.

5 Separate the phases in a microcentrifuge for 10 min.

6 Transfer the upper phase to a fresh microcentrifuge tube and add 70 μl *iso*-propanol.

7 Mix well and centrifuge immediately at full speed in a microcentrifuge for 5 min.

8 Decant the supernatant and wash the pellet with 70% ethanol.

9 Drain and air-dry the pellet.

10 Dissolve in 15 μl TE buffer.

This provides approximately 0.5 μg good quality DNA readily amplifiable by PCR.

Protocol 8

Large-scale isolation of genomic DNA

This method can be used to isolate DNA suitable for Southern Blot analysis. The same basic-procedure is used as described in Protocol 7.

Equipment

- Extraction buffer: 0.1 M Tris-Cl, pH 8.0; 2% CTAB; 1.42 M NaCl; 20 mM Na_2EDTA; 2% PVP-40. Autoclave and store at room temperature. Immediately prior to use, add 7 μl β-mercaptoethanol, and 10 mg ascorbic acid to 10 ml buffer stock. Prewarm at 65 °C.

- Small mortar and pestle

- 10 mg/ml RNase A (pre-boiled for 10 min to denature any contaminating DNase)
- Chloroform-*iso*-amyl alcohol (24:1)
- TE buffer: 10 mM Tris-Cl, pH 8,1 mM Na_2EDTA

Method

1 Harvest one plate of tissue as shown in *Figure 2* and *Protocol 6*.

2 Freeze the tissue in liquid N_2 and grind with a small mortar and pestle.[a] Add 1 ml extraction buffer and continue homogenizing to obtain a smooth paste.

3 Add a further 1 ml extraction buffer, stirring to obtain a uniform homogenate and transfer this to a suitable high-speed centrifuge tube (e.g. a 15 ml Corex tube). Wash out residual homogenate into the tube with another 1 ml aliquot of buffer.

4 Add 30 μl RNase A and incubate at 65 °C for 5 min.

5 Add 3 ml chloroform-*iso*-amyl alcohol and vortex to emulsify.

6 Separate the phases by centrifuging at 10,000 rpm for 10 min in a swing-out rotor (e.g. Sorvall HB-4).

Protocol 8 continued

7 Precipitate the DNA by adding 2.1 ml *iso*-propanol, mix well, and immediately centrifuge at 12 000 g for 5 min in a swing-out rotor.

8 Wash the pellet with 70% ethanol and air-dry.

9 Dissolve the pellet in 100 μl TE and transfer it to a 1.5 ml microcentrifuge tube, then wash the residue from the Corex tube with another 100 μl TE, and add to the first 100 μl in the microcentrifuge tube.

10 Centrifuge the resuspended DNA for 2 min at 12 000 g in a microcentrifuge.[b]

11 Recover the supernatant carefully and transfer it to a clean microcentrifuge tube for storage at −20 °C.

12 Typically, this provides about 15 μg DNA which is sufficient for five Southern blots.

[a] The liquid N_2 treatment facilitates effective rupture of the cells. It is not necessary to use a pre-chilled mortar. When using liquid nitrogen appropriate safety precautions should be taken.

[b] You may observe a translucent pellet of carbohydrate.

6.3 Digestion of DNA

Although the design of any particular experiment may require the use of a specific restriction enzyme for Southern blot analysis of *Physcomitrella* DNA, it should be noted that some enzymes cleave *Physcomitrella* DNA to a greater extent than others. This probably relates to the extent and distribution of methylation of the moss genome. This has been discussed by Krogan & Ashton (16) who demonstrated that enzymes with recognition sites subject to C-methylation at CG and CNG sequences were less effective in digesting *Physcomitrella* DNA. In our hands, *Hind*III routinely yields the most effective digests. The most heavily methylated sequences in *Physcomitrella* may lie outside the coding sequences. We have noted that in screening a lambda genomic library for numerous *Physcomitrella* genes, that in every case the gene of interest was located at the end of the inserted genomic fragment. This implies that the sites most accessible to the restriction enzyme *Sau*3A, used for partial digestion of the genomic DNA, were undermethylated sites within transcribed regions of the genome.

Protocol 9

RNA extraction

Physcomitrella protonemal tissue presents no great difficulties for the extraction of RNA, as long as the general preliminary procedures used in DNA isolation are followed (see *Protocol 6*). There is little endogenous ribonuclease activity, and the tissue is amenable to RNA isolation using most commercially available kits. However, we typically use an aqueous-SDS/phenol extraction procedure that has the merit of permitting simultaneous isolation of both DNA and RNA from the same sample. High-molecular weight RNA (including mRNA) is recovered by selective precipitation with NaCl. At 2.5 M NaCl, DNA and low molecular weight RNA remain in solution, while high molecular weight RNA species precipitate.

Protocol 9 continued

Equipment and reagents

- Extraction buffer: 0.1 M Tris–HCl, pH 9; 0.5% SDS; 5 mM Na$_2$EDTA.
- Autoclave and store at room temperature. Immediately prior to use, add 7 µl β-mercaptoethanol to 10 ml extraction buffer.
- A Kontes-type homogenizer (see *Protocol 7*).

- Tissue collected from half a 9-cm plate (see *Figure 2*)
- Phenol-chloroform-*iso*-amyl alcohol (25:24:1)
- 3 M Na-acetate pH 5.2
- 5 M NaCl.

Method

1 Harvest the tissue, as described in *Protocol 8*, step 1.

2 Drop the squeeze-dried tissue (approximately 50 mg) into a microcentrifuge tube and freeze the sample by brief immersion in liquid N$_2$. Grind to a powder using a Kontes-type homogenizer.

3 Add 250 µl extraction buffer and homogenize to produce a slurry.

4 Add 250 µl phenol-chloroform-*iso*-amyl alcohol (25:24:1), cap the tube, and vortex to produce an emulsion.

5 Centrifuge at 12 000 g for 2 min in a microcentrifuge and recover the upper (aqueous) phase, transferring it to a fresh tube.

6 Add 25 µl 3 M Na-acetate (pH 5.2) and 600 µl ethanol, and mix well. Plunge the tube into liquid nitrogen for a few seconds, then centrifuge at 12 000 g for 5 min in a microcentrifuge.

7 Discard the supernatant and wash the pellet by resuspension in 70% ethanol, followed by immediate re-centrifugation at 12 000 g for 2 min in a microcentrifuge.

8 Drain the pellet and allow it to air-dry briefly. The pellet contains DNA, RNA, and carbohydrate.

9 Dissolve the pellet in 100 µl TE and add 100 µl 5 M NaCl. Mix well and place on ice for 4 h.

10 To recover the RNA, centrifuge at 12 000 g for 10 min in a microcentrifuge.

11 Carefully aspirate the supernatant using a micropipette. Care is required, since the (RNA) pellet is relatively sloppy, having been centrifuged through a relatively viscous DNA-rich supernatant.

12 Retain the supernatant for DNA recovery (steps 16–19).

13 Meanwhile, resuspend the pellet by vortexing with 200 µl 2.5 M NaCl, and recentrifuge for 5 min at 12 000 g in a microcentrifuge.[b]

14 Discard the supernatant, and wash the pellet with 70% ethanol by resuspension and recentrifugation at 12 000 g for 5 min in a microcentrifuge. Drain the pellet and allow it to air-dry thoroughly.

15 Dissolve the RNA pellet in 50 µl sterile water and store at −20 °C.

16 Dilute the supernatant from step 12 with an equal volume of water, then add two volumes of ethanol.

17 Place at −20 °C for 5 min and centrifuge at 12 000 g for 5 min.

18 Wash the pellet by resuspension in 70% ethanol and centrifuge at 12 000 g for 5 min.[a]

19 Finally, air-dry the DNA pellet and dissolve it in 50 µl TE and store at −20 °C.

[a] This DNA typically contains residual RNA, but is of sufficient quality that it is readily amenable to restriction enzyme digestion and subsequent Southern blot analysis. However, this DNA has not proven to be very tractable to amplification by PCR, and we normally recommend the use of the CTAB extraction procedure (*Protocol 7*) for this purpose.

[b] This time, the pellet should pack tightly as residual DNA is washed out.

Typically, approximately 50 μg RNA is obtained by this method. Agarose gel electrophoresis reveals it to be substantially composed of cytoplasmic and chloroplast rRNA species; the latter appearing as a ladder of fragments resulting from the 'hidden breaks' within the molecules. (Chloroplast ribosomal RNAs contain breaks in the sugar-phosphate backbone at defined points. These do not affect the working of the ribosome, since the structure is held together by the secondary structure of the rRNA. The nicks only become apparent when RNA is analysed by denaturing gel electrophoresis.) Occasionally, a trace of residual DNA is apparent in such preparations. If this is a problem, this can be removed by repeating the 2.5 M NaCl wash step in the protocol. RNA obtained by this procedure can be used directly for northern blot analysis or for the subsequent enrichment of poly(A)-containing mRNA, by oligo-dT affinity chromatography. We have used RNA prepared in this way for translation, *in vitro*, and for the synthesis of cDNA in the construction of cDNA libraries.

6.4 PCR analysis

Using genomic DNA isolated by the method described in *Protocol 7* and suitably chosen pairs of primers, integration at the homologous sites of both the 5′ and 3′ sequences flanking the *npt*II selection cassette can be tested. We recommend the use of a forward and reverse pair of primers, homologous with sequences within the *npt*II coding sequence as an internal control (e.g. forward primer, 5′-AACAGACAATCGGCTGCTCTGATGC-3′, Tm = 73 °C, and reverse primer, 5′-TCCAGATCATCCTGATCGACAAGAC-3′, Tm 69.4 °C). The 5′ reverse primer should recognize the *npt*II sequence within the transforming plasmid (e.g. 5′-GCATCAGAGCAGC CGATTGTCTGTT-3′, Tm = 73 °C) and the 5′ forward primer recognize an upstream moss sequence, which is not represented in the transforming sequence. In this way, amplification of a fragment can only occur at the homologous site, thus confirming gene targeting. Lack of amplification with this primer set is indicative of integration at a non-homologous site. True gene disruption should demonstrate amplification of a fragment also at the 3′ site, using a set of primers in which the forward primer corresponds to the *npt*II gene in the transforming plasmid 5′-AACAGACAATCGGCTGCTCTGATGC-3′, Tm = 73 °C, and the reverse primer corresponds to a downstream moss sequence which is not present in the plasmid (see *Figure 5*).

7 Phenotype analysis

Having identified transgenics with molecular evidence for both gene targeting and for integration of the plasmid DNA at a non-homologous site, candidates from each of these classes can be tested for altered phenotypes. As moss development is affected by factors such as temperature and lighting conditions, it is best to compare a number of targeted transgenics with some non-targeted controls, under a range of conditions, to establish whether targeting leads to consistent changes in growth rate or development. For more details of normal development under standard conditions see ref. 1.

The next challenge for moss gene technology is to develop a gauntlet of phenotypic screens. Undoubtedly, many interested in using this technology will themselves be expert in

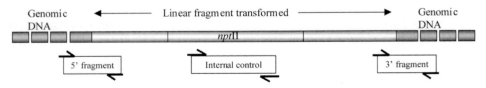

Figure 5 Location of test and control primer pairs for confirming gene targeting.

the analysis of specific gene functions; however, it will also be useful to know how the knock-out of a given gene affects a range of processes and this research is ongoing.

8 Future developments

The use of moss technology to study gene function is gaining interest world-wide. The community is, however, still small and communication, as always, remains an advantage. To assist this, the International Bryophyte Research Information Service (IBRIS) has been established and registration is possible by e.mail (d.j.cove@leeds.ac.uk). All future developments, such as improvements to the technology, major new resources, and training and service elements, will be advertised through this route.

References

1. Cove, D. J. (1992). In: *Development: the molecular genetic approach* (ed. V. E. A. Russo, S. Brody, D. Cove and S.Ottolenghi). Springer-Verlag, Berlin.
2. Reski, R. (1998). *Botanica Acta*, **111**, 1.
3. Schaefer, D. & Zryd, J-P. (1997). *Plant J.*, **11**, 1195.
4. Strepp, R., *et al.* (1998). *Proc. Natl Acad. Sci.*, **95**, 4368.
5. Schaefer, D. G. (2001). *Curr. Opin. Plant Biol.*, **4**, 143.
6. Machuka , J., *et al.* (1999). *Plant Cell Physiol.*, **40**, 98.
7. Reski, R. (1998). *Botanica Acta*, **111**, 141.
8. Kaul, S., *et al.* (2000). *Nature*, **408**, 816.
9. Insertional mutagenesis of *Arabidopsis* by T-DNA transposon insertion (SLAT lines) and Ds-launch pads (Available at: http://www.jic.bbsrc.ac.uk).
10. Cove, D. J., *et al.* (1997). *Trends Plant Sci.*, **2**, 99.
11. Grimsley, N. H. and Withers, L. A. (1983). *Cryolett.*, **4**, 251.
12. Croy, R. R. D. (1993). *Plant Molecular Biology Labfax*. Bios Scientific, Oxford.
13. Schaefer, D., *et al.* (1991). *Mol. Gen. Genet.*, **226**, 418.
14. Grimsley, N. H., *et al.* (1977a). *Mol. Gen. Genet.*, **154**, 97.
15. Grimsley, N. H., *et al.* (1977b). *Mol. Gen. Genet.*, **155**, 103.
16. Krogan, N. T. and Ashton, N. W. (1999). *J. Bryol.*, **21**, 289.

List of suppliers

Ambersil Limited, Wylds Road, Castlefield Industrial Estate, Bridgwater, Somerset TA6 4DD, UK

Amersham Pharmacia Biotech UK Ltd, Amersham Place, Little Chalfont, Buckinghamshire HP7 9NA, UK (see also Nycomed Amersham Imaging UK; Pharmacia)
Tel: 0800 515313
Fax: 0800 616927
URL: http//www.apbiotech.com/

Anderman and Co. Ltd, 145 London Road, Kingston-upon-Thames, Surrey KT2 6NH, UK
Tel: 0181 5410035
Fax: 0181 5410623

Autoquant Imaging Co., 877 25th St., Watervliet, NY 12189, USA
Tel. 001 5182762138
Fax 001 5182763069
sales@aqi.com; http://www.aqi.com

BD Biosciences Clontech
1020 East Meadow Circle, Palo Alto, CA 94303-4230, USA
Tel: 001 650 424 8222
Fax: 001 650 424 1352
URL: http://www.clontech.com

Beckman Coulter Inc., 4300 N. Harbor Boulevard, PO Box 3100, Fullerton, CA 92834–3100, USA
Tel: 001 714 8714848
Fax: 001 714 7738283
URL: http://www.beckman.com/

Beckman Coulter (UK) Ltd, Oakley Court, Kingsmead Business Park, London Road, High Wycombe, Buckinghamshire HP11 1JU, UK
Tel: 01494 441181   Fax: 01494 447558
URL: http://www.beckman.com/

Becton Dickinson and Co., 21 Between Towns Road, Cowley, Oxford OX4 3LY, UK
Tel: 01865 748844   Fax: 01865 781627
URL: http://www.bd.com/
Becton Dickinson and Co., 1 Becton Drive, Franklin Lakes, NJ 07417–1883, USA
Tel: 001 201 8476800
URL: http://www.bd.com/

Bio 101 Inc., c/o Anachem Ltd, Anachem House, 20 Charles Street, Luton, Bedfordshire LU2 0EB, UK
Tel: 01582 456666 Fax: 01582 391768
URL: http://www.anachem.co.uk/
Bio 101 Inc., PO Box 2284, La Jolla, CA 92038–2284, USA
Tel: 001 760 5987299 Fax: 001 760 5980116
URL: http://www.bio101.com/

Bio-Rad Laboratories Ltd, Bio-Rad House, Maylands Avenue, Hemel Hempstead, Hertfordshire HP2 7TD, UK
Tel: 0181 3282000 Fax: 0181 3282550
URL: http://www.bio-rad.com/
Bio-Rad Laboratories Ltd, Division Headquarters, 1000 Alfred Noble Drive, Hercules, CA 94547, USA
Tel: 001 510 7247000 Fax: 001 510 7415817
URL: http://www.bio-rad.com/

Boehringer-Mannheim (see Roche)

Branson Ultrasonics Corp., Calibron Instruments Div., 41 Eagle Rd., Danbury, CT 06813–1961, USA

Calbiochem-Novabiochem GmbH, Postfach 1167, D-65796 Bad Soden/Ts. Germany
Calbiochem-Novabiochem (UK) Ltd., Boulevard Industrial Park, Padge Road, Beeston, Nottingham NG9 2JR, UK
Calbiochem-Novabiochem Corporation, PO Box 12087, La Jolla, CA 92039–2087, USA

CBS Scientific Company Inc., PO Box 856, Del Mar, CA 92014, USA
For world-wide distributers, see www.cbss-ci.com

Clontech Laboratories, Inc., (see BD Biosciences Clontech)

Costar Corning Inc., Science Products Div., 45 Nagog Park, Acton, MA 01720, USA

CP Instrument Co. Ltd, PO Box 22, Bishop Stortford, Hertfordshire CM23 3DX, UK
Tel: 01279 757711
Fax: 01279 755785
URL: http//:www.cpinstrument.co.uk/

Difco Laboratories, Division of Becton Dickinson and Company Sparks, Maryland 21152, USA
Tel: 001 410 316 4000
URL: http://www.bd.com

Dupont Co. (Biotechnology Systems Division), PO Box 80024, Wilmington, DE 19880–002, USA
Tel: 001 302 7741000 Fax: 001 302 7747321
URL: http://www.dupont.com/
Dupont (UK) Ltd, Industrial Products Division, Wedgwood Way, Stevenage, Hertfordshire SG1 4QN, UK
Tel: 01438 734000   Fax: 01438 734382
URL: http://www.dupont.com/

Dynal, Inc., 5 Delaware Drive, Lake Success, NY 11042, USA
Dynal (UK) Ltd, 26 Grove Street, New Ferry, Wirral, L62 5AZ, UK.

Eastman Chemical Co., 100 North Eastman Road, PO Box 511, Kingsport, TN 37662–5075, USA
Tel: 001 423 2292000
URL: http//:www.eastman.com/

Electromedical Instruments, PO Box 8, Pangbourne, Reading RG8 7HU, UK
Tel: +44 (0)118 984 3888
Fax: +44 (0)118 984 5374
clark@cwcom.net
http://www.clark.mcmail.com/

Eppendorf-Fisher Scientific UK, Bishop Meadow Road, Loughborough, Leicestershire LE11 0RG, UK
Tel. +44–1509–231166
Fax +44–1509–231893
andrear@fisher.co.uk
http://www.eppendorf.com/

Fastnacht Laborbedarf GmbH 53119, Bonn, Germany
Tel: +49(0)228 988740
Fax: +49(0)228 9887450

Fisher Scientific UK Ltd, Bishop Meadow Road, Loughborough, Leicestershire LE11 5RG, UK
Tel: 01509 231166
Fax: 01509 231893
URL: http://www.fisher.co.uk/
Fisher Scientific, Fisher Research, 2761 Walnut Avenue, Tustin, CA 92780, USA
Tel: 001 714 6694600
Fax: 001 714 6691613
URL: http://www.fishersci.com/

Fluka, PO Box 2060, Milwaukee, WI 53201, USA
Tel: 001 414 2735013
Fax: 001 414 2734979
URL: http://www.sigma-aldrich.com/

Fluka Chemical Co. Ltd, PO Box 260, CH-9471, Buchs, Switzerland
Tel: 0041 81 7452828
Fax: 0041 81 7565449
URL: http://www.sigma-aldrich.com/

FMC, Flowgen, Lynn Lane, Shenstone, Staffs, WS14 0EE, UK
FMC Marine Colloids, Bioproducts Department, 5 Maple Street, Rockland, ME 04841, USA

Gensiphere 3DNA, 3 Fir Court, Oakland, NJ, 07436, USA
URL: http://www.genisphere.com

Geno Technology, Inc. 3047 Bartold Ave., Maplewood, MO 63143, USA

Gibco-BRL (see Life Technologies)

Glen Spectra, 2–4 Wigton Gardens, Stanmore, Middlesex HA7 1BG, UK
Tel: 020 8204 9517
Fax: 020 8204 5189
gs@isa-gs.co.uk
http://www.isa-gs.co.uk/

HT Biotechnology Ltd, Unit 4, 61 Ditton Walk, Cambridge CB5 8QD, UK

Hybaid Ltd, Action Court, Ashford Road, Ashford, Middlesex TW15 1XB, UK
Tel: 01784 425000
Fax: 01784 248085
URL: http://www.hybaid.com/
Hybaid US, 8 East Forge Parkway, Franklin, MA 02038, USA
Tel: 001 508 5416918
Fax: 001 508 5413041
URL: http://www.hybaid.com/

HyClone Laboratories, 1725 South HyClone Road, Logan, UT 84321, USA
Tel: 001 435 7534584
Fax: 001 435 7534589
URL: http//:www.hyclone.com/

IDEXX Laboratories, Inc., One IDEXX Drive, Westbrook, ME 04092, USA

Invitrogen BV, PO Box 2312, 9704 CH Groningen, The Netherlands
Tel: 00800 53455345
Fax: 00800 78907890
URL: http://www.invitrogen.com/

Invitrogen Corp., 1600 Faraday Avenue, Carlsbad, CA 92008, USA
Tel: 001 760 6037200   Fax: 001 760 6037201
URL: http://www.invitrogen.com/

Jencons-PLS, Cherrycourt Way Ind. Est., Stambridge Road, Leighton Buzzard, Beds, LU7 8UA, UK

Kinematica AG, Luzernerstrasse 147a, CH-6014 Littau-Lucerne, Switzerland

Levington, The Scotts Company UK Ltd., Paper Mill Lane, Bramford, Ipswich IP8 4BZ, UK

Life Technologies Inc., 9800 Medical Center Drive, Rockville, MD 20850, USA
Tel: 001 301 6108000
URL: http://www.lifetech.com/
Life Technologies Ltd, PO Box 35, 3 Free Fountain Drive, Inchinnan Business Park, Paisley PA4 9RF, UK
Tel: 0800 269210
Fax: 0800 243485
URL: http://www.lifetech.com/

Ludl Electronic Products Ltd., 200 Brady Avenue, Hawthorne, NY 10532, USA
Tel (914) 769 6111   Fax (914) 769 4759
lbonomini@ludl.com
http://www.ludl.com

Media Cybernetics, L.P., Frambozenweg 139,2321 KA Leiden The Netherlands
Tel: + 31 715–730–639
Fax: + 31 715–730–640
paul@mediacy.com
http://www.mediacy.com/mediahm.htm

Menzel-Gläser, Menzel, Gerhard Glasbearbeitungswerk GmbH + Co. KG, Braunschweig, Postfach 3157, Germany
Tel: +49–531–59008–0

Merck Sharp & Dohme Research Laboratories, Neuroscience Research Centre, Terlings Park, Harlow, Essex CM20 2QR, UK
URL: http://www.msd-nrc.co.uk/

MSD Sharp and Dohme GmbH, Lindenplatz 1, D-85540, Haar, Germany
URL: http://www.msd-deutschland.com/

Millipore Corp., 80 Ashby Road, Bedford, MA 01730, USA
Tel: 001 800 6455476 Fax: 001 800 6455439
URL: http://www.millipore.com/
Millipore (UK) Ltd, The Boulevard, Blackmoor Lane, Watford, Hertfordshire WD1 8YW, UK
Tel: 01923 816375 Fax: 01923 818297
URL: http://www.millipore.com/local/UKhtm/

MJ Research, Inc., Waltham, MA, 02451, USA

Molecular Dynamics, 928 East Arques Ave., Sunnyvale, CA 94086–4520, USA

Molecular Probes, Inc., 4849 Pitchford Avenue, Eugene, OR 97402–9165, PO Box 2210, Eugene, OR 97402–0469, USA
Molecular Probes Europe BV, Poortgebouw, Rijnsburgerweg 10, 2333 AA Leiden, The Netherlands
http://www. probes.com

Nalge Nunc International. 75 Panorama Creek Drive, P.O. Box 20365, Rochester, NY 14602–0365, USA

Narishige International Ltd., Unit 7, Willow Business Park, Willow Way, London SE26 4QP, UK
Tel: +44 (0) 2086999696
Fax: +44 (0) 2082919678
eurosales@narishige.co.uk
http://www.narishige.co.jp/order/staff.htm

National Diagnostics, 305 Patton Drive, Atlanta, Georgia 30336, USA
National Diagnostics, Itlings Lane, Hessle, Hull Hu139LX, UK

New England Biolabs, 32 Tozer Road, Beverley, MA 01915–5510, USA
Tel: 001 978 9275054

Nickerson-Zwaan Ltd., Rothwell, Market Rasen, Lincoln LN7 6DT, UK
Nickerson-Zwaan bv., Postbus 19, 2990 AA, Barendrecht, Netherlands

Nikon Corp., Fuji Building, 2–3, 3-chome, Marunouchi, Chiyoda-ku, Tokyo 100, Japan
Tel: 00813 32145311 Fax: 00813 32015856
URL: http://www.nikon.co.jp/main/index_e.htm/
Nikon Inc., 1300 Walt Whitman Road, Melville, NY 11747–3064, USA
Tel: 001 516 5474200 Fax: 001 516 5470299
URL: http://www.nikonusa.com/

Nycomed Amersham, 101 Carnegie Center, Princeton, NJ 08540, USA
Tel: 001 609 5146000
URL: http://www.amersham.co.uk/
Nycomed Amersham Imaging, Amersham Labs, White Lion Rd, Amersham, Buckinghamshire HP7 9LL, UK
Tel: 0800 558822 (or 01494 544000)
Fax: 0800 669933 (or 01494 542266)
URL: http//:www.amersham.co.uk/

Olympus Optical Co. (Europa) GmbH, Wendenstr. 14–16, 20097 Hamburg Germany
Tel: +49–40–237730
Fax: +49–40–23773 653
andrea.jensen@olympus-europa.com
http://www.olympus-europa.com/home.htm

Omega Optical, Inc., 3 Grove St Brattleboro, VT, USA 05301
Tel. (802) 254–2690
Fax (802) 254 3937
info@omegafilters.com
http://www.omegafilters.com/index.html

Perkin Elmer Ltd, Post Office Lane, Beaconsfield, Buckinghamshire HP9 1QA, UK
Tel: 01494 676161
URL: http//:www.perkin-elmer.com/
Perkin Elmer (Applied Biosystems Division), Kelvin Close, Birchwood Science Park North, Warrington, Cheshire WA3 7PB, UK

Photonic Science Ltd, Millham, Mountfield, Robertsbridge, East Sussex, TN32 5LA, UK
Tel: +44 (0) 1580 881199

Fax: +44 (0) 1580 880910
infow@photonic-science.ltd.uk
http://www.photonic-science.ltd.uk

Pierce, 3747 N. Meridian Road, P.O.Box 117, Rockford, IL 61105, USA
Pierce Chemical Co., PO Box 117, 3747 N. Meridian Rd, Rockford, IL 61105, USA
cs@piercenet.com

Pierce & Warriner (UK) Ltd., 44, Upper Northgate St, Chester, Cheshire CH1 4EF, UK
PWTECH@piewar.com

Pharmacia, Davy Avenue, Knowlhill, Milton Keynes, Buckinghamshire MK5 8PH, UK (also see Amersham Pharmacia Biotech)
Tel: 01908 661101
Fax: 01908 690091
URL: http//www.eu.pnu.com/

Phenix Research Products, 3540 Arden Road, Hayward, CA, USA

Promega Corp., 2800 Woods Hollow Road, Madison, WI 53711-5399, USA
Tel: 001 608 2744330
Fax: 001 608 2772516
URL: http://www.promega.com/
Promega UK Ltd, Delta House, Chilworth Research Centre, Southampton SO16 7NS, UK
Tel: 0800 378994 Fax: 0800 181037
URL: http://www.promega.com/

Qiagen Inc., 28159 Avenue Stanford, Valencia, CA 91355, USA
Tel: 001 800 4268157 Fax: 001 800 7182056
URL: http://www.qiagen.com/
Qiagen UK Ltd, Boundary Court, Gatwick Road, Crawley, West Sussex RH10 2AX, UK
Tel: 01293 422911   Fax: 01293 422922
URL: http://www.qiagen.com/

Roche Diagnostics Corp., 9115 Hague Road, PO Box 50457, Indianapolis, IN 46256, USA
Tel: 001 317 8452358
Fax: 001 317 5762126

URL: http://www.roche.com/
Roche Diagnostics GmbH, Sandhoferstrasse 116, 68305 Mannheim, Germany
Tel: 0049 621 7594747
Fax: 0049 621 7594002
URL: http://www.roche.com/
Roche Diagnostics Ltd, Bell Lane, Lewes, East Sussex BN7 1LG, UK
Tel: 0808 1009998 (or 01273 480044)
Fax: 0808 1001920 (01273 480266)
URL: http://www.roche.com/

Schleicher and Schuell Inc., Keene, NH 03431A, USA
Tel: 001 603 3572398

Scientific Laboratory Supplies Ltd., Unit 26/27, Wilford ind. Estate, Ruddington Lane, Nottingham NG11 7EP, UK

Shandon Scientific Ltd, 93–96 Chadwick Road, Astmoor, Runcorn, Cheshire WA7 1PR, UK
Tel: 01928 566611
URL: http//www.shandon.com/

Sigma-Aldrich Co. Ltd, The Old Brickyard, New Road, Gillingham, Dorset SP8 4XT, UK
Tel: 0800 717181 (or 01747 822211)
Fax: 0800 378538 (or 01747 823779)
URL: http://www.sigma-aldrich.com/

Sigma Chemical Co., PO Box 14508, St Louis, MO 63178, USA
Tel: 001 314 7715765
Fax: 001 314 7715757
URL: http://www.sigma-aldrich.com/

Stratagene Europe, Gebouw California, Hogehilweg 15, 1101 CB Amsterdam Zuidoost, The Netherlands
Tel: 00800 91009100
URL: http://www.stratagene.com/
Stratagene Inc., 11011 North Torrey Pines Road, La Jolla, CA 92037, USA
Tel: 001 858 5355400
URL: http://www.stratagene.com/

Sutter Instrument Co., 51 Digital Drive.
Novato, CA 94949, USA
Tel: (415) 883–0128
Fax: (415) 883–0572
info@sutter.com
http://www.sutter.com/

Tepnel Life Sciences, Scotscroft Building,
Towers Business park,
856 Wilmslow Road, Didsbury,
Manchester, M20 2RY, UK

TWP Inc., 2831 Tenth Street, Berkeley, CA
94710, USA

United States Biochemical (USB), PO Box
22400, Cleveland, OH 44122, USA
Tel: 001 216 4649277

Yakult Honsha Co., Japan
13-5 Shinbashi S-chome, Minatoku,
Tokyo 105-0004 Japan
Tel: 03 5470 8911   Fax: 03 5470 8921
URL: http://yakult.co.jp/ypi/english/index.html

Index

Page numbers in bold refer to this volume, bracketed page numbers in italics refer to volume 1.

frameshift mutation (5) (*see also* mutation)
French press **115–116**
FRET (see fluorescence resonance energy transfer)
Freund's adjuvant **200** (*see also* antibody)
Fura-2 **240, 251** (*see also* calcium-sensitive fluorescent dyes)
fusion protein
β-galactosidase (207)
cAMP-dependent kinase site **113**
cellulose binding domain **113**
denaturation/renaturation of, in South-western screening (216–217)
epitope tag **113, 116–117, 121, 175, 177, 231**
glutathione-S-transferase (GST) **113–114, 116–117, 121**
histidine tag **113, 116–117, 121, 231–232**
maltose binding protein **113**
myc tag **113**
PKA site **113**
pull down assay **121**
S-tag **113**
T7-tag **113**
thioredoxin **113**

G418 **293–295** (*see also* antibiotics and selectable markers for plant transformation)
GA1 locus (83–84, 95)
GAL1 promoter **175–176** (*see also* promoters)
GAL4–based yeast two-hybrid **101–102, 176, 178, 192–195** (*see also* yeast two hybrid)
GalactonStar™ **183, 185–186** (*see also* β-galactosidase)
gametes (3, 102)
gametophore **286, 290** (*see also* Physcomitrella patens)
gametophyte **286–287** (105) (*see also* Physcomitrella patens)
GCG (187, 189) (*see also* bioinformatics software)

gel retardation assay (see electrophoretic mobility shift assay)
gel shift assay (see electrophoretic mobility shift assay)
gene fusion (44)
gene mapping (101–136, 159)
gene surgery **285** (*see also* Physcomitrella patens)
gene tagging
activation tagging (33)
enhancer trap (33, 44, 56, 68)
exon trap (56, 58)
insertional mutagenesis (33, 41 184)
promoter trap (56)
target selected gene inactivation (75)
T-DNA (33–52)
transposon tagging (53–82)
gene transfer (see plant transformation and Agrobacterium tumefaciens)
genetic background (4)
genetic distances (102)
genetic map (101–103, 106, 110–111, 148) (*see also* linkage analysis, map based cloning and physical map)
genetic map unit (see centiMorgan)
genetically effective cell number (GECN) (2–3)
Genisphere 3DNA software **69, 71** (*see also* bioinformatics software)
genome mapping (148)
genome painting (see chromosome painting)
GenomeWalker™ (185–186, 189, 193–195) (*see also* PCR)
genome-wide approaches (see systems approaches)
genomic subtraction (5, 83–99)
genotype (4)
gentamycin (20) (*see also* antibiotics)
germinal excision frequency (56, 58)
germinal reversion (69–70) (*see also* reversion frequency)
germline (2–3, 8)
germline mutation (2) (*see also* mutation)

GFP (see green fluorescent protein)
GFP spectral variants (see green fluorescent protein)
glucocorticoid-regulated promoter **100–106** (*see also* promoters)
glucosinolates **124**
glutaraldehyde **43, 212**
glutathione beads **116–117, 120**
glutathione-S-transferase (GST) **113–114, 116–117, 121** (*see also* fusion protein)
glyoxysome **152**
Golgi
isolation of **149, 160**
marker enzymes for **163–164, 169**
gradient mixer **160–161**
green fluorescent protein
advantages and disadvantages **276–278**
Aequorea victoria **254, 261, 275**
applications of **100, 102, 261–262, 278–282**
cameleon **262 281**, (*see also* fluorescence resonance energy transfer (FRET) and calcium-sensitive fluorescent dyes)
construction of GFP-tagged proteins **278**
destabilised GFP **281**
drFP583 **275**
dsRed **275**
enhanced GFP **277–278, 281**
enhancer trap reporter (44)
lifetime **276**
optical filters for **278** (*see also* optical filters for fluorescence microscopy)
photobleaching **276, 280**
properties **275–276**
spectral variants **276–278, 282**
green safelight **153**
gridding robot **70** (*see also* DNA microarrays)
GST (see glutathione-S-transferase)
guanidinium hydrochloride for RNA extraction **9–10**
GUS (see β-glucuronidase)